Recent Trends in Modelling of Environmental Contaminants

Debashish Sengupta
Editor

Recent Trends in Modelling of Environmental Contaminants

Springer

Editor
Debashish Sengupta
Geology and Geophysics
Indian Institute of Technology
Kharagpur
West Bengal
India

ISBN 978-81-322-1782-4 ISBN 978-81-322-1783-1 (eBook)
DOI 10.1007/978-81-322-1783-1
Springer New Delhi Heidelberg New York Dordrecht London

Library of Congress Control Number: 2014940089

Printed on acid-free paper

Springer is part of Springer Science+Business Media (www.springer.com)

Dedicated to Ma, Baba and Gogol for the immense affection, perseverance and faith in me.

Preface

The present book titled "Recent Trends in Modelling of Environmental Contaminants" has been based on the vast experience of various eminent Professors and Scientists working in the field of Earth and Environmental sciences. It encompasses the significant results from extensive research work undertaken by different research groups including ours, both in India and abroad. Various chapters have been immensely upgraded and major recent research work reported by internationally renowned researchers has been incorporated. I am thankful to my wife, Sonali and son, Arnab, who have immensely supported to the effort without any complaints.

Acknowledgments

The present work has evolved from a long series of active interaction with my colleagues, both in India and abroad, especially, during the past one decade, on "Modeling of Environmental Contaminants and Remediation Methodologies."

The present book deals with specific and instructive case history from various parts of the world which has important ramifications and assumes significance as a learning exercise, in terms of modelling of the contaminant migration and rapidly evolving remediation techniques, during the recent past. The book showcases state of the art experimental methodology and modeling aspects, related to contaminant modelling. The present study also stresses on the significance of exploration, production and remediation strategy (EPR) for long term research studies and applications based on economic minerals, energy resources and environmental impact.

The book has an extensive literature survey and all pains have been taken to take care of proper citation at the requisite places. Quite a lot of work reported has been undertaken as joint collaboration with other Institutes and Labs. I would personally like to thank them on behalf of me and the authors of other chapters, respectively. Any inadvertent error/omission in this regard is sincerely regretted. In addition, I also thank all the authors of the respective chapters, who have contributed to the same with their valuable time, effort and expertise in the respective area of research/study, as provided in each chapter.

Contents

Radiometric Studies and Base Line Calibrations for NORM and TENORM Studies 1
Debashish Sengupta, Saurabh Mittal and Kajori Parial

Arsenic Fate and Transport in the Groundwater-Soil-Plant System: An Understanding of Suitable Rice Paddy Cultivation in Arsenic Enriched Areas 21
P. K. Sahoo and A. Mukherjee

Geophysical Signatures for Detection of Fresh Water and Saline Water Zones 45
Kalyan Kumar Roy

Prevention and Treatment of Acid Mine Drainage: An Overview 95
D. P. Tripathy

Very Low-Frequency Electromagnetic Method: A Shallow Subsurface Investigation Technique for Geophysical Applications 119
S. P. Sharma, Arkoprovo Biswas and V. C. Baranwal

Treatment of Industrial Wastewater 143
A. K. Gupta and C. Sahoo

Environmental Effects of the Fukushima Daiichi Nuclear Power Plant Accident 167
Katsumi Hirose

Applications of Remote Sensing, Geographic Information System and Geostatistics in the Study of Arsenic Contamination in Groundwater 197
A. R. Ghosh and Kajori Parial

Modeling Radon Behavior for Characterizing and Forecasting Geophysical Variables at the Atmosphere–Soil Interface 213
Antonello Pasini, Roberto Salzano and Alessandro Attanasio

Index 239

Contributors

Alessandro Attanasio CNR—Institute of Atmospheric Pollution Research, Rome, Italy

V. C. Baranwal Geological Survey of Norway (NGU), Trondheim, Norway

Arkoprovo Biswas Department of Geology and Geophysics, IIT Kharagpur, Kharagpur, West Bengal, India

A. R. Ghosh Department of Science & Technology, Government of West Bengal, Kolkata, India

A. K. Gupta Environmental Engineering Division, Department of Civil Engineering, Indian Institute of Technology, Kharagpur, India

Katsumi Hirose Department of Materials and Life Sciences, Faculty of Science and Technology, Sophia University, Tokyo, Japan

Geosphere Research Institute, Saitama University, Saitama, Japan

Saurabh Mittal Department of Geology and Geophysics, Indian Institute of Technology Kharagpur, Kharagpur, West Bengal, India

A. Mukherjee Department of Geology & Geophysics, Indian Institute of Technology Kharagpur, Kharagpur, West Bengal, India

Kajori Parial Department of Geology and Geophysics, Indian Institute of Technology Kharagpur, Kharagpur, West Bengal, India

Kajori Parial Department of Geology and Geophysics, Indian Institute of Technology Kharagpur, Kharagpur, India

Antonello Pasini CNR—Institute of Atmospheric Pollution Research, Rome, Italy

Kalyan Kumar Roy Golf Garden, Tollygunge, Kolkata, India

Indian Institute of Engineering, Science and Technology, Bengal Engineering and Science University, Howrah, West Bengal, India

C. Sahoo Department of Civil Engineering, Indira Gandhi Institute of Technology, Sarang, Odisha, India

P. K. Sahoo Vale Institute of Technology, Belém, Pará, Brazil

Roberto Salzano CNR—Institute of Atmospheric Pollution Research, Rome, Italy

Debashish Sengupta Department of Geology and Geophysics, Indian Institute of Technology Kharagpur, Kharagpur, West Bengal, India

S. P. Sharma Department of Geology and Geophysics, IIT Kharagpur, Kharagpur, West Bengal, India

D. P. Tripathy Department of Mining Engineering, National Institute of Technology, Rourkela, India

About the Editor

Dr. Debashish Sengupta is a Professor, Higher Administrative Grade and Head of the Department of Geology and Geophysics in Indian Institute of Technology Kharagpur, West Bengal, India. Dr. Sengupta has more than 25 years of teaching and research experience. He has completed his PhD in 1987 in Applied Geophysics. His areas of interests are nuclear geophysics including petroleum logging using subsurface nuclear data, radioactive methods and geochronology, radon emanometry and its applications, applications of isotopes and radionuclides in earth and environmental geosciences, heat flow and geothermics. Dr. Sengupta has 73 research publications in international journals and more than 50 papers in Conference Proceedings apart from a large number of Invited Talks delivered both in India and abroad. The research work has been seminal and resulted in the formulation of Environmental Regulation policies in various countries both in India and countries like USA, South America and the European Union. Dr. Sengupta had also been a Visiting Professor at the University of Sao Paulo, Brazil and as Senior Visiting Professor at the University of Salamanca, Spain, earlier, while on a sabbatical leave from the institute. He has received Society of Geoscientists and Allied Technologists (SGAT's) Award of Excellence in Earth Sciences for the year 2003. Till date, 9 students have been awarded PhD degrees under his supervision.

Radiometric Studies and Base Line Calibrations for NORM and TENORM Studies

Debashish Sengupta, Saurabh Mittal and Kajori Parial

Abstract Natural ambient radiations are present everywhere due to the spontaneous emissions by radioactive elements. There are two sources of radiations in the earth's environment: cosmic and terrestrial. Cosmic radiation originates from high-energy cosmic ray particle interaction with the Earth's atmosphere, and terrestrial radiation comes from the naturally occurring radioactive materials (NORM) present in the Earth's crust. Radiometric studies provide a suitable tool for detecting and measuring the natural radiation. Radioactivity measurements are also referred to as baseline measurements when the radioactivity is only from NORM. Baseline measurements provide the background measurements for comparative studies, since background radiation is not the same everywhere. There are some regions in the world with higher levels of radiation, known as high background radiation areas (HBRAs); a few places even have much higher background radiation, and are known as very high background radiation areas (VHBRAs). In this chapter, applications of radioactivity measurements are demonstrated with case studies for various studies, namely characterization of the area as background or HBRAs or VHBRAs, environmental studies for assessing the hazard due to radiation and locating hot springs. Mining and production of uranium also cause environmental radioactivity, termed as technologically enhanced naturally occurring radioactive material (TENORM). Radiometric studies also provide classification of activity under NORM and TENORM. Some of the recent studies undertaken primarily by us in this regard are discussed along with their implications.

The studies presented in the Chapter reflect the significance of research related to integrated near-surface geophysics for various applications, namely radiometric measurements, contamination studies primarily due to anthropogenic causes, shallow subsurface exploration for economic deposits like shale gas, coal bed methane, and associated rare earths.

Keywords NORM and TENORM · Near surface geophysics · High background radiation areas (HBRAs) · Radioactivity measurements · Uranium exploration

---

D. Sengupta (✉) · S. Mittal · K. Parial
Department of Geology and Geophysics, Indian Institute of Technology Kharagpur, Kharagpur, West Bengal 721302, India
e-mail: dsgg@gg.iitkgp.ernet.in

D. Sengupta (ed.), *Recent Trends in Modelling of Environmental Contaminants*,
DOI 10.1007/978-81-322-1783-1_1, © Springer India 2014

1 Introduction

Naturally occurring radionuclides are present all over the Earth's crust in varying quantities depending on the ambient geological and geochemical environment. The considerable concentration of radionuclides in soil, water, air, and living organisms is due to the presence of naturally occurring radioactive elements like ^{232}Th, ^{238}U, and ^{40}K that occur mostly in minerals such as monazite and zircon. The reason for their presence today since the origin of Earth is their sufficiently long half-lives. Their half-lives are comparable to the age of the Earth and therefore these are called primordial radionuclides. The major contribution to environmental radiation is from ^{232}Th, ^{238}U, ^{40}K and their decay products; other radionuclides, namely ^{235}U, ^{87}Rb, ^{138}La, ^{147}Sm, etc. also exist in nature but in such low quantities that their contribution to the ambient environmental radiation can be neglected.

Natural radiation, also known as ambient radiation or background radiation, is quantified by the natural radiation environment. There are two sources of background radiation in the environment: cosmic and terrestrial. Cosmic radiation originates from high-energy cosmic ray particle interaction with the Earth's atmosphere. Terrestrial radiation comes from the naturally occurring radioactive materials (NORM) present in the Earth's crust. NORM include primordial radionuclides that are naturally present in the rocks and minerals of the earth's crust and cosmogenic radionuclides produced by interactions of cosmic nucleons with target atoms in the atmosphere and in the earth. NORM does not include natural radioactive material from anthropogenic sources, such as those produced by nuclear power and used in nuclear medicine. Therefore radioactive materials can be classified into (a) man-made and (b) NORM. Man-made radionuclides include Cobalt-60, Strontium-90, and Cesium-137, which are produced in nuclear reactors by nuclear fusion or nuclear fission.

TENORM is an acronym used for technologically enhanced NORM, and is produced when naturally occurring radionuclides are concentrated or their activities and exposure is increased compared with the unaltered situation by anthropogenic activities.

Terrestrial radiation affects us both internally and externally. External exposures primarily include gamma radiation. Internal radiation exposure is due to intake of radionuclides by inhalation or ingestion. The important contributor in internal exposure is radon (from uranium decay series) and its decay products. Therefore, the total exposure to radiation comes from gamma radiation (32.2 %) and inhalation of radon (51.9 %) (UNSCEAR 2000). Figure 1a and b shows the schematic of a radiation counter operating to detect radiation from an area of 2π and $>2\pi$, respectively.

Radiometric studies coupled with other geophysical techniques provide a valuable tool for detecting and measuring the natural radiation. This also helps in the baseline calibration for the average radiation background. Table 1 represents an outline of the geophysical techniques and the attributes or parameters measured through it that can be applied along with radiometric studies for a better understanding of the ambient geology. Radioactivity measurements are also called baseline measurements when the radioactivity is from NORM and not from any other

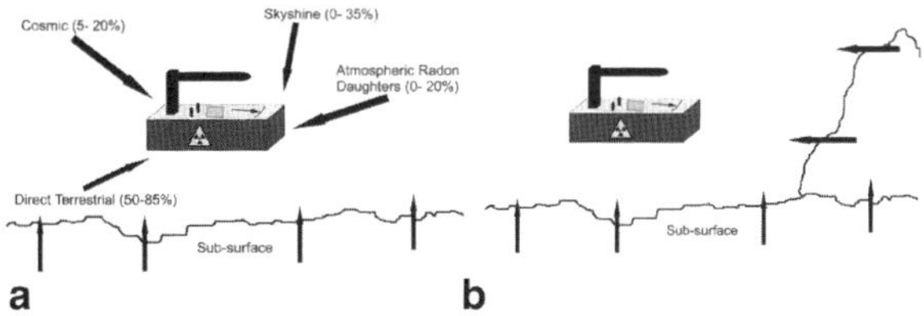

Fig. 1 Schematic of radiation detection. **a** Within a solid angle 2π. **b** Within solid angle $>2\pi$

Table 1 Geophysical techniques employed and corresponding attributes or parameters measured to identify various geological features

Relevant aspects	Geophysical methods	Output
Resistivity/Conductivity anomaly	Resistivity, VLF-EM	Faults and fractures
High or low conductivity	Resistivity, VLF-EM, radiometrics	Alteration
High conductivity, potassium content	VLF-EM, radiometrics	Hydrothermal alteration
High conductivity contrast, potassium content	VLF-EM, radiometrics	Silicification (dyke-like feature)
Weak conductor, low and patchy mineralization	Resistivity, VLF-EM	Disseminated deposit
High over low conductivity	Resistivity, VLF-EM, radiometrics	Weathering

source or by any kind of anthropogenic activity (such as mining and oil and gas production). Baseline measurements aid in the identification and quantification of naturally occurring radionuclides. It also provides the background measurements for comparative studies, since background radiation is not the same everywhere. There are some regions in the world that have higher levels of radiation, known as high background radiation areas (HBRAs). A few places in the world have an even greater level of background radiation, and are known as very high background radiation areas (VHBRAs). VHBRAs include: Guarapari, in the coastal region of Espirito Santo, Brazil (Cullen 1977; UNSCEAR 1993); Yangjiang, China (Wei et al. 1993; Wei and Sugahara 2000); the southwest coast of India (Sunta et al. 1982; Paul et al. 1998); Ramsar and Mahallat, Iran (Sohrabi 1993; Ghiassi-nejad et al. 2002); and places in some other countries (UNSCEAR 2000). Sources of background radiation include radon, monazite sand deposits, and radium. The eastern coast of Orissa, India has also been delineated as an HBRA (Mohanty et al. 2004a, b). Further details of the same have been incorporated as a case study subsequently.

Apart from characterization of the area as background or HBRA or VHBRA, radiometric studies are helpful in environmental studies for assessing the hazard due to radiation and locating hot springs. Mining and production of uranium also cause environmental radioactivity. Such radioactivity is caused by materials formed due to anthropogenic activities like mining and oil and gas production and is termed TENORM. Baseline measurement of natural radioactivity helps in the classification of activity under NORM or TENORM. A case study explaining the migration of radon from a uranium tailing pond is also presented in this chapter.

Besides this, radiometric studies are particularly useful in the exploration of radioactive mineral deposits. An integrated geophysical survey with radiometric surveys is quite useful in providing details about the presence or absence of radioactive minerals, geometry, and the lateral extent of ore bodies. The case study provided in this chapter illustrates the advantage of radiometric studies in exploration. Radiometric data were correlated with very low frequency (VLF) electromagnetic (EM) data to investigate the possibility of shallow economic uranium deposit along the South Purulia Shear Zone (SPSZ).

The present study amply reflects the significance of research related to integrated near-surface geophysics for various applications, right from radiometric measurements and contamination studies to shallow-depth subsurface exploration for non-radioactive minerals like shale gas, coal bed methane, etc.

1.1 Ionization Chambers

Geiger–Müller (GM) counters

A Geiger–Müller (GM) counter is a gas-filled cylindrical tube (cathode) with a thin metal wire (anode) (Fig. 2). When a γ-ray photon interacts with the wall of the cylinder an energetic electron is produced that enters the tube and ionizes the gas, creating an avalanche of electrons that moves towards the central wire. Due to high voltage within the tube a strong electric field is produced within the tube. The electrons move towards the positive electrode and the positive atoms move towards the negative electrode. A current pulse is generated as a result of this and is measured as a count. Once detected or counted the ions become neutral, restoring the GM counter to normal state, ready for count detection. The operation and efficiency of a GM counter are highly dependent on the operating voltage. Lower voltages will not record any pulse and higher voltages result in constant discharge, resulting in erroneous count. The operating voltage of a GM counter is to be strictly maintained in order to ensure maximum efficiency and the health of the detector. The simple yet efficient counting capability of GM counters makes it a reliable, cost-effective, and portable instrument for detecting total environmental radiation counts. It is mostly used for ground reconnaissance survey for identifying radioactive zones.

Scintillation Counters

Scintillation counters are made up of a luminescent material connected to a device, mostly a photomultiplier tube (PMT), which detects the γ-ray-induced photon emission. When the incident γ-ray interacts with the scintillator the ionized atoms emits a photon of light to attain a lower energy level. This light energy is detected and amplified by the photo multiplier tube.

The amount of scintillation produced is proportional to the γ-ray energy as the entire energy is used to excite the atom. An electric pulse is generated by the PMT

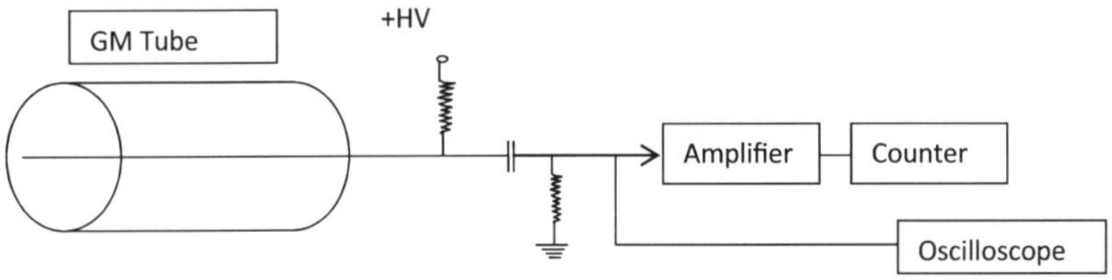

Fig. 2 Schematic diagram of Geiger–Müller (GM) counter (After Smith and Lucas 1991)

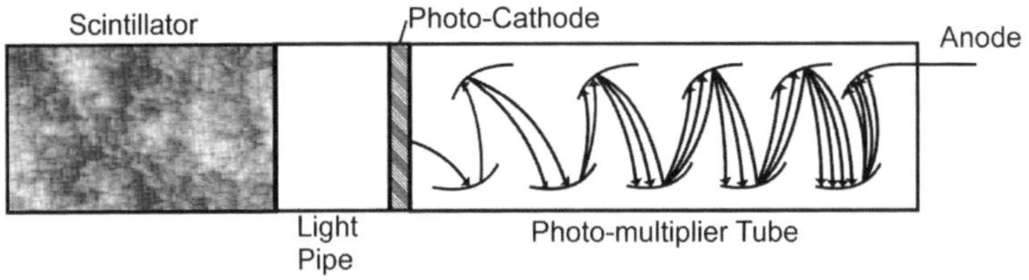

Fig. 3 Schematic diagram of Scintillation counter (After Smith and Lucas 1991)

proportional to the amount the scintillation and the pulse height correlates with the incident γ-radiation energy. The scintillation detectors can be either organic (anthracene, plastics, etc.) or inorganic (NaI(Tl), BGO, ZnS, etc.). The scintillation counters are advantageous compared to the GM counters in terms of efficiency; however, they costlier than GM counters and are not portable enough for ground survey.

1.2 γ-Ray Spectroscopy:

γ-rays are detected based on their interaction with the charged electrons of atoms of matter. During ionization the γ-ray photon gives part or the whole of its energy to the electron. These electrons in turn collide with other electrons and release more electrons. The charge can be collected through GM counters (as mentioned earlier) or through the solid-state detectors. Natural radioactive sources yields a complex spectrum in contrast with mono energetic sources. The energy of the photopeak is considered to be equal to the incident γ-ray. Germanium (Ge) is found to be the most efficient element for solid-state semiconductor detectors. Initially, lithium-drafted Ge was used as a detector element. However, in recent times very high purity Ge is used as the detector material.

When γ-ray is incident on the Ge crystal an electron–hole pair is created that helps in detecting the γ-radiation energy. When the γ-ray is incident on the detector with energy greater than the band-gap energy, an electron–hole pair is created. These electrons allow the conduction of electricity, generating an electric pulse. The greater the incident energy of the γ-photon, the greater the pair production, thus

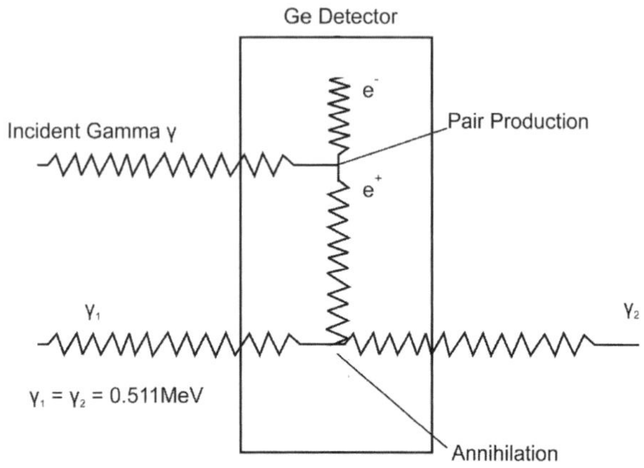

Fig. 4 Process of pair production (After Khandaker 2011)

generating a higher pulse. Figure 4 shows a schematic view of the pair production generation.

The activity (in Bq/Kg) of each radionuclide in the samples is determined using the total net counts under the selected photo-peaks after subtracting the background counts, and applying factor for photo-peak efficiency γ-intensity of the radionuclide, the weight of the sample and the time of counting. The relationship or the formula used to calculate activity concentration is given by

$$A = (dps/w) \times 1000 \quad (1)$$

Where, A = activity concentration (in Bq/kg),
w = weight of the sample (in g), and
dps = disintegrations per second

$$dps = P/(t \times \varepsilon \times I) \quad (2)$$

Here, P = net peak area,
t = time for which counting is done (second)
ε = efficiency of photo-peak, and
I = γ-intensity of a particular radionuclide

2 Case Studies

2.1 Erasama Beach, Orissa, India (HBRA)

Terrestrial background radiation from naturally occurring radionuclides like ^{238}U, ^{232}Th, and ^{40}K, is a major contributor towards humankind's exposure to radiation (UNSCEAR 2000). However, there are certain local geological factors that enhance

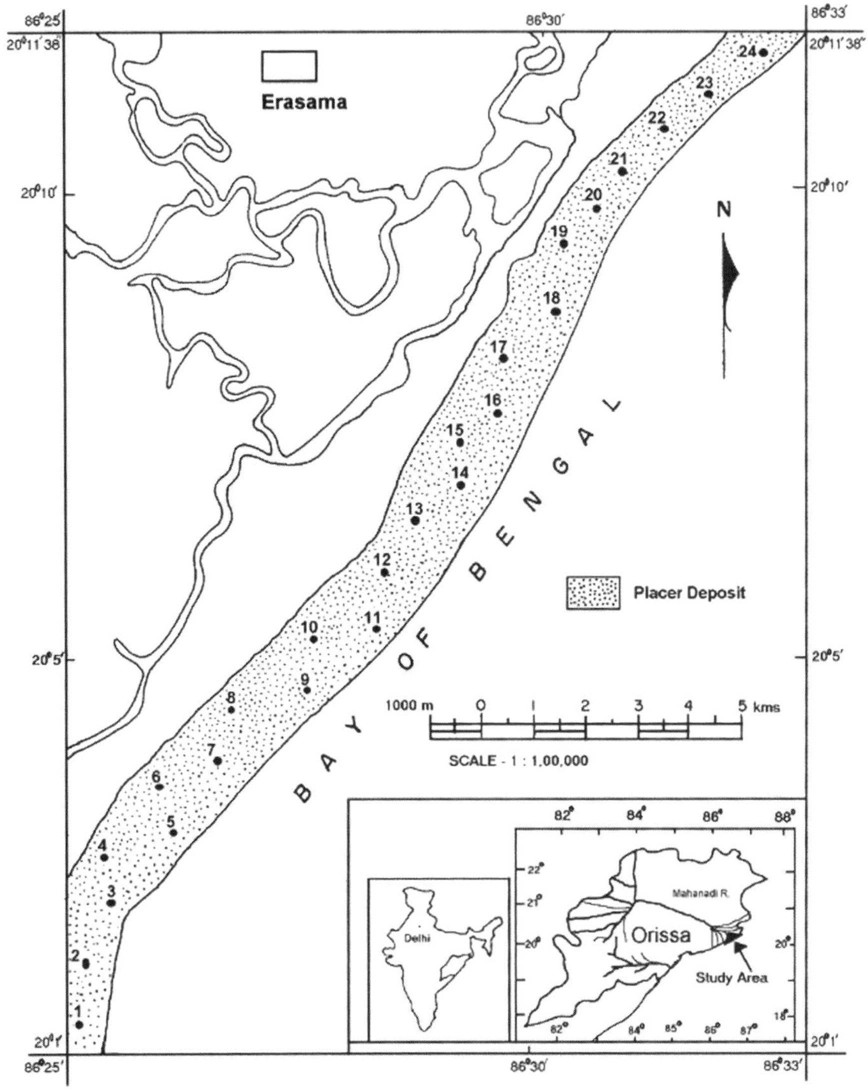

Fig. 5 Map showing Erasama beach (After Mohanty et al. 2004a)

this radiation significantly above normal (or background); these regions are known as HBRAs. In India, the Erasama beach region, on the eastern coast of Orissa, is one such HBRA. Erasama beach is a coastal strip (Lat. 86°25′–86°33′N, Long. 20°1′–20°11′E) on the eastern coast of Orissa state, India (Fig. 5). The area of study considered by Mohanty et al. (2004a) is a flat terrain with a gentle slope towards the Bay of Bengal, and consists of recent alluvium and coastal deltaic plain sediments. Extensive radiometric studies were carried by Mohanty et al. (2004a) out in this region to measure the ambient radiation in terms of activity concentration of radionuclides.

Radiometric Analysis

Standard procedure was followed for sampling in the field and sample preparation in the laboratory. 24 bulk sand samples and different individual fraction of mineral sands were subjected to radiometric analysis. Individual mineral fractions were separated and only zircon, sillimanite, and rutile fractions were considered for this study. Samples were dried and sealed for some time for attaining secular equilibrium between ^{232}Th, ^{238}U, and their daughter products. The activity of the samples was then analysed using γ-ray spectrometric analysis. It was carried out at the Radiochemistry Divison, Variable Energy Cyclotron Centre, BARC, Kolkata, using a coaxial HPGe detector (EG & G, ORTEC) with 15 % relative efficiency.^{152}Eu liquid source (Amersham Company, UK) of known activity was used to determine the efficiency of detector, since ^{152}Eu has large range of energies (122–1,408 keV) with emission abundances of 3–29 % (Firestone and Shirley 1998; Grigorescu et al. 2002).

In the absence of any interference, ^{40}K is measured directly by its own γ-rays; ^{232}Th and ^{238}U are indirectly measured from the γ-rays of their decay products. Decay products for ^{238}U (^{214}Pb: 295 and 352 keV; and ^{214}Bi: 609 keV) and ^{232}Th (^{228}Ac: 209, 338, and 911 keV; ^{212}Pb: 239 keV; ^{212}Bi: 727 keV; and ^{208}Tl: 583 keV) were used with the assumption that the decay series is in secular equilibrium. Mass of the sample, branching ratio of the γ-decay, time of counting, and efficiency of the detector were duly considered during estimation of activity concentration. After calculation of activity concentration from intensity of several γ-rays, they were grouped together to compute a weighted average activity per nuclide. Absorbed gamma dose rates were also calculated based on the guidelines of UNSCEAR (1993, 2000).

Results and Discussion

A total of 24 samples were analysed. Activity concentration of ^{232}Th ranged from 900–4,700 Bq kg^{-1} with an average of 2,825 ± 50 Bq kg^{-1}. However, ^{238}U activity concentration ranged from 150–500 Bq kg^{-1}, with an average of 350 ± 20 Bq kg^{-1} and that of ^{40}K ranged from 100–250 Bq kg^{-1} with an average of 180 ± 25 Bq kg^{-1}. ^{238}U and ^{232}Th was correlated in sand samples indicating the presence of monazite and zircon rich U and Th respectively. The higher concentration of monazite could be attributed to its significant supply of fine grain sediments from the hinterland areas by the Mahanadi River system and its tributaries (Mohanty et al. 2004a).

The absorbed γ dose rates in air for the Erasama beach region, on the eastern coast of Orissa was found to vary from 650 to 3,150 nGyh^{-1} with a mean (± SD)

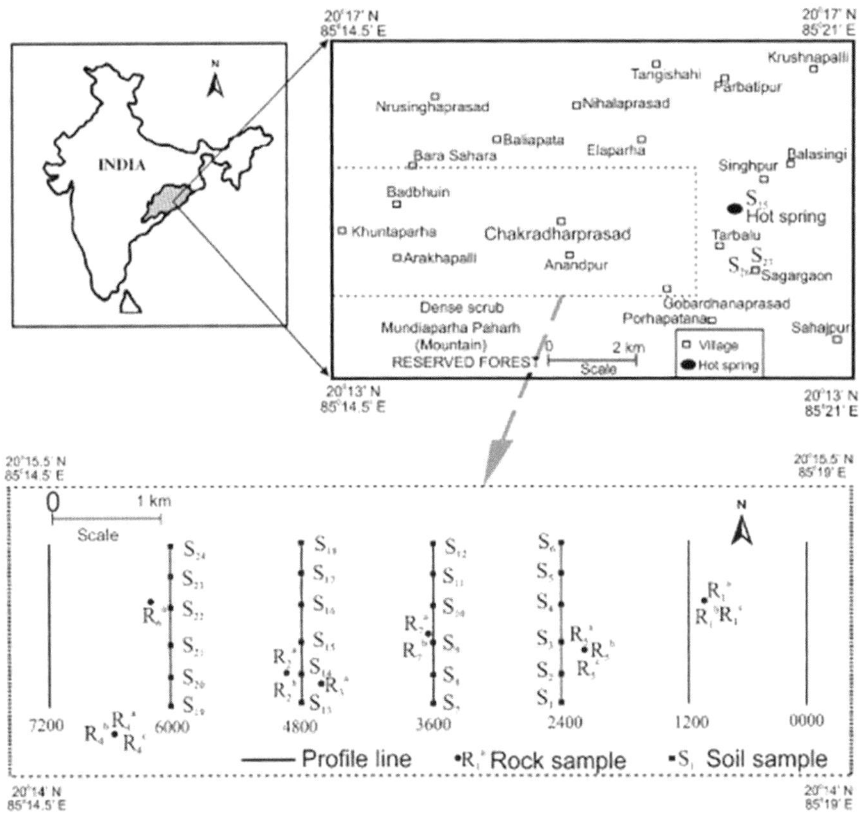

Fig. 6 Survey area and sample locations in Tarbalu region, Orissa (After Baranwal et al. 2006)

value of 1,925 ± 718 $nGyh^{-1}$ and the annual external effective equivalent dose rate of 2.0 ± 1.5 mSv yr^{-1}. Based on the higher levels of natural radioactivity and gamma-absorbed dose rates in air, compared to other monazite rich areas, this region can be considered an HBRA and is comparable to other monazite sand-bearing HBRAs in southern and southwestern coastal regions of India (Mohanty et al. 2004a).

2.2 *Tarbalu Hot Spring, Orissa, India*

Presence of radioactive mineral deposits enhances the background radiations in the area and this helps in the detection of such deposits. However, they also produce heat, called radiogenic heat, which is a source of heat for hot springs. Tarbalu is one such hot spring in the geothermal region of the Eastern Ghats metamorphic province in Orissa, India (Fig. 6). The study area extends from Lat. 20°14′– 20°15.5′N,

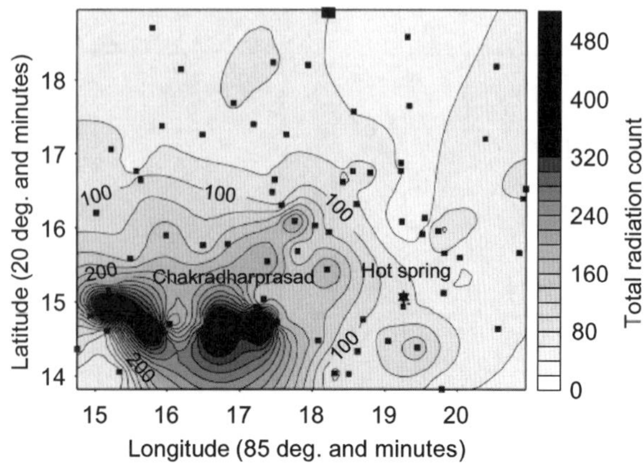

Fig. 7 Isorad map of the study area (value in total counts per 100 s). *Solid squares* represents data point locations (Taken from Baranwal and Sharma 2005)

and Long. 85°14.5′–85°19′E. The study area is characterized by high-grade metamorphic rocks of granulite facies such as khondalite, charnockite, and enderbite, with interlayered mafic horizons (Kundu et al. 2002). It is worth mentioning that deformed granites were found coinciding with high-radioactivity zones, and these high-radioactivity zones are aligned in the east–west direction.

A radiometric reconnaissance survey was carried out in the adjoining geothermal region by Baranwal et al. (2006) using a portable GM counter to identify a relatively high natural background radiation region. Soil and rock samples were collected from these HBRA regions for further measurements in the laboratory to determine the relative concentration of radioactive elements, ^{238}U, ^{232}Th, and ^{40}K, responsible for high radiation level. Soil and rock sampling locations and the corresponding profiles in the high-radiation area are shown in Fig. 6. The samples were analyzed by γ-ray spectrometry (GRS) using NaI(Tl) and HPGe detector.

Radiometric Studies

The entire region near the hot spring was surveyed using a portable pulsed GM counter. The contour map over the area is shown in Fig. 7. Soil and rock samples were collected along four profiles (Fig. 6) based on high-radiation areas identified (Fig. 7). The profile-wise detailed radiation survey results are provided in Fig. 5. For detailed measurements, counts were measured at an interval of 25 m along all profiles and profile distance was kept at 1200 m to cover the area of study. The sensor of GM counter was kept approximately at 1 foot above the ground surface to have the appropriate geometry for subsequent sample collection and device was operated for 100 s. From the data obtained in the survey a high background radiation zone was identified (shaded area in Fig. 8) (Baranwal et al. 2006).

Total radiation counts measured over all the profiles are shown as linear plots (Fig. 8 a–g) for simplicity and quick correlation among them to identify the high background radiation zones. Soil samples were collected at 200 m intervals from

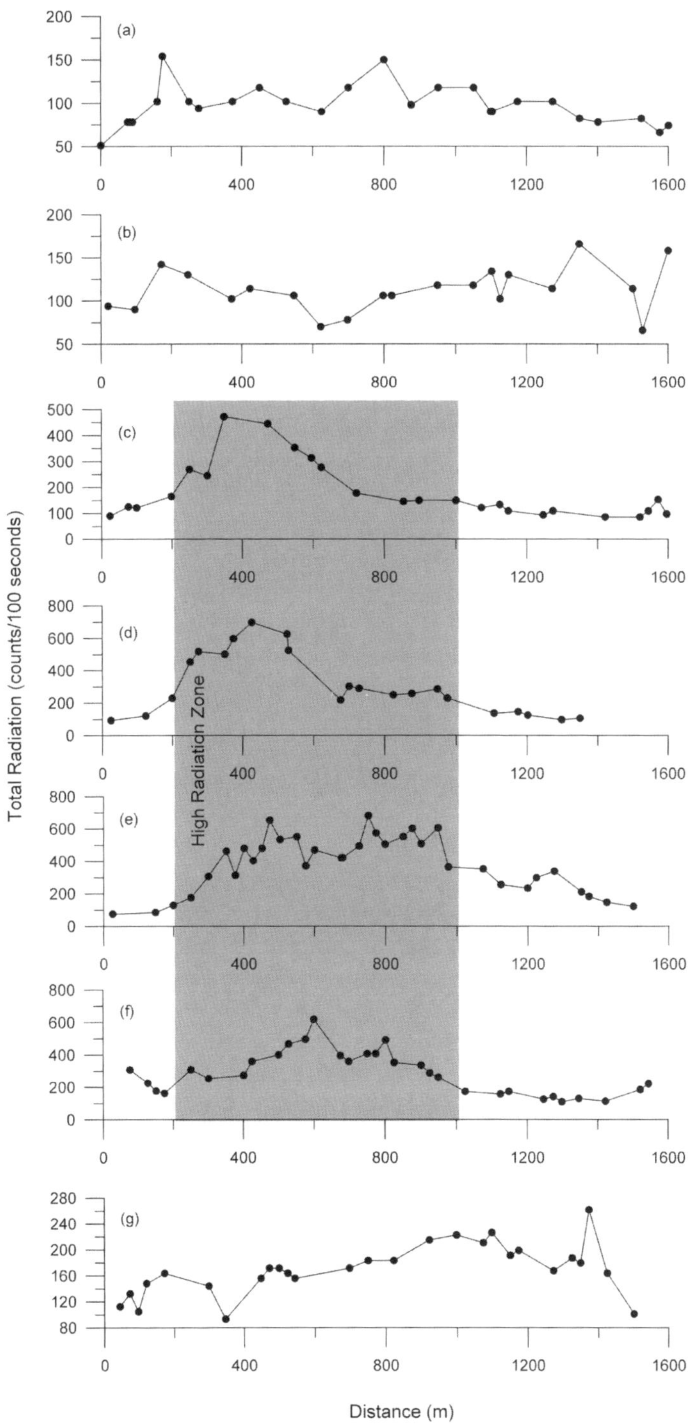

Fig. 8 Total radiation (counts per 100 s) for different profiles. **a** 0000. **b** 1200. **c** 2400. **d** 3600. **e** 4800. **f** 6000. **g** 7200. *Shaded region* represents high-radiation zones (After Baranwal et al. 2006)

four profiles (2400, 3600, 4800 and 6000) exhibiting the higher radiations. A few rock samples were also collected wherever outcrop was found exposed along these profiles. Subsequently, rock and soil samples were crushed and powdered for gamma ray spectrometry (GRS) analysis and dose rate calculations.

γ-ray spectrometric analysis was carried out using NaI(Tl) detector and HPGe detector to estimate the actual concentration of radionuclides. Since radioactive elements emit various energies, high energies were utilized for analysis using a NaI(Tl) detector and low energies using an HPGe detector, respectively. HPGe detectors have better resolution than NaI(Tl) detectors for low energies and so they show sharp peaks for corresponding energies of its decay products, whereas, efficiency of NaI(Tl) detector is higher. The concentration of thorium, equivalent radium, and potassium were estimated. The actual concentration of uranium was also estimated using β–γ technique (Eichholz et al. 1953).

Absorbed gamma dose rates were also calculated based on the guidelines of UNSCEAR (1993, 2000). The expression to calculate dose rate is

$$D = \left(0.692C_{Th} + 0.462C_{U} + 0.0417C_{K}\right) \text{nGy/h} \tag{3}$$

where, 0.692 nGy/h is the conversion factor for Th, 0.462 nGy/h is the conversion factor for uranium and 0.0417 nGy/h is the conversion factor for potassium. C_{Th},C_{U}, and C_{K} are average activity concentrations (in Bq/kg) of thorium, uranium, and potassium, respectively.

Results and Discussion

High-radiation zones were identified along four profiles (Fig. 8 c–f) with a maximum value of 650 counts/100 s. The width of these zones varies from 200–1,400 m with widest zone along profile 4800. Much of this radiation is due to thorium, since concentration of thorium is very high as compared to uranium and potassium. The maximum concentration of thorium was observed to be 1,194 and 1,066 ppm in sample from profile 3600 and 4800, respectively. Granitic rocks exhibited higher thorium content than other rocks found in the area. Activity concentration of the radionuclides in soil samples also showed high thorium concentration. The presence of thorium was also confirmed by analysis of rock samples using HPGe detector, since the result shows distinct energy peaks corresponding to various isotopes in the decay chain of thorium. The absorbed gamma dose rate in air and external annual dose rate of the high-radiation zone were estimated to be 2431 nGyh^{-1} and 2.99 mSvy^{-1}, respectively, which is 10 times greater than outside the high-radiation zones. The high concentration of thorium in the samples also supports the possibility of radiogenic heating to be the source of subsurface heat flow and hence, that of the hot springs (Baranwal et al. 2006).

2.3 *Radiometric Study of Coastal, Beach, and the Monazite Placer Deposits*

Beach placer deposits are reported worldwide to be the major source of economically extractable heavy minerals, mostly monazites, zircon, ilmenite, rutile, garnet, sillimanite (Sulekha Rao and Misra 2009). Globally, high-resolution radiometric methods were employed effectively in heavy mineral deposit exploration and associated studies (Grosz 1983; Noakes et al.1989; de Meijer et al. 1997). Mohanty et al. (2004a, b), Sengupta et al. (2005) and Sulekha Rao et al. (2009) studied the beach deposits at Chhatrapur, Erasama, Gopalpur, and Rushikulya beach on the Orissa coast.

Radiometric Studies

Samples were collected and subjected to the standard procedure of sample preparation as mentioned earlier. The samples were sealed in plastic vials to attain secular equilibrium. The γ-spectrometric analysis was done using HPGe (EG & G, ORTEC) to determine the activity concentration of the radionuclides. Absorbed γ-dose rate, effective dose rate, radium equivalent activity, and level index were calculated from the activity concentration based on standard equations provided by UNSCEAR and NEA-OECD (Sulekha Rao et al. 2009).

Results and Discussion

For the Gopalpur samples, the average activity concentrations of ^{232}Th was 1,670 ± 340 Bq kg^{-1}, for ^{238}U an average activity of 200 ± 40 Bq kg^{-1} and for ^{40}K an average activity of 650 ± 110 Bq kg^{-1} was found. For Rushikulya samples, the average activity concentrations of ^{232}Th, ^{238}U, and ^{40}K was 990 ± 250 Bq kg^{-1}, 100 ± 25 Bq kg^{-1} and 775 ± 100 Bq kg^{-1}, respectively. For Chhatrapur samples, the average activity concentrations of ^{232}Th, ^{238}U, and ^{40}K was 2500 ± 1850 Bq kg^{-1}, 220 ± 135 Bq kg^{-1}, and 120 ± 80 Bq kg^{-1}, respectively (Mohanty et al. 2004b). However, apart from the activity and calculated dose rates the radiometric studies revealed various other significant implications. High radioactivity in the sand dune samples indicated these regions as potential zones for radiogenic minerals. Higher (40–67 times) mean activity than world average of ^{232}Th suggests predominance of thorium rich monazites (Sulekha Rao et al. 2009). The thorium uranium ratio calculated from the activity also offers valuable insight into the host environment. A higher Th/U concentration ratio suggests enrichment of heavy minerals in the rocks as observed in Gopalpur, Rushikulya beach deposits. The eTh/eU ratio calculated provides information about the geochemical facies, the oxidizing/reducing conditions of the environment. In Gopalpur and Rushikulya beach the average eTh/eU

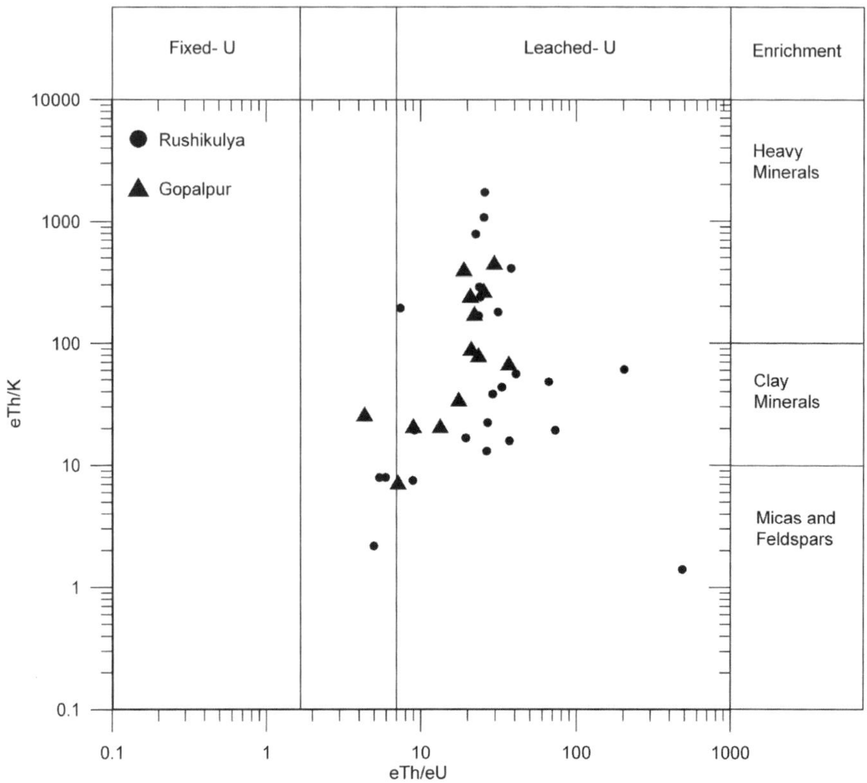

Fig. 9 Cross plot of eTh/eU vs. eTh/K on sand samples of Gopalpur and Rushikulya beach placer deposits, Orissa, India, plotted on the discrimination diagram of Anjos et al. 2006 (After Sulekha Rao et al. 2009)

ratio suggests uranium mobilization through weathering and/or leaching conditions indicating an oxidizing terrestrial environment. The cross plot between eTh/eU and eTh/k (fig. 9)showed that majority of Gopalpur falls in heavy mineral field and Rushikulya samples in clay mineral field(Sulekha Rao et al. 2009).

2.4 Radon Migration Study From Uranium Tailings Pond

Uranium tailings are a major source for anthropogenic radon. The tailings are wastes generated from uranium mining and milling activities and are reported to contain ^{226}Ra activity similar to ^{238}U present in the ore before extraction. The contamination from the tailings pond can be through dispersion in air and water (Banerjee 2011). Radon is emitted from the tailings either in form of emanation or exhalation (Sahoo et al. 2010). The study by Sahoo et al. (2010) investigated the seasonal variation of Radon flux over the U tailings pond and generated preliminary data for modelling radon flux from the tailings surface. Figure 10 shows the study area.

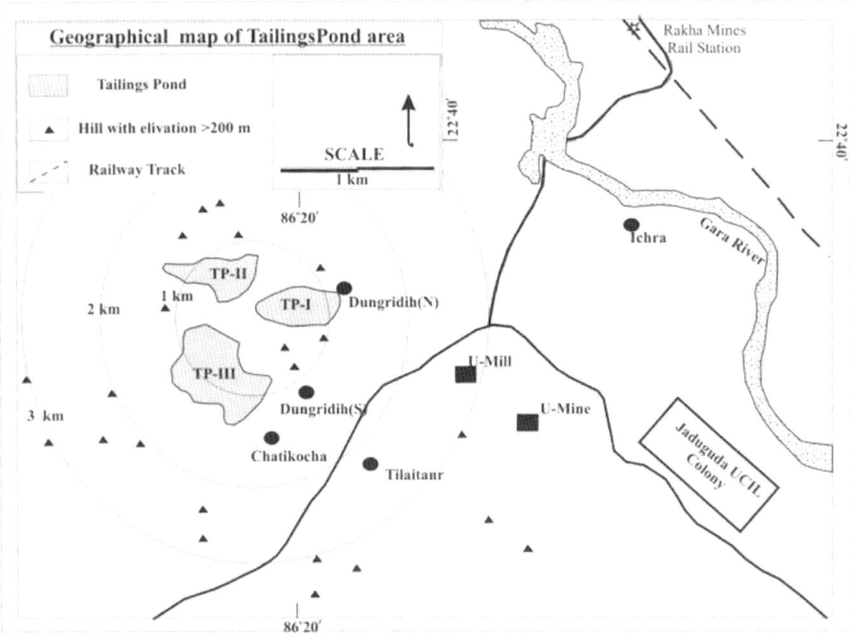

Fig. 10 Geographical map of U tailings pile region at Jaduguda, Jharkhand, India. (After Sahoo et al. 2010)

Radiometric Studies

Based on the accumulator technique developed by Mayya (2004), in situ radon flux was measured. The accumulator, of 30 cm diameter and 30 cm height, was attached to a continuous radon monitor (AlphaGuard). The set up was installed on the tailings surface and was made air-tight (Fig. 11a). The radon concentrations were recorded at an interval of 10 min for 2–3 h starting from the deployment time considering the saturation state. Equation 4 (Aldenkamp et al. (1992), Sahoo et al. 2010) was used to fit the concentrations data with elapsed time, after necessary corrections:

$$C(t) = \frac{J_s^m A}{V \lambda_e}(1 - e^{-\lambda_e t}) + C_0 e^{-\lambda_e t} \quad (4)$$

Where $\mathbf{J_s^m}$ is the radon flux from tailings surface (Bq m^{-2} s^{-1}), V is the accumulator volume (m^3), λ_e denotes the effective decay constant (s^{-1}) of radon for the given set up and is given by sum of radon decay constant, leakage rates (if any) and back-diffusion rate for the set up. A is the soil surface area under the accumulator (m^2) and C_0 is initial radon concentration inside the accumulator at $t=0$.

Theoretical radon fluxes were calculated from the in-situ radon emanation values using UNSCEAR 2000 formula (Sahoo et al. 2010):

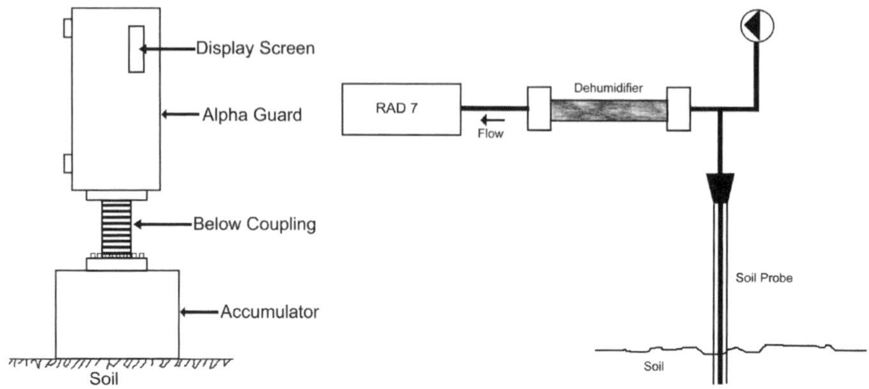

Fig. 11 **a** Experimental setup for in-situ radon flux measurement (After Sahoo et al. 2010). **b** Experimental setup for the measurement of radon concentration at different depth in U tailings (After Sahoo et al. 2010)

$$J_s^e = \lambda l_s Q_{Ra} \rho_b f \tag{5}$$

Where the soil radium content (Bq kg^{-1}) is given by Q_{Ra},ρ_b denotes the soil bulk density (kg m^{-3}), f is radon emanation factor measured in situ, l_s (m) is the radon diffusion length in the tailings and the decay constant for ^{222}Rn (s^{-1})is given by λ.

Tailings parameters were calculated from samples collected by introducing a 1000 cc stainless steel cylinder into the tailings. The samples were processed followed by oven drying at 110°C to attain constant weight and sealed in 300 cc plastic containers for gamma spectrometric analysis. ^{226}Ra in the tailings (Q) (after Shukla et al. 2001, c. i. in Sahoo et al. 2010), radon mass exhalation rate, radon emanation factor in-situ (f) and in dry conditions (f_0) (after Sahoo et al. 2007) was measured. Soil porosity (ϵ) and S was calculated using the relations given by Rogers and Nielson (1991) in Sahoo et al. (2010) (Eq. 6):

$$(\epsilon) = [1\text{-}\rho_b / \rho_g] \tag{6}$$

(where ρ_a is the average specific gravity of soil (2700 kg m^{-3}) and ρ_b is the dry bulk density (kg m^{-3}) of tailings) and $100S = \rho_b M_w / \rho_w \epsilon$ where M_w is soil water content (dry weight percent) and ρ_w is the density of water (kg m^{-3}), respectively. The temperature at 10 cm depth from tailings surface was measured in situ using a temperature sensor. Depth profile of radon concentration in the U tailings pile was used to estimate the in situ radon diffusion lengths (l_s). Online radon monitor with silicon alpha detector (RAD 7, Durridge Company Inc, Bedford) was used to measure radon concentrations in the tailings pond. Readings were taken by air sample collection at the rate of 0.5 L min^{-1} through a probe (Fig. 11b) at an interval of 20 cm up to a depth of 120 cm. The radon concentration vs. soil depth was plotted and fitted to the concentration depth profile using equation (Eq. 7) (Nazaroff and Nero 1988 c.i. Sahoo et al. 2010).

$$C(z) = C_\infty(1 - \exp(-z / l_s) \quad (7)$$

where, $C(z)$ is the ^{222}Rn concentration (Bq m^{-3}) at a depth z (cm), C_∞ is the ^{222}Rn concentration at deeper depth (Bq m^{-3}) and l_s (cm) is the diffusion length of ^{222}Rn in soil.

Results and Discussion

The radon fluxes at different locations of U tailings pile at Jaduguda, India showed a weak seasonal variation (Sahoo et al. 2010). The mean values of radon fluxes obtained were 3.74 ± 2.18, 3.27 ± 1.82, and 4.24 ± 1.83 Bq m^{-2} s^{-1} in winter, summer, and rainy seasons, respectively (Sahoo et al. 2010). From this data, the radon source term contributed from this area was estimated as 8.1 ± 10^{13} Bq y^{-}—^{1} (Sahoo et al. 2010). A good linear corelation was observed between ^{226}Ra contents and seasonal averaged radon fluxes in tailings (Sahoo et al. 2010).The slope of value $(8.3 \pm 0.4) \times 10^{-4}$ kg m^{-2} s^{-1} was used as a conversion factor for the quick estimation of radon flux (Bq m^{-2} s^{-1}) using ^{226}Ra (Bq kg^{-1}) in U tailings or vice versa (Sahoo et al. 2010).

2.5 *Radiometric Studies for Shallow Uranium Deposits*

In nature the ambient geological and geochemical conditions (oxidation, reduction, Eh, pH, etc.) determines the concentration of radionuclide in minerals and rocks. A lithological unit, primary and unaltered, in a geological region can be characterized by a certain range of radioactivity (Savosin and Sinitsyn 1965). However, subjected to successive geological episodes there can be considerable changes in the radioactivity of the same. The radionuclides and the decay products of some of them emit highly energetic penetrating radiations like γ-rays. When analyzed, different γ-rays with different energies gives out a spectrum composed of various photo peaks. This γ-ray spectrum can be used to obtain the activity concentration of the respective radionuclides present in the source material. In a similar study by Mittal et al. (2013), the nature of uranium mineralization in SPSZ as well as the efficacy of this zone as an economic uranium prospect was investigated. The study area lies around SPSZ in Purulia district of West Bengal, India. The shear zone is an arc type curvilinear tectonic zone extending from Tamar in the west to Khatra in the east with a strike extension of 120–150 km and an average width of ~4–5 km. To the north of it lies the Chhotanagpur Granite Gneissic Complex (CGGC) and in south it is bounded by the Singhbhum Group.

The area is mainly made up of meta-sedimentary rock suites (Bhattacharya et al. 1991) and the major rock types include Tuffaceous-phyllite, Schist, Quartzite, Pyroxenite, Carbonatite, Quartz-magnetite-apatite, Tourmalinite, Quartz breccias, Alkali granites, Alkaline ultramafics (Chakrabarty et al. 2009). Moderate-to-steep dip foliations, ductile to brittle-ductile characters are also found in the rocks comprising the study area. Initial survey was done using portable GM counter for identifying the sampling locations. Samples were then collected from those sites exhibiting higher radioactivity than background, for γ–ray spectrometric study. All the samples (soil and rock) were crushed, dried, and cooled to the room temperature, and followed by sieving to a particular size, coning, and quartering in order to achieve a homogeneous representation of the bulk samples. Known amounts of the soil and rock samples were sealed in plastic vials, labelled, and kept for a minimum duration of 40 days to attain the secular equilibrium. Subsequently, the samples were then subjected to γ-ray spectrometric analysis for ^{238}U, ^{232}Th, and ^{40}K measurements using HPGe γ-ray spectrometry system (EG & G, ORTEC) with 50 % relative efficiency.

The activity concentration of ^{238}U in the samples ranges from 61.5 ± 2.3 to 276.7 ± 2.6 Bq/kg, ^{232}Th ranges from 66.3 ± 18.9 to 246.9 ± 3.2 Bq/kg, and ^{40}K ranges from 2354.2 ± 3.8 to 6329.7 ± 15.8 Bq/kg, respectively. Uranium content showed a decreasing trend towards the periphery and highest concentration at the central part of the shear zone. It is important to note that the central part of the zone represents a zone of intense shearing; this implies that shearing has an active control on distribution of radionuclides. The uranium activity also showed a similar trend of decrement away from the central part of shear zone. This is because the local soil is the weathered product of rocks and the distribution of radioactive elements is significantly affected by the geological processes like solution activity, weathering, erosion, leaching, etc. (Åkerblom and Mellander 1997). Therefore, soil samples showed a higher radiation level compared to the rocks. The U : Th ratio indicated an enrichment of Uranium in the central part that decreased radially outward. As compared to average Indian values, the concentration of uranium is found higher in the study area. This suggests the possibility of shallow uranium deposits in the area along the shear zone.

References

Åkerblom G, Mellander H (1997) Geology and radon. In: Durrani SA, Ilic´ R (eds) Radon measurements by etched track detectors: applications to radiation protection, earth sciences and the environment. World Scientific press, Singapore, pp 21–49

Aldenkamp FJ, de Meijer RJ, Put LW, Stoop P, (1992) An assessment of in-situ radon exhalation measurements and the relation between free and bound exhalation rate, Radiation Protection Dosimetry 45:449–453

Banerjee KS (2011) Measurement of source term for environmental radioactivity and study of geologic control on radon release in a uranium mineralized zone, Jharkhand, India. Unpublished doctoral dissertation thesis, Indian Institute of Technology Kharagpur, Kharagpur, India, 61

Baranwal VC, Sharma SP (2005) Integrated geophysical studies in the East-Indian geothermal province. Pure Appl Geophys 163:209–227

Baranwal VC, Sharma SP, Sengupta D, Sandilya MK, Bhaumik BK, Guin R, Saha SK (2006) A new high background radiation area in the geothermal region of Eastern Ghats Mobile Belt (EGMB) of Orissa, India. Radiat Meas 41:602–610

Bhattacharya C, Chakraborty A, Banerjee PK (1991) Petrology-geochemistry of apatite-magnetite amphibolites and their role in phosphate mineralisation along the southern shear zone, Puruliya District, West Bengal. Indian J Earth Sci 18:94–109

Chakrabarty A, Sen AK, Ghosh TK (2009) Amphibole—a key indicator mineral for petrogenesis of the Puruliacarbonatite, West Bengal, India. Mineral Petrol 95(1–2):105–112

Cullen TL (1977) A review of Brazilian investigations in areas of high natural radioactivity, Part I: radiometric and dosimetric studies. In: Cullen TL, Penna Franca E (eds) Proceedings of international symposium on areas of high natural radioactivity, 1975. Academia Brasileria De Ciencias, Rio de Janeiro, pp 49–64

de Meijer RJ, Stapel C, Jones DG, Roberts PD, Rozendaal A, Macdonald WG (1997) Improved and new uses of natural radioactivity in mineral exploration and processing. Explor Min Geol 6:105–117

Eichholz GG, Hilborn JW, McMahon C (1953) The determination of uranium and thorium in ores. Can J Phys 31:613–628

Firestone RB, Shirley VS (1998) Table of isotopes, 8th edn. Wiley, New York

Ghiassi-nejad M, Mortazavi SMJ, Cameron JR, Niroomand-rad A, Karam PA (2002) Very high background radiation areas of Ramsar, Iran: preliminary biological studies. Health Phys 82:87–93

Grigorescu EL, Razdolescu AC, Sahagia M, Luca A, Ivan C, Tanase G (2002) Standardization of 152Eu. Appl Radiat Isot 56:435–439

Grosz AE (1983) Applications of total count aeroradiometric maps to the exploration for heavy-mineral deposits in the coastal plain of Virginia. US Geological Survey Prof. Paper 1263

Khandaker MU (2011) High purity germanium detector in gamma-ray spectrometry. Int J Fundam Phys Sci 1(2):42–46

Kundu N, Panigrahi MK, Sharma SP, Tripathy S (2002) Delineation of fluoride contaminated groundwater around a hot spring in Nayagarh, Orissa, India using geochemical and resistivity studies. Environ Geol 43:228–235

Mayya YS (2004) Theory of radon flux into accumulators placed at the soil-atmosphere interface. Radiat Dosim 111:305–318

Mittal S, Guin R, Sharma SP, Sengupta D (2013) Estimation of 238U, 232Th and 40K Concentrations in Rock and Soil Samples Around South Purulia Shear Zone, India. International Journal of Low Radiation 9:110–118

Mohanty AK, Sengupta D, Das SK, Vijayan V, Saha SK (2004a) Natural radioactivity in the newly discovered high background radiation area on the eastern coast of Orissa, India. Radiat Meas 38:153–165

Mohanty AK, Sengupta D, Das SK, Saha SK, Van KV (2004b) Natural radioactivity and radiation exposure in the high background area at Chhatrapur beach placer deposit of Orissa, India, J. Environ. Radioact 75:15–33

Nazaroff W W and Nero Jr. A V (1988). In: John Wileyand Sons Inc. (Ed.), Radon and its decay products in indoor air, pp 57–112, New York.

Noakes JE, Harding JL, Staehle CM (1989) Survey of nuclear methods applied to marine mineral exploration. Mar Min 8:215–244

Paul AC, Pillai PMB, Haridasan P, Radhakrishnan S, Krishnamony S (1998) Population exposure to airborne thorium at the high natural radiation areas in India. J Environ Radioactiv 40:251–259

Rogers VC, Nielson KK (1991) Multiphase radon generation and transport in porous materials. Health Phys 60:807

Sahoo BK, Nathwani D, Eappen KP, Ramachandran TV, Gaware JJ, Mayya YS (2007) Estimation of radon emanation factor in Indian building materials. Radiat Meas 42:1422–1425

Sahoo BK, Mayya YS, Sapra BK, Gaware JJ, Banerjee KS, Kushwaha HS (2010) Radon exhalation studies in an Indian uranium tailings pile. Radiat Meas 45:237–241

Savosin SI, Sinitsyn VI (1965) Application of methods of nuclear geophysics in ore prospecting, exploration, and development. Translated from Atomnaya ~nergiya, Vol. 18, 1, pp 81–84

Sengupta D, Mohanty A K, Das S K and Saha S K (2005) Natural radioactivity in the high background radiation area at Erasama beach placer deposit of Orissa, India, International Congress Series, 1276, pp. 210–211

Shukla VK, Sartendel SJ, Ramachandran TV (2001) Natural radioactivity level in soil from high background radiation areas of Kerala. Radiat Prot Environ 24:437–439

Smith HA Jr, Lucas M (1991) Gamma-ray detectors. In: Reilly D, Ensslin N, Smith Hastings Jr (eds) Passive nondestructive assay of nuclear materials. NuREG/cR-5550, LA-UR-90-732., pp 43–64. http://www.lanl.gov/orgs/n/n1/panda/00326398.pdf. Accessed 24 Sept 2013

Sohrabi M (1993) Recent radiological studies of high level natural radiation areas of Ramsar. In: Proceedings of the International Conference on High Levels of Natural Radiation Areas, 1990, Ramsar, Iran. IAEA Publication Series, IAEA, Vienna

Sengupta D, Mohanty AK, Das SK, Saha SK (2005) Natural radioactivity in the high background radiation area at Erasama beach placer deposit of Orissa, India, International Congress Series, 1276, pp 210–211

Sulekha Rao N, Misra S (2009) Sources of Monazite Sand in Southern Orissa Beach Placer, Eastern India, Journal of Geological Society of India, Vol. 74, September 2009, pp 357–362

Sunta CM, David M, Abani MC, Basu AS, Nambi KSV (1982) Analysis of dosimetry data of high natural radioactivity areas of southwest coast of India. In: Vohra KG et al (ed) The natural radiation environment. Wiley Eastern Limited, India, pp 35–42

UNSCEAR (1993) Sources and effects of ionizing radiation. New York, United Nations Scientific Committee on the Effect of Atomic Radiation

UNSCEAR (2000) Sources and effects of ionizing radiation. New York, United Nations Scientific Committee on the Effect of Atomic Radiation

Wei L, Sugahara T (2000) An introductory overview of the epidemiological study on the population at the high background radiation areas in Yangjiang, China. J Radiat Res 41(Suppl):1–7

Wei L, Zha Y, Tao Z, He W, Chen D, Yuan Y (1993) Epidemiological investigation in high background radiation areas in Yangjiang, China. In: Proceedings of the international conference on high levels of natural radiation areas, Ramsar, Iran, 1990. IAEA Publication Series, IAEA, Vienna, pp 523–547

Arsenic Fate and Transport in the Groundwater-Soil-Plant System: An Understanding of Suitable Rice Paddy Cultivation in Arsenic Enriched Areas

P. K. Sahoo and A. Mukherjee

Abstract Irrigation with arsenic (As)-rich groundwater in agricultural soil is one of the major causes for As contamination in rice. This can be observed in Bangladesh where the highest levels of As were reported in rice grains. However, As risk assessment in rice based on the As contaminated groundwater and soil can be misleading, since As accumulation in plant is controlled by a number of factors in the soil–rhizosphere system. In this chapter, we have discussed the impact of As-rich groundwater on rice cultivation, and the major factors which control the fate of As in the soil–rhizosphere–plant system. Soil parameters such as pH, redox potential, and concentrations of phosphorous, sulfur, silica, and iron affect the availability of As in the soil–rhizosphere zone, thus, should be considered in the assessment of As toxicity for rice. Concentration of silica and phosphate compete with arsenite and arsenate respectively during uptake by plants. Environmental factor such as soil flooding condition is also one of the potential factors influencing arsenic accumulation in rice by increasing As mobility in the soil–rhizosphere. In biological factors, rice genotype, which controls the root aeration, is significant in affecting accumulation of As. Therefore, selection of suitable rice genotype besides growing rice in aerobic conditions would be important to minimize As accumulation in rice. The translocation of As from roots to above-ground parts is dependent on As speciation. Although, organic As is more readily translocated, its uptake is much lower compared to inorganic As; thus, inorganic As predominates in grains.

Keywords Arsenic · Rice · Bioavailability · Soil–Rhizosphere system · As uptake mechanism

P. K. Sahoo (✉)
Vale Institute of Technology, Boaventura da Silva, Belém, Pará 66055-090 Brazil
e-mail: prafulla.iitkgp@gmail.com; prafulla.sahoo@vale.com

A. Mukherjee
Department of Geology & Geophysics, Indian Institute of Technology Kharagpur, Kharagpur, West Bengal 721302, India

D. Sengupta (ed.), *Recent Trends in Modelling of Environmental Contaminants*,
DOI 10.1007/978-81-322-1783-1_2, © Springer India 2014

1 Introduction

Arsenic (As) is a ubiquitous element that ranks 20th among the most abundant elements in the earth's crust (Cullen and Reimer 1989). Its atomic number is 33, and the atomic weight is 74.92. Arsenic is a redox-sensitive element, which exists mainly in four oxidation states, + V (arsenate), + III (arsenite), 0 (arsenic), and − III (arsine). Unlike other toxic elements (Pb, Cu, and Co), As occurs as cation (positively charged ion) and oxyanion (negatively charged ions with oxygen) species. It can occur in the environment both in organic and inorganic forms. Inorganic forms usually associate with many minerals and other elements, especially oxygen, sulfur, and chlorine, while organic forms associate with carbon and hydrogen. The most important inorganic species are arsenate As(V) and arsenite As(III), while the most important organic species are monomethylarsonic acid (MMA) and dimethylarsinic acid (DMA). The toxicity of As is mainly dependent on the forms of arsenic. Inorganic As species are more toxic to living organisms than organic forms (Meharg and Whitaker 2002). Arsenic toxicity in humans can cause severe health problems ranging from skin lesions to cancers of the brain, liver, kidney, and stomach, and many other cardiovascular and neurological diseases (Smith et al. 1992; Ng et al. 2003; Martinez et al. 2011).

Arsenic naturally occurs in many environmental media, such as rocks, soil, sediments, and surface/groundwater, and it can further be released into the aquatic and terrestrial ecosystem via natural and anthropogenic activities (Lombi et al. 2000). According to the World Health Organization (WHO) guideline, the baseline limit of As concentration in safe drinking water is 10 μg/L, which is similar to the European Union (EU), US Environmental Protection Agency (USEPA), Netherlands, and Germany limits of As in drinking water (Ahsan and DelValls 2011). However, in Bangladesh and many other developing countries, such as India, China, and Nepal, it is set to 50 μg/L (World Bank 2005). Currently, Australia has set a strict national standard of 7 μg/L for acceptable As concentration in drinking water (World Bank 2005). Arsenic poisoning via drinking of As contaminated groundwater is well documented (Smedley and Kinniburgh 2002); however, recent investigation revealed that rice, which is the staple food for nearly half of the world's population, is another potential As exposure pathway to human health (William et al. 2005; Zavala et al. 2008). It can add significant amounts of dietary intake of As than most crops because rice plants have the special ability to soak up As from soil compared to other cereals (Heikens 2006), and use of arsenic-contaminated groundwater for irrigation of rice cultivation (Heikens et al. 2007). However, accumulation of As in rice is influenced by various environmental, geochemical, and biological factors which controls As solubility, bioavailability, and uptake in the soil–rhizosphere–plant system (Walter and Wenzel 2002; Zhao et al. 2010).

This chapter presents the As levels in rice from different parts of the world, assesses the impact of irrigation with As rich-groundwater on soil and paddy rice, and discusses the major factors controlling As behavior in the soil–rhizosphere–plant systems, along with As toxicity to plants. Understanding of these aspects may be helpful to reduce As accumulation in rice grains.

2 Arsenic Concentration in Rice

Several surveys of As concentration in rice grains have been carried out in various countries (Table 1). It was found that As concentration in rice varied widely. The highest level of As, up to 1.85 mg/kg (0.05–1.85 mg/kg), was found in Bangladesh (Meharg and Rahman 2003). Other Asian countries also reported high levels (ranges) of arsenic in rice: China, 0.46–1.18 mg/kg (Sun et al. 2008); Taiwan, 0.1–0.63 mg/kg (Lin et al. 2004); Thailand, 0.06–0.5 mg/kg (Adamako et al. 2011); Japan, 0.07–0.42 mg/kg (Meharg et al. 2009); Korea, 0.24–0.72 mg/kg (Lee et al. 2008); and India, particularly from West Bengal, 0.085–0.66 mg/kg (Roychowdhury et al. 2002). Besides Asian rice, significant arsenic levels were reported from Venezuela (0.19–0.46 mg/kg) by Schoof et al. (1998), France (0.12–0.61 mg/kg) by Williams et al. (2007a), and USA (0.16–0.71 mg/kg) by Zavala et al. (2008).

Rice As values (ranges) from Europe, 0.13–20 mg/kg (Williams et al. 2005); Spain, 0.07–0.21 mg/kg (Williams et al. 2007a); Italy, 0.19–0.22 mg/kg (Williams et al. 2005); and Ghana <0.01–0.15 mg/kg (Adomako et al. 2011) fall into intermediate category. Some parts of the Asian rice also fall into this category, as follows: Philippines, 0–0.25 mg/kg (Williams et al. 2006); Vietnam, 0.09–0.17 mg/kg (Adomako et al. 2011); and Pakistan, 0.01–0.2 mg/kg (Adamako et al. 2011). On contrary, rice from Australia, 0.02–0.04 mg/kg (Williams et al. 2006); Canada, 0.02 mg/kg (Williams et al. 2005); and Egypt (0.02–0.08 mg/kg) (Meharg et al. 2007) is found to contain the lowest levels of As contamination (Table 1). The overall range found among all samples, the lowest grain As (0.01 mg/kg) is nearly two order magnitude difference with that of the highest concentration (1.835 mg/kg). According to the information available to us, in terms of As contamination levels in rice, there is no limit set by EU and US (Francesconi 2007). However, China has regulated the maximum contamination level (MCL) of As in rice as 0.15 mg/kg (Zhu et al. 2008). On comparison with Table 1, it was noticed that the mean values of most of the Asian, USA, and France rice were exceeded the MCL. However, higher enrichment in Asian rice, particularly in Bangladesh, which can be the significant source of dietary As for population of this area and population from other countries import rice from this region.

3 Arsenic in Groundwater

Arsenic contamination of groundwater has been indentified in many countries, including Argentina, Bangladesh, China, Cambodia, India, Japan, Mongolia, Mexico, Nepal, Taiwan, Thailand, Vietnam, and USA (Mandal and Suzuki 2002; Smedley and Kinniburgh 2002). There are a number of people are at risk of chronic arsenic poisoning from drinking As-contaminated water in several parts of the world, however highest number of people at risk from Ganges Delta (Bangladesh and West Bengal, India) (Fig. 1) (Smedley and Kinniburgh 2005; Garelick and Jones 2008).

Table 1 Arsenic concentration (in mg/kg) in rice grains from different countries

Country	Maximum	Minimum	Average	No. of samples	Study (Author/year)
Bangladesh	1.835	0.058	0.49	13	Meharg and Rehman (2003)
China	1.18	0.46	0.82	2	Sun et al. (2008)
India	0.66	0.085	0.3	10	Roychowdhury et al. (2002)
Taiwan	0.63	0.1	0.1	5	Lin et al. (2004)
Thailand	0.5	0.06	0.15	53	Adomako et al. (2011)
Korea	0.72	0.24	0.41	–	Lee et al. (2008)
Japan	0.42	0.07	0.19	26	Meharg et al. (2009)
Vietanam	0.17	0.09	0.12	18	Adomako et al. (2011)
Philippines	0.25	0	0.07	–	Williams et al. (2006)
Pakistan	0.2	0.01	0.09	5	Adomako et al. (2011)
Venezuela	0.46	0.19	0.3	4	Schoof et al. (1998)
Egypt	0.08	0.02	0.05	108	Meharg et al. (2007)
Spain	0.21	0.07	0.13	10	Williams et al. (2007a)
Europe	0.2	0.13	0.15		Williams et al. (2005)
France	0.61	0.12	0.32	22	Williams et al. (2007a)
USA	0.71	0.16	0.29	24	Zavela et al. (2008)
Italy	0.22	0.19	0.14	–	Williams et al. (2005)
Ghana	0.15	<0.01	–	7	Adomako et al. (2011)
Australia	0.04	0.02	0.03	5	Williams et al. (2006)
Canada	–	–	0.02	–	Williams et al. (2005)
Lebanon	0.07	0.01	0.04	11	Adomako et al. (2011)

"–": not avilable

Scores of people from China, Vietnam, Taiwan, Chile, Argentina, and Mexico are likely at risk as well (Fig. 1).

3.1 Arsenic Mobility in Groundwater

Based on the As geochemistry, there are three major mechanisms controlling As mobility in the groundwater, which have been reported by various workers. These are:

Oxidation of Pyrite Mobilization of As due to oxidation of arsenic-rich pyrite minerals associated with sediment and rock materials have been identified by many workers (Williams et al. 1996; Kim et al. 2012) (Table 2). Insoluble As-bearing minerals such as arsenopyrite (FeAsS) are rapidly oxidized when in contact with oxygen, releasing soluble As(III) and Fe(II) (Smedley and Kinniburgh 2002). The oxidation of the As-bearing minerals is dependent on availability of oxygen. High oxygen in pyritiferrous system suffers by the excess pumping and water-table draw-down. Chakraborti et al. (2001) and Sikdar et al. (2001) observed that water table declined in the Bengal basin because of the agricultural pumping that began in

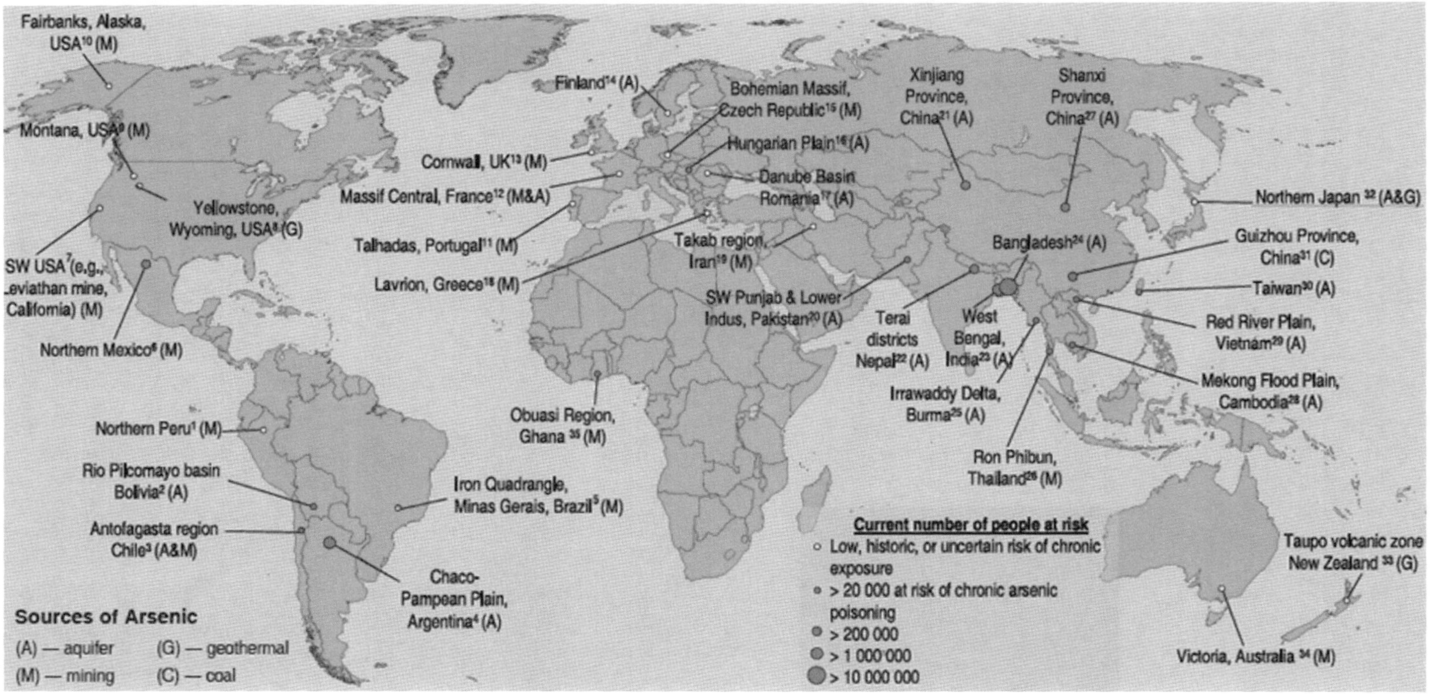

Fig. 1 Worldwide distribution of arsenic contaminated regions, illustrating numbers of people at risk of chronic As exposure. (Adapted from Garelick and Jones 2008)

Table 2 As concentration in groundwater (μg/L) relations with major release mechanism reported from different countries

Country	As in groundwater	Major release mechanism	References
USA	Up to 12,000	Oxidation of As-rich sulfides	Schreiber et al. (2000)
Thailand	1.25–5114	Oxidation of As-rich sulfides	Williams et al. (1996)
Korea	23–178	Oxidation of As-rich sulfides	Kim et al. (2012)
Bangladesh	Up to 640	Reductive dissolution of Fe or Mn (hydro)oxides	Harvey et al. (2002)
Bangladesh	6–934	Reductive dissolution of Fe or Mn (hydro)oxides	Bibi et al. (2008)
India	0.77–1059	Reductive dissolution of Fe or Mn (hydro)oxides	Bhattacharyya et al. (2008)
Nepal	<10–740	Reductive dissolution of Fe or Mn (hydro)oxides	Gurung et al. (2005)
China	0.6–572	Reductive dissolution of Fe or Mn (hydro)oxides	Guo et al. (2008)
Taiwan	Up to 12,000	Reductive dissolution of Fe or Mn (hydro)oxides	Chen and Liu (2007)
Vietnam	1–741	Reductive dissolution of Fe or Mn (hydro)oxides, oxidation of As-rich sulfides[a]	Naguyen and Itoi (2009)
USA	11–2620	Reductive dissolution of Fe or Mn (hydro)oxides, Desorption from Fe (hydro)oxides	Welch and Lico (1998)
India	0.82–538	Reductive dissolution of Fe or Mn (hydro)oxides, Desorption from Fe (hydro)oxides[a]	Kar et al. (2010)
Argentina	7–14,969	Desorption from Fe (hydro)oxides	Bhattacharya et al. (2006)

[a] Minor mechanism

the 1970s. Subsequently, in 1978, first As contamination in the Bengal Basin was detected. The rate of oxidation of pyrite is dependent on the redox potential (Eh) and pH. The release of Fe from pyrite oxidation can form Fe oxides that can immobilize As.

Reductive Dissolution of Fe Oxyhydroxides Arsenic in Fe-oxyhydroxides is mobilized when the environment becomes anaerobic (Smedley and Kinniburgh 2002). Bhattacharya et al. (1997) first proposed the reductive dissolution of Fe oxyhydroxides process for arsenic release. After that, many other studies have been reported this mechanism (Welch and Lico 1998; Harvey et al. 2002; Bibi et al. 2008; Gurung et al. 2005; Chen and Liu 2007; Bhattacharyya et al. 2008; Guo et al. 2008; Kar et al. 2010) (Table 2). McArthur et al. (2001) stated that the good correlation between As and HCO_3^- in groundwater is an indication of reduction. Other studies from Bengal basin reported that low Eh, O_2, NO_3^-, and SO_4^{2-}, and high Fe and Mn in groundwater are the indication of reducing conditions (Bhattacharya et al. 1997; Dowling et al. 2002). Similarly, Kim et al. (2009) and Sahoo et al. (2013) stated that the higher arsenic concentration in groundwater associated with lower Eh, NO_3, and

SO_4^{2-}, and higher alkalinity, Fe, and PO_4 is the indication of reductive dissolution of Fe (hydr)oxides.

Desorption of Fe-Oxydroxides Besides anaerobic conditions for releasing As from Fe oxyhydroxides, there is also effect of pH, particularly at pH > 8.5, which can cause mobilization of As from Fe-oxides. This is indentified by some workers (Bhattacharya et al. 2006; Kar et al. 2010) (Table 2).

4 Impact of As-Rich Groundwater on Paddy Cultivation

Arsenic concentration in paddy fields can occur naturally via weathering processes, and/or by anthropogenic activities such as mining (Zhu et al. 2008), pesticide use, fertilizer application (Williams et al. 2007b), and irrigation with As contaminated groundwater (Mehrag and Rahman 2003; Williams et al. 2006). However, irrigation with As contaminated groundwater plays a dominant role, especially in Asian countries where groundwater-irrigated rice cultivation has increased dramatically over the past few decades to maintain the country's food security. In the long term, use of this As-contaminated irrigation water can cause elevated As concentration in paddy soils and crops (Heikens et al. 2007; Hossain et al. 2008; Baig et al. 2011). Several field studies from Bangladesh and West Bengal showed that irrigation with As-rich groundwater increases the As levels in the top soil (Meharg and Rahman 2003; Norra et al. 2005; Williams et al. 2006; Roychowdhury et al. 2002). Studies from Argentina, Chile, and Taiwan (Chen et al. 2002; Tu and Ma 2004) also reported that As contents in agricultural soils are significantly higher than the normal levels due to irrigation with As-rich groundwater.

There are a number of studies that investigated the impact of irrigation with As-rich groundwater on soil and its subsequent impact on rice grains (Mehrag and Rahman 2003; Williams et al. 2006; Dittmar et al. 2010). Meharg and Rahman (2003) reported the positive relations among As in irrigation water, soil, and rice from Bangladesh. Williams et al. (2006) did an extensive sampling of rice throughout Bangladesh, and found that high As concentration in rice is associated with As contaminated irrigation water. In another study from Murshidabad district of Bangladesh, Roychowdhury et al. (2005) reported the positive relations among As concentrations in irrigated water, soil, and plants. Recently, Singh et al. (2010) observed the positive relations between irrigation water and rice. However, this relationship is conflicting in some cases. Norra et al. (2005) found no significant relationship between rice grains from West Bengal, India, where plants are irrigated with low and high concentrations of As containing water. Van Geen et al. (2006) reported that there was no evidence of a proportional transfer As from soil to rice grains, despite the soil irrigated with As rich-groundwater. Furthermore, Williams et al. (2007a) reported elevated As concentration in rice grains in background paddy soil. These observations indicate that As content in rice is not directly dependent on total concentration of As in soil and irrigation water, but may be with other factors in the soil–rhizosphere–plant system, which are relevant.

5 Arsenic Behavior in Soil–Plant System

5.1 Soil

Arsenic is naturally present in soil. Its average concentration in non-contaminated soils is ~5 mg/kg; however, the concentration can increase up to 27,000 mg/kg or more than that in contaminated soil (Mclaren et al. 1996). The solubility and bioavailability of As in soil is influenced by various factors (Fitz and Wenzel 2002; Zhao et al. 2009), which are discussed below.

As Speciation

Arsenic in soil exists in both inorganic and organic forms. The most common inorganic As species in soil are As(V) and As(III), while organic forms are MMA and DMA (Zhao et al. 2009). As(III) is more toxic than As(V), and also much more toxic than MMA or DMA (Zhao et al. 2010). Inorganic As species are dominating in paddy soils, whereas organic species are in low quantities (Fitz and Wenzel 2002). Inorganic species can be converted to organic form by methylation linked with microbial action in paddy soil (Takamatsu et al. 1982). Each species has different solubility and bioavailability. Marine et al. (1992) reported that the As availability to rice plants followed the order As(III) > MMA > As(V) > DMA. Meharg and Whitaker (2002) also observed that both As(III) and MMA are more available to rice plants. This indicates that speciation of As in soil environment is essential to assess As toxicity to plants.

Redox Potential

The balance of reducing and oxidizing condition (redox status) in the soil is important because it controls arsenic mobility and speciation (Fitz and Wenzel 2002). Under oxidizing conditions (aerobic), As(V) usually dominates, which has strong affinity for soil minerals such as Fe-oxhydroxides, leading to decrease As solubility and bioavailability to plants (Takahashi et al. 2004; Xu et al. 2008). However, under reducing conditions, such as soil flooding, As(III) is the most common inorganic species, and its mobility is sharply increased in the soil-rhizopshere (Takahashi et al. 2004). This is due to the reductive dissolution of Fe-oxyhydroxides, and relatively high abundance of Fe reducing bacteria and algae in reduced soil, which help in As solubility via reduction of As(V) and methyl As species to more soluble As(III) species (Horneman et al. 2004; Mahimairaja et al. 2005). Therefore, in flooded soil, As is readily available for plant uptake. This may be the cause for higher As content in lowland crops usually growing in a reducing environment, such as paddy rice, than upland cereal crops (Williams et al. 2007a). However, changing irrigation practice can manage As uptake in plants (Das et al. 2008). Somennahally et al. (2011) conducted an experiment by changing the irrigation process, and found that rice takes up less As in intermittently flooding than continuous flooding. Other studies also supported that irrigation practice can significantly control As accumulation in plants (Arao et al. 2009; Roberts et al. 2011; Rahaman et al. 2011; Talukdar et al. 2011).

pH

The sorption of As to metal-oxyhydroxides is generally affected by soil pH (Fitz and Wenzel 2002; Quazi et al. 2011), but there is not any universal agreement on this issue. Arsenic (AsV) tends to be sorbed by Fe–Al oxyhydroxides at near neutral to acidic pH (Mukherjee et al. 2009). However, at extremely acidic pH ($pH<3$), arsenic mobility can increase due to the dissolution of arsenic binding species, such as Fe and Al oxycompounds (Signes-Pastor et al. 2007). Furthermore, the mobility of As in soil can increase at higher soil pH, above pH 8.5, because at high pH, mineral surfaces (mainly Fe-oxides) become progressively negatively charged that facilitate desorption of As from Fe-oxides (Marine et al. 1992; Streat et al. 2008), and increases labile As in the rhizosphere, leading to more As accumulation in plants (Fitz and Wenzel 2002). Campbell et al. (1985) observed that uptake of arsenic in plants increases at higher soil pH. In a recent work, Ahmed et al. (2011) also observed the positive relationship between grain As and soil pH.

Organic Matter

Organic matter (OM) can have a profound effect on As solubility in the soil since it tends to form insoluble and soluble complexes with As (Mukhaopadhyay and Sanyal 2004). Pikaray et al. (2005) reported that organic matter has a greater potential for As sorption due to formation of organo-As complex. Thus, high OM containing soil can reduce As availability to plants. Considering this effect, Rahaman et al. (2011) conducted an experiment, and observed that organic amendments significantly reduce arsenic content in rice grains. This can be supported by Fu et al. (2011) who found the negative correlation between As concentration in rice grain and soil OM.

On the contrary, organic matter can increase arsenic mobility in soil by forming water-soluble complexes with As, leading to inhibit As sorption on mineral surfaces (Bauer and Blodau 2006). Dissolved organic matter contains negative charge, which has high potential to compete with As for sorption sites in the soil, thus, it can increase As mobility (Lin et al. 2008). Furthermore, dissolved organic matter can promote As release by changing the redox chemistry of site surfaces and As species (Wang and Mulligan 2006). Selim Reza et al. (2010) reported that As is released by the reductive dissolution of Fe-Ox hydroxides linked to organic matter. Others studies also supported that As solubility in soil is positively related with organic matter (Hunag et al. 2006; Harvey et al. 2002; Rowland et al. 2006). This may be the cause for the positive relationship between rice and OM reported by Bhattacharya et al. (2010a), and elevated grain As in paddy soil even at background soil (Williams et al. 2011).

Soil Texture

Soil texture is another important factor that can influence arsenic mobility (Fitz and Wenzel 2002; Quazi et al. 2011). In general, clay or finer texture soils have much more surface area than coarse or sandy soils. In addition, Fe oxides are mainly present in the clay size soil fraction; therefore, clayey soils can have a higher As retention potential compared to sandy soils (Fitz and Wenzel 2002; Mahimairaja et al. 2005; Heikens et al. 2007), and soils with clayey texture are supposed to be less toxic of arsenic to plants compared to sandy soils (Heikens et al. 2007). Sheppard (1992) observed that phytotoxicity of As is five times more in sands and loams than clay soils. Similar studies have reported that plant uptake of As is high in loamy sand and low in silty clay loam (Woolson 1973; O'Neil 1995).

As Bound to Fe–Mn Oxides

Oxides of Fe and Mn are common in soils, and these phases are very efficient in sorbing As due to their high sorption capacity (Dudas 1987). However, their sorption property is strongly dependent on the environmental conditions (Mukherjee et al. 2009). Under aerobic conditions, the affinity of the oxyhydroxides is very high for As. However, under flooding conditions, oxyhydroxides release As from soil by the reductive dissolution of Fe oxydroxides, leading to more As available for plant uptake (Fitz and Wenzel 2002; Takahashi et al. 2004). Fu et al. (2011) found the positive relations between Fe–Mn bound As and grain As. Moreover, amorphous or poorly crystalline Fe oxides are much more readily dissolved than the crystalline Fe oxides under flooding conditions (Biswas et al. 2003). Ahmed et al. (2011) observed the significant positive relations between As in Fe-amorous fraction and rice. This indicates that As content in amorous/poorly crystalline Fe-oxides in soil can be an important factor for rice As.

Phosphate

Phosphate (PO_4) is an important parameter in agricultural field not only as an important nutrient, but also it controls As phytoavailability (Woolson et al. 1973; Peryea and Kammereck 1995; Bodgan and Schenk 2009). Generally, phosphate is a chemical analog of AsV (Fitz and Wenzel 2002); thus, it competes with As(V) to occupy the same sorption sites in soils, leading to increase As solubility (Smith et al. 2002; Fitz and Wenzel 2002). Many studies reported that addition of PO_4 increases As mobility in soil (Peryea and Kammereck 1995; Smith et al. 2002; Campos 2002; Cao et al. 2003; Signes-Pastor et al. 2007), which can subsequently increase As uptake by plants (Bogdan and Schenk 2009; Islam and Jahiruddin 2010). However, it is conflicting at some point, since, PO_4 also competes with As(V) at the same transport during uptake, and can reduce arsenate accumulation in plants (Fitz and Wenzel 2002). This is supported by several works (Abedin et al. 2002a; Tu and Ma 2003; Lihong and Guilan 2009).

Sulfur

Sulfate has a high sorption affinity for arsenic. Under reducing conditions, it forms insoluble As-sulphide minerals, thus less As in the soil–rhizosphere (Signes-Pastor et al. 2006). In addition, sulfur containing thiol (SH) compounds make chelation of As in plant roots, that can inhibit As transfer to grains (Raab et al. 2005). Zhang et al. (2011) highlights the important role of sulfur on arsenic accumulation in rice. They reported that increase of sulfur content reduced the As content in rice. Hu et al. (2007) in a pot experiment also found that the addition of sulfur to soil considerably decreased the As content of rice shoots. This result hypothesized that application of S fertilizers in paddy fields can obtain high yields, and additionally benefit of restricting As uptake by plants. However, this needs to be studied under field conditions.

Fe-Plaque

In flooded paddy fields, rice plants release oxygen to their root zones through aerenchyma, which leads to oxidized Fe(II) concentrations, and form Fe-oxhydroxide coating, termed as Fe-plaque, on the root surface (Fig. 2). This plaque is known to control As uptake by rice plants (Chen et al. 2005; Zhao et al. 2010). Numerous studies have reported that Fe-plaque scavenges significant amounts of As in root zones and decreases As uptake by plants (Hossain et al. 2009; Sefferth et al. 2010; Garnier et al. 2010). However, the formation of Fe-plaque is strongly influenced by soil properties. Hu et al. (2005) observed that the formation of Fe-plaque is decreased on the rice roots after the application of phosphate. Liu et al. (2004) found that phosphorous (P) sufficiency had little iron plaque on the rice roots while P-deprivation in solution enhanced the formation of iron plaque on the roots. This may be because the low P in plant tissue increases oxygen release from rice roots to the rhizosphere (Kirk and Du 1997), which stimulates the formation of iron plaque. Nitrate concentrations also affect Fe-plaque formation in the rhizosphere (Chen et al. 2008), because it is an oxidizer, which reduces Fe(II) concentration in soil solution leading to decrease Fe-plaque formation in the rhizosphere (Chen et al. 2008). However, in spite of decreasing Fe-plaque, it reduces As uptake by rice (Chen et al. 2008). This is due to NO_3^- stimulates Fe(II) oxidation and/or inhibits Fe(III) reduction, which sequesters mobile As on Fe(III) minerals in soil and subsequently make unavailable for plants. On the other hand, the amendment of NH_4 enhances Fe(III) reduction, which increases Fe(II) concentration in the rhizosphere, leading to more Fe-plaque formation on the root surface (Chen et al. 2008).

Nevertheless, accumulation of As on Fe plaque can be a source for As for plants (Voegelin et al. 2007) depending on the localized soil condition (Somennahally et al. 2011). Further, Seyfferth et al. (2010) found that Fe-plaque does not coat many of the young roots or the younger portion of mature roots, which are important for solute uptake. Thus, Fe-plaque does not directly prevent As supply and uptake by rice, but rather serves as a bulk scavenger of As at the root base.

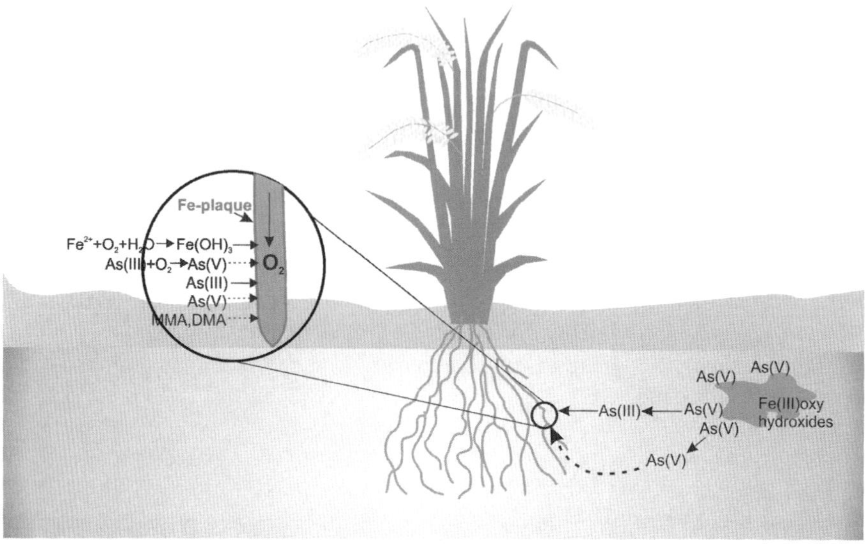

Fig. 2 Formation of Fe-plaque on rice roots in the soil–rhizosphere, and the arsenic uptake by rice plants. *Arrows* with solid and broken lines indicate major and minor processes, respectively. (Modified after Zhao et al. 2010)

5.2 Plants

The pathway of arsenic uptake, accumulation, and transformation can vary within and among species (Heikens 2006), which are discussed below.

Rice Varieties

Arsenic uptake by plants is varied with respect to rice varieties (Table 3). Meharg and Rehman (2003) found the significant variation of As accumulation among the rice varieties; BR11 was the highest accumulator, collected from Bangladesh. In a study from West Bengal, Bhattacharya et al. (2010b) reported that among the rice varieties, IR 50, White Minikit, and Red Minikit were the higher accumulators of As compared to Nayanmani, Jaya, Ratna, Ganga-kaveri, and Lal Sanna. Ahmed et al. (2011) also recently reported the As variation among the rice varieties. Though, there are a number of factors that can influence this variation, Zhang and Duan (2008) stated that rice genotype is the one of the dominating ones. In this regard, Norton et al. (2009a) conducted a field based experiment in Bangladesh by selecting 76 cultivars, and noticed 4–5-fold variations in grain As among genotypes, higher accumulation in landraces cultivars compared to BRRI. The same authors conducted another study in a multiple environment (Norton et al. 2009b). They cultivated 13 cultivars at 2 field sites each in Bangladesh, India, and China, and found 4–5-fold variations in grain As concentration among 7 rice cultivars grown

Table 3 Arsenic variations in rice grains (mg/kg) with respect to rice varieties

Bangladesh (Meharg et al. 2003)	
Rice variety	Grain As (mg/kg)
Khatobada	0.058
Pajam	0.082
Swarna	0.096
BR11	0.203
BR11	1.830
BR11	1.740
BR11	1.770
BR8	0.078
BR11	0.185
Gocha	0.075
Kalla	0.117
Swarna	0.096
WB, India (Bhattacharya et al. 2010b)	
IR 50	0.29
Ratna	0.20
Red minikit	0.24
White minikit	0.31
Ganga-kaveri	0.19
Jaya	0.14
Lal Sanna	0.20
Nayanmani	0.16
Barisal, Bangladesh (Ahmed et al. 2011)	
BR3	0.078
BR4	0.068
BR5	0.091
BR10	0.095
BR11	0.086
BR22	0.105
BR23	0.068
BR25	0.130
BR32	0.140
BR33	0.215
BR34	0.110
BR39	0.139
BR40	0.109
BR41	0.124

in 2 paddy fields in Bangladesh, and a significant correlation between 2 different As contamination sites in Bangladesh and India among the common genotypes. Chaturvedi (2006) and Ahmed et al. (2011) also reported that genotype is one of the major causes for As variation in rice plants. This is may be because genotype controls the physiological or morphological attributes of the root systems, which influence the root aeration capacity to form iron plaque on roots (Wu et al. 2011;

Dwivedi et al. 2011). Therefore, selection of suitable rice genotype based on root aeration capacity is one of the important factors to control As uptake in rice.

Uptake Mechanisms

All As species can be taken up by rice roots; however, inorganic species are taken up in a much higher rate than organic species (Abedin et al. 2002a). In inorganic As, As(V) takes up in the plasma membrane via phosphate transport pathway (Meharg 2004). This mechanism involves cotransport of phosphate or arsenate and protons, with with stoichiometry of at least 2 H^+ for each $H_2PO_4^-$ or H_2AsO_4 (Ullrich-Ebirius et al. 1989). Plants contain a number of phosphate transporters, most of are in roots (Zhao et al. 2009). Due to the competition of both ions in the same transport, it has been hypothesized that increasing phosphate supply can decrease As uptake (Zhao et al. 2009). This is supported by several studies (Abedin et al. 2002a; Tu and Ma 2003; Lihong and Guilan 2009). However, phosphate is not an analog of As(III), thus the presence of phosphate in anaerobic soils is probably less relevant in terms of As behavior (Takahashi et al. 2004). In a pot experiment, Abedin et al. (2002b) observed that no response of the effect phosphate application on As uptake in rice under flooding condition. This may be due to the conversion of soil As(V) to As(III) in reducing environment, and subsequent uptake of As(III) by plant, which leads to lack of effect of phosphate. This is consistent with other studies (Wang et al. 2002; Hu et al. 2005). However, there are conflicts in certain cases since some studies showed that addition of phosphorus enhances the As accumulation in the grains of rice in flooded condition (Jahiruddin et al. 2004; Hossain et al. 2009). In addition, it is not known if arsenite can be oxidized to arsenate in the root rhizosphere, since the rice root is known to develop oxidized conditions proximal to the root zones (Armstrong 1967), and competing with phosphate for plant uptake. Therefore, it cannot be excluded that phosphate does not influence arsenite behavior in the rhizosphere where aerobic conditions can occur under flooded conditions. However, a detail investigation is still required in this aspect.

As(III) is actively taken up by roots via aquaporin channels, which is used for silica transport in the roots (Ma et al. 2008). Since both ions compete during uptake at the same transport (Zhao et al. 2009), the high silica availability in the soil solution reduced the arsenite uptake by rice (Bogdan and Schenk 2008; Guo et al. 2007, 2009). This implies that Si amendment of paddy soils could possibly reduce As(III) uptake in rice plants. Reason for the inhibiting effect of Si is probably because the supply of Si reduces the expression of Si transporters, Lsi1 and Lsi2, which mediate arsenite uptake as well (Ma et al. 2006, 2007, 2008). However, these mechanisms need to be studied further. The ratio of the plant-available Si to arsenite in pore-water may be useful to predict the arsenic acumination in rice.

Translocation and Accumulation of As

Several studies (Table 4) reported that the translocation of As from the roots to the above-ground parts is significantly limited. Bhattacharya et al. (2009) conducted a micro level study from West Bengal, India, and noticed the As concentration decreasing in plant parts following the order: root > stem > leaves > grain, regardless of the sampling locations (Table 4). Similar observations were made by Bhattacharya et al. (2010b) and Biswas and Santra (2011) from the Gangetic West Bengal, India (Table 4). Das et al. (2004) reported that in rice plant, 73 % of total As is deposited in roots, while in stems and grain, it is reduced to 22 and 4 %, respectively. This decrease in As transfer may be influenced by the plant physiological processes (Abedin et al. 2002b; Williams et al. 2007a).

Among all species, organic As is more readily translocated, but, its uptake is very lower compared to inorganic As (Carbonell et al. 1998). Therefore, inorganic As dominates in rice grains (Zavala et al. 2008). According to Heikens (2006), about 50 % of the total As in grains is inorganic, though it is varied from 10 to 90 %. Compared to various countries, rice from India and Bangladesh contains highest percentage of inorganic As (Adomako et al. 2009), implying rice from these regions are at a higher risk than other parts of world, as this is the main carcinogenic component (NRC 2001). William et al. (2005) reported that the average percentage of inorganic As is near about 80 % in Dhaka. With the information on As speciation in rice, the potential risk of the rice to the human diet can be estimated.

Phytotoxicity

The As exposure to rice plants leads to phytotoxicity, often resulting in straighthead disease, which is a physiological disorder of rice (Zhao et al. 2009; Van Geen et al. 2006). Yan et al. (2005) and Williams (2003) reported "straighthead" disease is considered to be an indicator of As toxicity in the USA and Australia. Recently, this disease has been noticed in rice from Faridpur, Bangladesh (Duxbury and Panaullah 2007). The symptom of this disease includes shortened plant height, delayed heading, reduction of panicle formation and panicle length, sterile florets, and reduced grain yields (Williams 2003; Rahman et al. 2008). However, this disease may be influenced by soil type and cultivation practices. Frans et al. (1988) found that straighthead disease is frequently noticed on sandy loam soils but rarely on clay soils, and often in fields where undecayed vegetation had been ploughed into soils just before planting rice. Williams (2003) reported that though straighthead disease was observed over a wide range of Australian soils, the toxicity was worse in soils with low nitrogen, and in those after pasture or where previous crop stubble had been incorporated in soils.

Table 4 Arsenic accumulation (mg/kg) in different parts of the rice plant

Regions	No. of samples	Root	Straw	Husk	Grain
Bhattacharya et al. (2009)					
Tatla-1	3	8.32	3.58	0.96	0.52
Tatla-2	3	10.48	1.9	0.94	0.59
Ghetugachi	4	7.74	1.88	0.74	0.41
Dubra	3	10.42	2.96	1.35	0.73
Dewli	4	9.02	1.17	0.67	0.25
Hingrara	4	8.93	3.05	0.56	0.47
Shilinda-1	4	12.22	1.94	0.71	0.58
Shilinda-2	3	7.19	2.62	0.79	0.3
Chanduria-1	4	10.09	1.47	0.89	0.72
Chanduria-2	3	18.63	1.41	1.34	0.6
Madanpur-1	3	10.89	1.17	0.99	0.66
Madanpur-2	3	16.83	3.09	1.28	0.69
Sarati	4	11.85	3.35	0.91	0.42
Saguna	4	9.5	1.79	0.74	0.34
Simurali	3	17.06	4.15	0.74	0.48
Routari	3	14.09	1.17	1.05	0.54
Kancharpara	3	9.33	2.13	0.78	0.65
Bhattacharya et al. (2010a) (Boro rice)					
Chakdaha	18	10.45	1.41	0.67	0.23
Ranaghat-I	12	18.93	1.47	0.79	0.3
Shantipur	12	28.63	2.13	0.99	0.4
Krishnagar	12	11.85	1.79	0.74	0.24
Haringhata	9	14.09	1.34	1.05	0.54
Contai	5	0.44	0.08	0.07	0.05
Biswas and Santra (2011)					
Chakdah	105	11.07	5.24	0.96	0.59
Haringhata	105	15.01	3.82	1.67	0.81
Ranaghat	105	17.61	4.24	0.71	0.49
Tehatta	105	13.01	5.98	1.13	0.4
Karimpur	105	11.92	3.57	1.79	0.52
Krisnanagar	105	9.79	2.52	0.97	0.1
Hanskhali	105	13.67	5.97	0.72	0.51
Control	105	2.39	1.9	0.02	0.03

6 Conclusions

Arsenic contaminated groundwater is not only posing a risk via drinking of water but also a serious threat to sustainable agriculture. This chapter discussed the As levels in rice in several countries, effect of As-rich groundwater on paddy soils and rice grains, and the factors controlling As mobilization and uptake in the soil–rhizosphere system, along with the toxic effects of As on rice plants.

Arsenic levels in rice significantly varied among countries; higher enrichment in Asian rice, particularly in Bangladesh. Therefore, rice from this region could add significant levels of As in the food stuffs. Arsenic contamination in rice is caused due to irrigation with As-rich groundwater, and subsequent contamination of soil. However, arsenic risk assessment in rice based on the total As concentration in soil and irrigation water can be misleading, since As accumulation in plants is controlled by a number of factors within the soil–rhizosphere–plant system. Soil parameters such as soil pH, Eh, texture and concentrations of P, Si, S and Fe and arsenic speciations should be considered in the assessment of soil As for potential rice toxicity. Environmental factors such as soil flooding is also one of the potential factors influencing arsenic accumulation in rice by increasing As mobility in the soil–rhizosphere. Therefore, growing rice in aerobic conditions may be a feasible strategy to mitigate this problem. In addition, rice genotype, which controls the root aeration, is significant in affecting the accumulation of As. Therefore, selection of suitable rice genotype is important to minimize As accumulation in rice.

Arsenic is taken up by roots, and then transferred to the edible portion of plant, which leads to phytotoxicity, often resulting straighthead disease in rice plants. However, the translocation of As from the roots to the above parts is significantly limited. Among all species, organic As is more readily translocated, but, its uptake is very low compared to inorganic As; thus, inorganic As predominates in grains. Uptake of arsenite in rice occurs through the silicon transport pathway; therefore, application of silica fertilizers in soil is suggested to decrease the transfer of arsenic from soil to rice. Further, phosphate fertilization is also suggested to lower the arsenate uptake in plants, since both enter to rice through the same transporter. However, there are arguments in certain cases, since in the flooding condition As is present as arsenite which cannot compete phosphate, and further, phosphate increases As mobility since it competes with arsenate for adsorption site on Fe-oxides/hydroxides. In addition, accumulation of As in the presence Fe-plaque and other soil parameters is still confusing. More work needed, especially on a close monitoring of soil, irrigation water, and rice in both lab and field based experiments to understand the role of Fe-plaque, phosphorous, silica, and sulfur in influencing As mobility and solubility in soil-rhizospehere and its uptake mechanisms by plants.

References

Abedin MJ, Feldmann J, Mehrag AA (2002a) Uptake kinetics of arsenic species in rice plants. Plant Physiol 128(3):1120–1128

Abedin MJ, Howells JC, Mehrag AA (2002b) Arsenic uptake and accumulation in rice (*Oryza sativa* L.) irrigated with contaminated water. Plant Soil 240:311–319

Adomako EE, Solaiman ARM, Williams PN, Deacon C, Rahman GKMM, Mehrag AA (2009) Enhanced transfer of arsenic to grain for Bangladesh grown rice compared to US and EU. Environ Int 35:476–479

Adomako EE, Williams PN, Deacon C, Mehrag AA (2011) Inorganic arsenic and trace elements in Ghanaian grain staples. Environ pollut 159:2435–2442

Ahmed UZ, Panaullah GM, Jr HG, McCouch SR, Tyagi W, Kabir MS, Duxbury JM (2011) Genotype and environment effect rice (*Oryza sativa* L.) grain arsenic concentration in Bangladesh. Plant Soil 338:367–382

Ahsan DA, DelValls TA (2011) Impact of arsenic contaminated irrigation water in food chain: an overview from Bangladesh. Int J Environ Res 5(3):627–638

Arao T, Kawasaki A, Baba K, Mori S, Matsumoto S (2009) Effects of water management on cadmium and arsenic accumulation and dimethyl arsenic acid concentrations in Japanese rice. Environ Sci Technol 43:9361–9367

Armstrong W (1967) The use of polarography in the assay of oxygen diffusing from roots in anaerobic media. Physiol Plant 20:540–553

Baig JA, Kazi TG, Shah AQ, Afridi HI, Kandhro GA, Khan S, Kolachi NF, Wadhwa SK, Shah F, Arain MB, Jamali MK (2011) Evaluation of arsenic levels in grain crops samples, irrigated by tube well and canal water. Food Chem Toxicol 49:265–270

Bauer M, Blodau C (2006) Mobilization of arsenic by dissolved organic matter from iron oxides, soils and sediment. Sci Total Environ 354:179–190

Bhattacharya P, Chatterjee D, Jacks G (1997) Occurrence of arsenic-contaminated groundwater in alluvial aquifers from delta plains, eastern India: options for safe drinking water supply. Int J Water Resour Dev 13(1):79–92

Bhattacharya P, Claesson M, Bundschuh J, Sracek O, Fagerberg J, Jacks G, Martin RA, Storniolo AR, Thir JM (2006) Distribution and mobility of arsenic in the Río Dulce alluvial aquifers in Santiago del Estero Province, Argentina. Sci Total Environ 358:97–120

Bhattacharyya P, Tripathy S, Kim K, Kim SH (2008) Arsenic fractions and enzyme activities in arsenic-contaminated soils by groundwater irrigation in West Bengal. Ecotoxicol Environ Saf 71:149–156

Bhattacharya P, Samal AC, Majumdar J, Santra SC (2009) Transfer of arsenic from groundwater and paddy soil to rice plant (*Oryza sativa* L.): a micro level study in West bengal, India. World J Agric Sci 5(4):425–431

Bhattacharya P, Samal AC, Majumdar J, Santra SC (2010a) Accumulation of arsenic and its distribution in rice plant (*Oryza sativa* L.) in Gangetic West Bengal, India. Paddy Water Environ 8:63–70

Bhattacharya P, Samal AC, Majumdar J, Santra SC (2010b) Uptake of arsenic in rice plant varieties cultivated with arsenic rich groundwater. Environ Asia 3:34–37

Bibi MH, Ahmed F, Ishiga H (2008) Geochemical study of arsenic concentrations in groundwater of the Meghna River Delta, Bangladesh. J Geochem Explor 97:43–58

Biswas A, Santra SC (2011) Arsenic distribution in winter rice and vetable crops-in vivo micro level study in a contaminated region. Int Res J Agric Sci Soil Sci 1(6):205–210

Biswas BK, Loeppert RH, Hossain MB, Rahman GKMM, Jahiruddin M, Miah MAM, Farid TM, Panaullah GM, Meisner CM, Duxbury JM (2003) Impact of soil oxides on retention of arsenic in Bangladesh rice-producing soils. In: Proceedings of 7th international conference on the biogeochemistry of trace elements, Uppasala, Sweden, June 15–19, 2003.

Bogdan K, Schenk MK (2008) Arsenic in rice (*Oryza sativa* L.) related to dynamic of arsenic and silicic acid in paddy soils. Environ Sci Technol 42:7885–7890

Bogdan K, Schenk MK (2009) Evaluation of soil characteristics potentially affecting arsenic concentration in paddy rice (*Oryza sativa* L.). Environ Pollut 157:2617–2621

Campbell JA, Stark JH, Carlton-Smith CH (1985) International symposium on heavy metals in the environment (vol 1). CEP Consultants, Athens

Campos V (2002) Arsenic in groundwater affected by phosphate fertilizers at São Paulo, Brazil. Environ Geol 42:83–87

Cao X, Ma LQ, Shiralipour AS (2003) Effect of compost and phosphate amendments on arsenic mobility in soils and arsenic uptake by the hyperaccumulator, *Pteris vittata* L. Environ Pollut 126:157–167

Carbonell AA, Aarabi MA, DeLaune RD, Gambrell RP, Patrick JWH (1998) Arsenic in wetland vegetation: availability, phytotoxicity, uptake and effects on plant growth and nutrition. Sci Total Environ 217:189–199

Chakraborti D, Basu GK, Biswas BK (2001) Characterization of arsenic-bearing sediments in the Gangetic delta of West Bengal, India. In: Chappell WR, Abernathy CO, Calderon RL (eds) Arsenic exposure and health effects IV. Elsevier, Oxford

Chaturvedi I (2006) Effect of arsenic concentrations on growth and arsenic uptake and accumulation by rice (*Oryza Sativa* L.) genotype. Electron J Environ Agric Food Chem 5:1546–1552

Chen KY, Liu TK (2007) Major factors controlling arsenic occurrence in the groundwater and sediments of the Chianan coastal plain, SW Taiwan. Terr Atmos Ocean Sci 18:975–994

Chen TB, Fan ZL, Lei MEI (2002) Effect of phosphorus on arsenic accumulation in As-hyperaccummulator: *Pteris vittata* L. and its implication. Chin Sci Bull 47:1876–1879

Chen Z, Zhu YG, Liu WJ, Meharg AA (2005) Direct evidence showing the effect of root surface iron plaque on arsenite and arsenate uptake into rice (*Oryza sativa*). New Phytol 165:91–97

Chen XP, Zhu YG, Hong MN, Kappler A, Xu YX (2008) Effects of different forms of nitrogen fertilizers on arsenic uptake by rice plants. Environ Toxicol Chem 27:881–887

Cullen WR, Reimer KJ (1989) Arsenic speciation in the environment. Chem Rev 89:173–764

Das HK, Mitra AK, Sengupta PK, Hossain A, Islam F, Rabbani GH (2004) Arsenic concentrations in rice, vegetables and fish in Bangladesh: a preliminary study. Environ Int 30(3):383–387

Das DK, Sur P, Das K (2008) Mobilization of arsenic in soils and in rice (*Oryza sativa* L.) plants affected by organic matter and zinc application in irrigation water contaminated with arsenic. Plant Soil Environ 54:30–37

Dittmar J, Voegelin A, Maurer F, Roberts LC, Hug SJ, Saha GC, Ali MA, Badruzzaman ABM, Kretzschmar R (2010a) Arsenic in soil and irrigation water affects arsenic uptake by rice: complementary insights from field and pot studies. Environ Sci Technol 44:8842–8848

Dowling CB, Poreda RJ, Basu AR et al (2002) Geochemical study of arsenic release mechanisms in the Bengal Basin groundwater. Water Resour Res 38(9):12.1–12.18

Dudas MJ (1987) Accumulation of native arsenic in acid sulphate soils in Alberta. Can J Soil Sci 67:317–331

Duxbury JM, Panaullah G (2007) Remediation of arsenic for agriculture sustainability, food security and health in Bangladesh. Working paper, Water Service, FAO (2007), Rome. http://www.fao.org/nr/water/docs/FAOWATER_ARSENIC.pdf

Dwivedi S, Tripathi RD, Srivastava S, Singh R, Kumar A, Tripathi P, Dave R, Rai UN, Chakrabarty D, Trivedi PK, Tuli R, Adhikari B, Bag MK (2010) Arsenic affects mineral nutrients in grain of various Indian rice (*Oryza sativa* L.) genotypes grown on arsenic-contaminated soils of West Bengal. Protoplasma 245:113–124

Fitz WJ, Wenzel WW (2002) Arsenic transformations in the soil–rhizosphere–plant system: fundamentals and potential application to phytoremediation. J Biotechnol 99:259–278

Frans R, Horton D, Burdette L (1988) Influence of MSMA on straight-head, arsenic uptake and growth responsible in rice. Res Ser 302: 1–12, Univ of Arkansan Agri Exp Sta, Fayetteville

Francesconi KA (2007) Toxic metal species and food regulations—making a healthy choice. Analyst 132:17–20

Fu Y, Chen M, Bi X, He Y, Ren L, Xiang W, Qiao S, Yan S, Li Z, Ma Z (2011) Occurrence of arsenic in brown rice and its relationship to soil properties from Hainan Island, China. Environ Pollut 159:1757–1762

Garelick H, Jones H (2008) Mitigating arsenic pollution: bridging the gap between knowledge and practice. Chem Int 30(4):7–12. (ISSN 0193–6484)

Garnier JM, Travassac F, Lenoble V, Rose J, Zheng Y, Hossain MS, Chowdhury SH, Biswas AK, Ahmed KM, Cheng Z, Geen A (2010) Temporal variations in arsenic uptake by rice plants in Bangladesh: the role of iron plaque in paddy fields irrigated with groundwater. Sci Total Environ 408:4185–4193

Guo W, Zhu YG, Liu WJ, Liang YC, Geng CN, Wang SG (2007) Is the effect of silicon on rice uptake of arsenate (AsV) related to internal silicon concentrations, iron plaque and Phosphate nutrition? Environ Pollut 148:251–257

Guo H, Yang S, Tang X, Li Y, Shen Z (2008) Groundwater geochemistry and its implications for arsenic mobilization in shallow aquifers of the Hetao Basin, Inner Mongolia. Sci Total Environ 393:131–144

Guo W, Zhang J, Teng M, Wang LH (2009) Arsenic uptake is suppressed in a rice mutant defective in silicon uptake. J Plant Nutr Soil Sci 172:867–874

Gurung JK, Ishiga H, Khadka MS (2005) Geological and geochemical examination of arsenic contamination in groundwater in the Holocene Terai Basin, Nepal. Environ Geol 49:98–113

Harvey CF, Swartz CH, Badruzzaman ABM, Keon-Blute N, Yu W, Ali MA, Jay J, Beckie R, Niedan V, Brabander D, Oates PM, Ashfaque KN, Islam S, Hemond HF, Ahmed MF (2002) Arsenic mobility and groundwater extraction in Bangladesh. Science 298:1602–1606

Heikens A (2006) Arsenic contamination of irrigation water, soil and crops in Bangladesh: risk implications for sustainable agriculture and food safety in Asia. Food and Agricultural Organization of the United Nations, Regional Office for Asia and the Pacific, Bangkok

Heikens A, Panaullah GM, Meharg AA (2007) Arsenic behavior from groundwater and soil to crops: impacts on agriculture and food safety. Rev Environ Contam Toxicol 189:43–87

Horneman A, VanGeen A, Kent DV, Mathe PE, Zheng Y, Dhar RK, O'Connell S, Hoque MA, Aziz Z, Shamsudduha M, Seddique AA, Ahmed KM (2004) Decoupling of As and Fe release to Bangladesh groundwater under reducing conditions. Part1: evidence from sediment profiles. Geochim et Cosmochim Acta 68:3459–3473

Hossian MB, Jahiruddin M, Panaullah GM, Loeppert RH, Islam MR, Duxbury JM (2008) Spatial variability of arsenic concentration in soils and plants, and its relationship with iron, manganese and phosphorus. Environ Pollut 56:739–744

Hossain MB, Jahiruddin M, Loeppert RH, Panaullah GM, Islam MR, Duxbury JM (2009) The effect of iron plaque and phosphorus on yield and arsenic accumulation in rice. Plant Soil 317:167–176

Hu Y, Li JH, Zhu YG, Huang YZ, Hu HQ, Christie P (2005) Sequestration of As by iron plaque on the roots of three rice (*Oryza sativa* L.) cultivars in a low-P soil with or without P fertilizer. Environ Geochem Health 27:169–176

Hu ZY, Zhu YG, Li M, Zhang LG, Cao ZH, Smith FA (2007) Sulphur(S)-induced enhancement of iron plaque formation in the rhizosphere reduces arsenic accumulation in rice (*Oryza sativa* L.) seedlings. Environ Pollut 147(17):387–393

Hunag RQ, Gao SF, Wang WL, Staunton S, Wang G (2006) Soil arsenic availability and the transfer of soil arsenic to crops in suburban areas in Fujian province, southeast China. Sci Total Environ 368:531–541

Islam MR, Jahiruddin M (2010) Effect of arsenic and its interaction with phosphorus on yield and arsenic accumulation in rice. 19th world congress of soil science, soil solutions for changing world, Brisbane, 1–6 Aug 2010

Jahiruddin M, Islam MA, Islam MR, Islam S (2004) Effects of arsenic contamination on rice crop. Environtropica 1:204–210

Kar S, Maity JP, Jean JS, Liu CC, Nath B, Yang HJ, Bundschuh J (2010) Arsenic-enriched aquifers: occurrences and mobilization of arsenic in groundwater of Ganges Delta Plain, Barasat, West Bengal, India. Appl Geochem 25(12):1805–1814

Kim K, Moon JT, Kim SH, Ko KS (2009) Importance of surface geologic condition in regulating As concentration of groundwater in the alluvial plain. Chemosphere 77:478–484

Kim K, Kim SH, Jeong GY, Kim RH (2012) Relations of As concentrations among groundwater, soil, and bedrock in Chungam, Korea: implications for As mobilization in groundwater according to the As-hosting mineral change. J Hazard Mater 15:25–35

Kirk GJD, Du VL (1997) Change in rice root architecture, porosity, and oxygen and proton release under phosphorus deficiency. New Phytol 135:191–200

Lee JS, Lee SW, Chon HT, Kim KW (2008) Evaluation of human exposure to arsenic due to rice ingestion in the vicinity abandoned Myungbong Au-Ag mine site, Korea. J Geochem Explor 96:231–235

Lihong W, Guilan D (2009) Effect of external and internal phosphate status on arsenic toxicity and accumulation in rice seedlings. J Environ Sci 21:346–351

Lin HT, Wong SS, Li GC (2004) Heavy metal content of rice and shell fish in Taiwan. J Food Drug Anal 12:67–74

Lin CF, Wu CH, Lai HT (2008) Dissolved organic matter and arsenic removal with coupled chitosan/UF operation. Sep Purif Technol 60:292–298

Liu WJ, Zhu YG, Smith FA, Smith SE (2004) Do phosphorus nutrition and iron plaque alter arsenate (As) uptake by rice seedlings in hydroponic culture? New Phytol 162:481–488
Lombi E, Sletten RS, Wenzel WW (2000) Sequentially extracted arsenic from different size fractions of contaminated soils. Water Air Soil Pollut 124:319–32
Ma JM, Tamai K, Yamaji N, Mitani N, Konishi S, Katsuhara M, Ishiguro M, Murata Y, Yano M (2006) A silicon transporter in rice. Nature 440:688–691
Ma JF, Yamaji N, Mitani N, Tamai K, Konishi S, Fujiwara T, Katsuhara M, Yano M (2007) An efflux transporter of silicon in rice. Nature 448:209–212
Ma JF, Yamaji N, Mitani N, Xu XY, Su YH, McGrath SP, Zhao FJ (2008) Transporters of arsenite in rice and their role in arsenic accumulation in rice grain. Proc Natl Acad Sci U S A 105(29):9931–9935
Mahimairaja S, Bolan NS, Adriano DC, Robinson B (2005) Arsenic contamination and its risk management in complex environmental settings. Adv Agron 86:1–82
Mandal BK, Suzuki KT (2002) Arsenic round the world: a review. Talanta 58(1):201–235
Marine AR, Masscheleyn PH, Patric WH (1992) The influence of chemical form and concentration of As on rice growth and tissue As concentration. Plant Soil 139:175–183
Martinez VD, Vucic EA, Becker-Santos DD, Gil L, Lam WL (2011) Arsenic exposure and the induction of human cancers. J Toxicol 2011:431287. doi:10.1155/2011/431287
McArthur JM, Ravenscroft P, Safiulla S, Thirlwall MF (2001) Arsenic in groundwater: testing pollution mechanisms for sedimentary aquifers in Bangladesh. Water Resour Res 37(1):109–17
Mclaren RG, Naidu R, Tiller KG (1996) Fractionation of arsenic in soils contaminated by cattle dip. 1st international conference: Contaminants and soil environment, Adelaide, pp 177–178, 1996
Meharg AA (2004) Arsenic in rice—understanding a new disaster for South-East Asia. Trends Plant Sci 9:415–417
Meharg AA, Rahman MDM (2003) Arsenic contamination of Bangladesh paddy field soils: implications for rice contribution to arsenic consumption. Environ Sci Technol 37:229–234
Meharg AA, Whitaker JH (2002) Arsenic uptake and metabolism in arsenic resistant and nonresistant plant species—review. New Phytol 154:29–43
Meharg AA, Adomako E, Lawgali Y, Deacon C, Williams Paul (2007) Arsenic in rice–a literature review. Food Standards Agency contract C101045
Meharg AA, Williams PN, Adomako E, Lawgali YY, Deacon C, Villada A, Cambell RCJ, Sun G, Zhu YG, Feldmann J, Raab A, Zhao FJ, Islam R, Hossain S, Yanai J (2009) Geographical variation in total and inorganic arsenic content of polished (white) rice. Environ Sci Technol 43:1612–1617
Mukherjee A, Fryar A, O'shea BM (2009) Major occurrences of elevated arsenic in groundwater and other natural waters. In: Henke K (ed) Arsenic: environment chemistry, health threats and waste treatment, Ch. 6. Wiley, New York
Mukhopadhyay D, Sanyal SK (2004) Complexation and release isotherm of arsenic in arsenic-humic/fulvic equilibrium study. Aust J Soil Res 42(7):815–24
Naguyen KP, Itoi R (2009) Source and release mechanism of arsenic in aquifers of the Mekong Delta, Vietnam. J Contam Hydrol 103:58–69
National Research Council (2001) Arsenic in drinking water—2001 Update. National Academy, Washington DC
Ng JC, Wang J, Shraim A (2003) A global health problems caused by arsenic from natural sources. Chemosphere 52:1353–1359
Norra S, Berner ZA, Agarwala P, Wagner F, Chandrasekharam D, Stuben D (2005) Impact of irrigation with As rich groundwater on soil and crops: ageochemical case study in West Bengal Delta Plain, India. Appl Geochem 20:1890–1906
Norton GJ, Islam MR, Deacon CM, Zhao FJ, Stroud JL, Mcgrath SP, Islam S, Jahiruddin M, Feldmann J, Price AH, Meharg AA (2009a) Identification of low inorganic and total grain arsenic rice cultivars from Bangladesh. Environ Sci Technol 43:6070–6075
Norton GJ, Duan G, Dasgupta T, Islam MR, Lei M, Zhu Y, Deacon CM, Moran AC, Islam S, Zhao FJ, Stroud JL, Mcgrath SP, Feldmann J, Feldmann J, Price AH, Mehagr AA (2009b) Environ-

mental and genetic control of arsenic accumulation and speciation in rice grain: comparing a range of common cultivars grown in contaminated sites across Bangladesh, china and India. Environ Sci Technol 43:8381–8386

O'Neil P (1995) Heavy metals in soils. In: Alloway BJ (ed) Arsenic. Blackie Academic and Professional, London, pp 105–121

Peryea FJ, Kammereck R (1995) Phosphate-enhanced move-ment of arsenic out of lead arsenate-contaminated top soil and through uncontaminated sub soil. Water Air Soil Pollut 93:243–254

Pikaray S, Banerjeem S, Mukherji S (2005) Sorption of arsenic onto Vindhyan shales: role of pyrite and organic carbon. Curr Sci 88:1580–1585

Quazi S, Datta R, Sarkar D (2011) Effect of soil types and forms of arsenical pesticide on rice growth and development. Int J Environ Sci Technol 8:45–460

Raab A, Schat H, Meharg AA, Feldmann J (2005) Uptake, translocation and transformation of arsenate and arsenite in sunflower (Helianthus annuus):formation of arsenic–phytochelatin complexes during exposure to high arsenic concentrations. New Phytol 168:551–558

Rahman MA, Hasegawa H, Rahman MM, Miah MAM, Tasmin A (2008) Straighthead disease of rice (*Oryza sativa* L.) induced by arsenic toxicity. Environ Exp Bot 62:54–59

Rahaman S, Sinha AC, Mukhopadhyay D (2011) Effect of water regimes and organic matters on transport of arsenic in summer rice (*Oryza sativa* L.). J Environ Sci 23:633–639

Roberts LC, Hug SJ, Voegelin A, Dittmar J, Kretzschmar R, Wehrli B, Saha GC, Badruzzaman ABM, Ali MA (2011) Arsenic dynamic in porewater of an intermittently irrigated paddy field in Bangladesh. Environ Sci Technol 45:971–976

Rowland HAL, Polya DA, Lioyd JR, Pancost RD (2006) Characterisation of organic matter in shallow, reducing, arsenic-rich aquifer, West Bengal. Org Geochem 37:1101–1114

Roychowdhury T, Uchino T, Tokunaga H, Ando M (2002) Survey of arsenic in food composites from anarsenic-affected area of West Bengal, India. Food Chem Toxicol 40:1611–21

Roychowdhury T, Tokunaga H, Uchino T, Ando M (2005) Effect of arsenic-contaminated irrigation water on agricultural land soil and plants in West Bengal, India. Chemosphere 58:799–810

Sahoo PK, Zhu W, Kim SH, Jung MC, Kim K (2013) Relations of arsenic concentrations among groundwater, soil and paddy from an alluvial plain of Korea. Geosci J 3:363–370

Schoof RA, Yost LJ, Crecelius E, Irgolic K, Goessler W, Guo HR, Greene H (1998) Dietary arsenic intake in Taiwanese districts with elevated arsenic in drinking water. Hum Ecol Risk Assess 4:117–135

Schreiber ME, Simo JA, Freiberg PG (2000) Stratigraphic and geochemical controls on naturally occurring arsenic in groundwater, eastern Wisconsin, USA. Hydrogeol J 8:161–176

Selim Reza AHM, Jena JS, Yang HJ, Lee MK, Woodall B, Liu CC, Lee JF, Luo SD (2010) Occurrence of arsenic in core sediments and groundwater in the Chapai-nawabgang District, northwestern Bangladesh. Water Res 44:2021–2037

Seyfferth AL, Webb SM, Andrews JC, Fendorf S (2010) Arsenic localization, speciation and co-occurrence with iron on rice (*Oryza sativa* L.) roots having variable Fe coatings. Environ Sci Technol 44:8108–8113

Sheppard SC (1992) Summary of phytotoxic levels of soil arsenic. Water Air Soil Pollut 64:539–550

Signes-Pastor A, Burló F, Mitra K, Carbonell-Barrachina AA (2006) Arsenic biogeochemistry as affected by phosphorus fertilizer addition, redox potential and pH in a west Bengal (India) soil. Geoderma 137(3–4):504–510

Signes-Pastor A, Burló F, Mitra K, Carbonell-Barrachina AA (2007) Arsenic biogeochemistry as affected by phosphorus fertilizer addition, redox potential and pH in a West Bengal (India) soil. Geoderma 137:504–510

Sikdar PK, Sarkar SS, Palchoudhury S (2001) Geochemical evolution of groundwater in the quaternary aquifer of Calcutta and Howrah, India. J Asian Earth Sci 19:579–94

Singh V, Brar MS, Sharma P, Malhi SS (2010) Arsenic in water, soil, and rice plants in the Indo-Gngetic plains of Northwestern India. Commun Soil Sci Plant Anal 41:1350–1360

Smedley PL, Kinniburgh DG (2002) A review of source, behavior and distribution of arsenic in natural water. Appl Geochem 17:517–568

Smedley P, Kinniburgh DG (2005) Arsenic in groundwater and the environment. In: Selinus O, Alloway B, Centeno JA, Finkelman RB, Fuge R, Lindh U, Smedley P (eds) Essentials of medical geology—impacts of the natural environment on public health. Academic, Waltham, pp 263–299

Smith AH, Hopenhayn-Rich C, Bates MN, Goeden HM, Hertz-Picciotto I, Duggan HM, Wood R, Kosnett MJ, Smith MT (1992) Cancer risks from arsenic in drinking water. Environ Health Perspect 97:259–267

Smith E, Naidu R, Alston AM (2002) Chemistry of arsenic in soils: II. Effect of phosphorous, sodium and calcium on arsenic sorption. J Environ Qual 31:557–563

Somenahally AC, Hollister EB, Yan W, Gentry TJ, Loeppert RH (2011) Water management impacts on arsenic speciation and iron-reducing bacteria in contrasting rice-rhizosphere compartments. Environ Sci Technol 45(19):8328–8335. doi:10.1021/es2012403

Streat M, Hellgardt K, Newton NLR (2008) Hydrous ferric oxide as an adsorbent in water treatment Part 3: Batch and mini-column adsorption of arsenic, phosphorus, fluorine and cadmium ions. Process Saf Environ Prot 86:21–30

Sun GX, Williams PN, Carey AM, Zhu YG, Deacon C, Raab A, Feldmann J, Islam RM, Meharg AA (2008) Inorganic arsenic in rice bran and its products are an order of magnitude higher than in bulk grain. Environ Sci Technol 42:7542–7546

Takahashi Y, Minamikawa R, Hattori KH, Kurishima K, Kihou N, Yuita K (2004) Arsenic behaviour in paddy fields during the cycle of flooded and non-flooded periods. Environ Sci Technol 38:1038–1044

Takamatsu T, Aoki H, Yoshida T (1982) Determination of arsenate, arsenite, monomethylarsonate, and dimethylarsinate in soil polluted with arsenic. Soil Sci 133:239–246

Talukder ASMHM, Meisner CA, Sarkar MAR Islam MS (2011) Effect of water management, tillage options and phosphorus status on arsenic uptake in rice. Ecotoxicol Environ Saf 74:834–839

Tu S, Ma LQ (2003) Interaction effect of pH, arsenic and phosphorous on uptake of As and P and growth of the arsenic hyperacumulation *Pteris vittata* L. under hydroponic conditions. Environ Exp Bot 50:243–251

Tu S, Ma LQ (2004) Comparison of arsenic and phosphate uptake and distribution in arsenic hyperaccumulating and nonhyperaccumulating Fern. J Plant Nutr 27:1227–1242

Ullrich-Eberius CI, Sanz A, Novacky AJ (1989) Evaluation of arsenate and vandate-associated changes of electrical membrane potential and phosphate transport in *Lemna gibba* GI. J Exp Bot 40:119–128

Van Geen A, Zheng Y, Cheng Z, He Y, Dhar RK, Garnier JM, Rose J, Siddique A, Hoque MA, Ahmed KM (2006) Impact of irrigating rice paddies with groundwater containing arsenic in Bangladesh. Sci Total Environ 367:769–777

Voegelin A, Weber FA, Kretzschmar R (2007) Distribution and speciation of arsenic around roots in a contaminated riparian flood plain soil: micro-XRF element mapping and EXAFS spectroscopy. Geochim et Cosmochim Acta 71:5804–5820

Walter JF, Wenzel WW (2002) Arsenic transformation in the soil–rhizosphere–plant system: fundamentals and potential application to phytoremediation. J Biotechnol 99:259–278

Wang J, Zhao FJ, Meharg AA, Raab A, Feldmann J, McGrath P (2002) Mechanisms of arsenic hyperaccumulation in *Pteris vittata*. uptake kinetics, interactions with phosphate, and arsenic speciation. Plant Physiol 130:1552–1561

Wang S, Mulligan CN (2006) Occurrence of arsenic contamination in Canada: sources, behavior, and distribution. Sci Total Environ 366:701–721

Welch AH, Lico MS (1998) Factors controlling As and U in shallow ground water, southern Carson Desert, Nevada. Appl Geochem 13:521–539

Williams P (2003) Investigating links between minerals in rice grain and straighthead. Final report. Leeton, Cooperative Research Centre for Sustainable Rice Production

Williams M, Fordyce F, Paijitprapapon A, Charoenchaisri P (1996) Arsenic contamination in surface drainage and groundwater in part of the southeast Asian tin belt, Nakhon Si Thammarat Province, southern Thailand. Environ Geol 27:16–33

Williams PN, Price AH, Raab A, Hossain SA, Feldmann J, Meharg AA (2005) Variation in arsenic speciation and concentration in paddy rice related to dietary exposure. Environ Sci Technol 39:5531–5540

Williams PN, Islam MR, Adomako EE, Raab A, Hossain SA, Zhu YG, Feldmann J, Meharg AA (2006) Increase in rice grain arsenic for regions of Bangladesh irrigating paddies with elevated arsenic in groundwater. Environ Sci Technol 40:4903–4908

Williams PN, Villada A, Deacon C, Raab A, Figuerola J, Green AJ, Feldmann J, Meharg AA (2007a) Greatly enhanced arsenic shoot assimilation in rice leads elevated grain levels compared to wheat and barley. Environ Sci Technol 41:6854–6859

Williams PN, Raab A, Feldmann J, Meharg AA (2007b) Market basket survey shows elevated levels of As in South Central U.S. processed rice compared to California: consequences for human dietary exposure, Environ Sci Technol 41:2178–2183

Williams PN, Zhang H, Davison W, Mehrag AA, Hossain M, Norton G, Brammer H, Islam MR (2011) Organic matter—solid phase interactions are critical for predicting arsenic release and plant uptake in Bangladesh paddy soils. Environ Sci Technol 45:6080–6087

Woolson EA, Axley JH, Kearney PC (1973) The chemistry and phytotoxicity of arsenic in soils. II. Effcets of time and phosphorus. Proc Soil Sci Soc Am 37(2):254–259

Woolson EA (1973) Arsenic Phytotoxicity and Uptake in Six Vegetable Crops. Weed Science. 21(6):524–527

World Bank (2005) Environment and social unit—South Asia region. Towards a more effective operational response. Arsenic contamination of groundwater in south and east Asian countries volume II technical report. Environmental Study (Report No 31303), The World Bank

Wu C, Ye Z, Shu W, Zhu Y, Wong M (2011) Arsenic accumulation and speciation in rice are affected by root aeration and variation of genotypes. J Exp Bot 62(8):2889–2898. doi:10.1093/jxb/erq462

Xu XY, McGrath SP, Meharg AA, Zhao FJ (2008) Growing rice aerobically decreases arsenic accumulation. Environ Sci Technol 42:5574–5579

Yan W, Dilday RH, Tai TH, Gibbons JW, McNew RW, Rutger JN (2005) Differential response of rice germplasm to straighthead disease by arsenic. Crop Sci 45:1223–1228

Zavala YJ, Gerads R, Gürleyük H, Duzbury JM (2008) Arsenic in rice: II. Arsenic speciation in USA and inplications for human health. Environ Sci Technol 42:3861–3866

Zhang J, Duan GL (2008) Genotypic difference in arsenic and cadmium accumulation by rice seedlings grown in hydroponics. J Plant Nutr 31:2168–2182

Zhang J, Zhao QZ, Duan GL, Huang YC (2011) Influence of sulfur on arsenic accumulation and metabolism in rice seedlings. Environ Exp Bot 72:34–40

Zhao FJ, Ma JF, Meharg AA, McGrath SP (2009) Arsenic uptake and metabolism in plants. New Phytol 181:777–794

Zhao FJ, McGrathMc SP, Mehrag AA (2010) Arsenic as a food chain contaminant: mechanisms of plant uptake and metabolism and mitigation strategies. Annu Rev Plant Biol 61:535–559

Zhu YG, Williams PN, Meharg AA (2008) Exposure to inorganic arsenic from rice: a global health issue? Environ Pollut 154:167–171

Geophysical Signatures for Detection of Fresh Water and Saline Water Zones

Kalyan Kumar Roy

Abstract Physical and physicochemical backgrounds behind possible geophysical signatures for the detection of saline water and fresh water zones are highlighted. Coastal areas generally encounter this type of problems. Self-potential (SP), direct current resistivity sounding and traversing (DC, VES), induced polarisation sounding and traversing (IP), electromagnetic geometric and frequency sounding (EM), electromagnetic transients (TEM), audio frequency magnetotellurics (AMT), controlled-source audio frequency magnetotellurics (CSAMT), ground-penetrating radar (GPR), remote sensing with Geographical Information System (GIS) and Global Positioning System (GPS), seismic reflection, seismic refraction and high-resolution seismic reflection are the surface geophysical tools needed for studying this problem. Borehole geophysical tools needed for studying saline water and fresh water problems are SP, resistivity, natural gamma ray, neutron–thermal neutron, neutron-capture gamma ray, gamma–gamma density and sonic logging. High flight imagery, viz. remote sensing, has wide application in groundwater industries. Low flight imagery, viz. aerial photos, airborne geomorphology and aeroelectromagnetics, has application in this problem.

Combination of geophysical tools needed for studying saline water and fresh water problems and discussed in this short chapter are (i) joint resistivity and IP sounding, (ii) central frequency sounding (CFS) and dipole frequency sounding (DFS), (iii) transient electromagnetic sounding, (iv) GPR, (v) remote sensing with GIS and GPS, (vi) high-resolution seismic reflection and GPR, (vii) seismic reflection or refraction jointly with DC resistivity sounding or electromagnetic frequency sounding, (vii) combined natural gamma ray, SP and DC resistivity logging, (viii) combined neutron–thermal neutron and neutron-capture gamma ray logging and (ix) combined sonic density and neutron logging will be useful geophysical tool combinations for (a) estimation of secondary porosity in all types of rocks, especially carbonate and hard rocks, and (b) accurate estimation of acoustic impedance and bed boundaries.

K. K. Roy (✉)
Flat No.1, 33, Anjumana Ara Begum Row, Golf Garden, Tollygunge,
Kolkata–700033, West Bengal, India
e-mail: kalyankrroy@gmail.com

Department of Earth Sciences, Indian Institute of Engineering, Science and Technology,
Shibpur, Howrah, West Bengal 711103, India

D. Sengupta (ed.), *Recent Trends in Modelling of Environmental Contaminants*,
DOI 10.1007/978-81-322-1783-1_3,

Basic reasons for combinations of these geophysical tools are as follows: (i) Both saline water sand and clay/shale beds can show the same order of resistivity, but IP effect for clay/shale will be higher than saline water sands. Therefore, combined resistivity and IP sounding can resolve the saline water–fresh water zone in a more effective way. (ii) Since DC resistivity sounding may not respond properly in a hard-rock area, electromagnetic sounding (CFS or DFS) can join hands with seismic reflection and a couple of porosity logs to resolve the issue. (iii) In transient electromagnetic sounding or time domain electromagnetics, the response signals are recorded in the absence of the primary field. Therefore, the signals are better and can be used for better detection of saline water aquifers at greater depth. (iv) Combination of GPR and a high-resolution seismics makes a very effective tools for mapping the first 10–15 m of the earth's surface with great details. Since dielectric constants of saline water and shale/clay are significantly different, GPR can differentiate between saline water sand and clay/shale. Electrical conductivity-dependent electromagnetic sounding may fail when the electrical conductivity is of the same order. (v) Since electrical conductivity of the water-saturated earth formation increases both with temperature as well as salinity, geoelectrical tools used for exploration of geothermal sources can also be used for exploration of deep-seated connate water pockets. Hence AMT, CSAMT, TEM, Long Offset Transient Electromagnetics (LOTEM) and deep EM and DC resistivity soundings can be used for detection of deep-seated saline water pockets. (vi) Remote sensors are instruments those measure electromagnetic signals sensing the surface or medium due to scattering or emission. Remote sensing is a very powerful tool for the detection of faults, fractures, fissures and lineaments in hard-rock areas. It is also widely used in soft-rock groundwater technology. It is an excellent tool for reconnaissance survey. (vii) Since shale and clay are better absorber of radionuclides, natural gamma ray logging along with SP and resistivity logs can differentiate between saline water sand and clay/shale. (ix) Nuclear capture cross-section of chlorine is much higher than that of common earth elements, viz. hydrogen, oxygen, silicon, etc.; therefore, saline water sand has higher capturing capacity of thermal neutrons than that of fresh water. As a result, combined neutron-thermal neutron and neutron-capture gamma ray logging can easily differentiate between saline water and fresh water. (x) SP reversal is an important qualitative diagnostics for identification of saline water and fresh water aquifers because one gets negative SP for saline water and positive SP for fresh water. Whether the salinity of the formation is more or less than the salinity of the borehole mud fluid dictates the direction of flow of SP current. This SP current is due to the formation of the liquid junction and membrane electrochemical cells and the direction of flow of current changes. (xi) Single-point resistance and normal and lateral resistivity logging tools along with SP are widely used geophysical tools for location of saline water-saturated beds. Appropriate geophysical tools best suited for different depth ranges are as follows: (i) GPR and high-resolution seismics will be for 1–10 m, (ii) DC resistivity, IP, CFS and DFS along with shallow seismic reflection or refraction are for 5–100 m or a little beyond, (iii) transient electromagnetic sounding, DC resistivity sounding/traversing using collinear dipole–dipole array and high-power low-frequency electromagnetic sounding jointly with seismic reflection sounding can go from 50 to 500 m, and (iv) deep resistivity sounding, AMT and CSAMT are for 100–2,000-m depths or a little beyond. Times are in (i) nanoseconds for GPR, (ii) microseconds to a few hundred milliseconds for TEM, (iii) milliseconds for time domain-induced polarisation and high-resolution seismics and (iv) fraction of a millisecond to 1 s for AMT, CSAMT, CFS, DFS and seismics. Longer-period signals go outside the periphery of groundwater geophysics. It is an important problem for the countries having a long coastline.

Keywords Saline and fresh water problems · Geophysical exploration tools

1 Introduction

Groundwater, in general, is subsurface water occupying most of the pore spaces or voids in a porous and permeable formation underground in a sedimentary argillaceous sand–shale/clay soft-rock terrain (Fig. 1). In the terrains of carbonate rocks occupied by limestone, dolomites and anhydrites, besides intergranular porosity, solution cavities and vugs are also present. Fractures, fissures, cracks, joints, faults, thrusts and lineaments are generally the locations of groundwater in a hard-rock (igneous and metamorphic) terrain, although fractures and cracks can be present in all forms of rocks. From permafrost region to dessert, from soft rock to hard rock, from coastal region to mountains, diverse geological problems do exist related to groundwater exploration. Only a few points will be discussed on saline water–fresh water problems in this short chapter. Porous and permeable formations that contain water are termed as aquifers. Electrolytic current conduction through geological medium takes place mostly through pores or void spaces filled with water (Fig. 1). Porosity of a rock can be primary or secondary. Intergranular porosity is a primary porosity. All other types of void spaces (just mentioned) come under the umbrella of secondary porosity. Both primary and secondary porosities control groundwater accumulation and flow. The porosity is defined as

$$\text{Porosity}\ \ \phi = \frac{\text{Pore volume}}{\text{Total volume}} \times 100 \tag{1}$$

It is a dimensionless fraction expressed in percentage. Secondary porosity comes from the solution cavities and vugs in carbonate rocks. Fractures, fissures, etc. are the sources of groundwater in granitic rocks because primary porosity can be as low as 1 % or even less (0.1 %). Sedimentary rocks can be highly porous. Clean fresh water coarse-grained sand can have a porosity as high as 35 %. In a water-saturated shale or clay, the porosity can be as high as 40 % or even more. Permeability, i.e. the ease with which a fluid can flow through a porous medium, has some contribution on current conduction path. For clean sandstones, the rock matrix is of silica grains, which are insulators. Current is mostly conducted through the pore fluids.

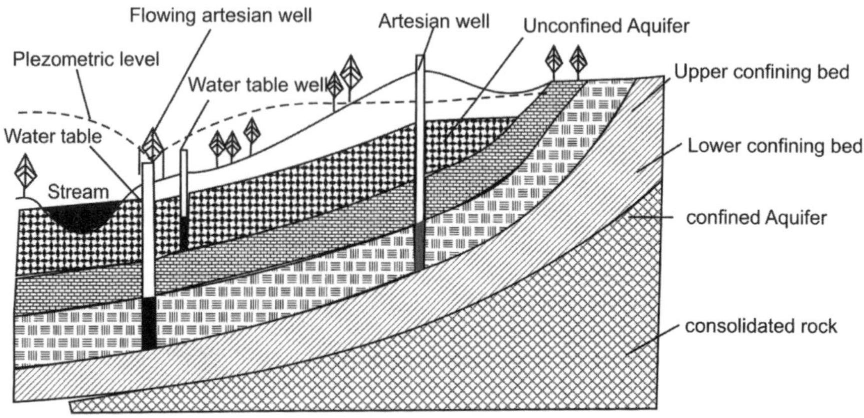

Fig. 1 The confined and unconfined aquifers, water table, confining layers, artesian wells, consolidated basement rock and streams

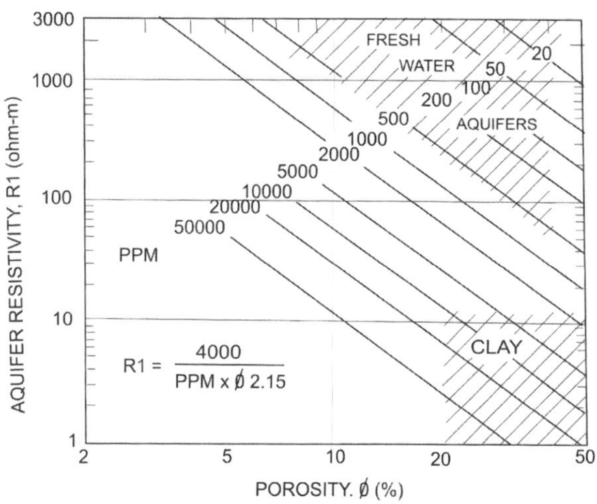

Fig. 2 The interconnectivity of porosity, resistivity and salinity of groundwater following Archie's law. (Taken from Hubert Guyod 1966)

Interconnectivity of these fractures in hard rocks dictates the permeability of a formation. Borehole geophysics contributes significantly in this front by providing (i) resistivity porosity ϕ_R, from combined self-potential (SP) and resistivity logging based on Archie's law, (ii) neutron porosity (ϕ_N) based on neutron logging, (iii) density porosity (ϕ_D) based on gamma–gamma density logging and (iv) sonic porosity (ϕ_S) based on sonic logging. Since sonic porosity responds to primary porosity and other porosities respond to total porosity, borehole geophysics contributes towards estimation of secondary porosities. Figure 2 shows the relation between, resistivity, salinity and porosity of a formation based on Archie's relation (1942).

The aquifers may be confined and unconfined. There may be multiple confined aquifers in a particular geological terrain. The confined aquifers are sandwiched between the two confining layers that are generally porous but composed of impermeable clay/shale layers as shown in Fig. 1. These confining layers are termed as aquicludes. In a sand shale sequence, the porous and permeable sand/sandstones are aquifers. In a carbonate area, porous and permeable limestones/dolomites are aquifers. Hard and compact limestones, dolomites/anhydrites or shale or clay may form the confining layers. In hard-rock areas, faulted, fractured weathered rocks and lineaments are the potential reservoirs for groundwater. Practically nonporous (porosity $<1\,\%$) and impermeable hard rocks (granitic/granodioritic/granulitic/basaltic terrain) become the confining layers. Layered earth situation is most unlikely to be present in granitic terrain. In flood basalt-covered areas, some of the intratrapean beds can become the potential aquifers. In the desert area, an aquifer may lie below the thick loose sand dunes. Aquifers may lie below the thick permafrosts in Arctic and Antarctic region. Buried underground flow channels can be the potential source of groundwater in many areas. Geophysical tools for their detection may be electrofiltration potential (SP), gravity, magnetic, IP, resistivity and electromagnetics (EM). Buried flow channels are not included in this discussion. Occurrence of groundwater itself created diverse geophysical problems. When saline water encroaches in these fresh water territories, special physical and chemical properties of the saline water fresh water contacts and the associated properties of the surrounding host rocks get added to these diverse geological problems. Depending on the geological situation, geophysical tools to be used are decided. Systematic studies of saline water–fresh water problems are needed because it is directly associated with water for domestic uses as well as for irrigation and industries keeping in mind that rain water harvesting is still not a very economically viable proposition for many countries.

Deep marine fossils at the top of the Himalayan mountain range indicate that a part of the lithosphere of the Indian plate was below the ocean. Since after the collision of the Indian plate with the Eurasian Plate, a part of the Indian plate subducted below the Tibetan plate and the upper part has gone up to build the Himalayan mountain, it is very likely that some deep-seated saline connate water pockets are formed at the foothills of Himalaya and below Indo-Gangetic alluviums and up to Narmoda–Son lineament during this collision process. Since India has a vast coastal shore line, the probability that saline water invaded the fresh water aquifers is quite high. Saline water has contaminated the coastal aquifers of West Bengal, Orissa, Tamil Nadu, Kerala, Maharashtra and Gujarat in India. In Gujarat, the saline water invasion has gone up to 12–14 km on shore behind the shore line (personal discussion with a Central Ground Water Board Scientist, CGWB). No significant and systematic geophysical survey has been done so far in India to handle this saline water invasion problem throughout the coastal areas. Some systematic discussions on the totality of the surface and borehole geophysical tools and the geological scenario basis prescriptions are available to handle the saline water–fresh water problem (Sabnabis and Patangay 1998). However, total saline water encroachment map of India based on geophysical survey is yet to appear. Two saline water/brackish water-encroached sections are shown as Indian examples from West Bengal. Direct current (DC) resistivity method has been used for this purpose. The first section is presented from Sonarpur area (Fig. 3) at the outskirts of Kolkata, a metropolitan city

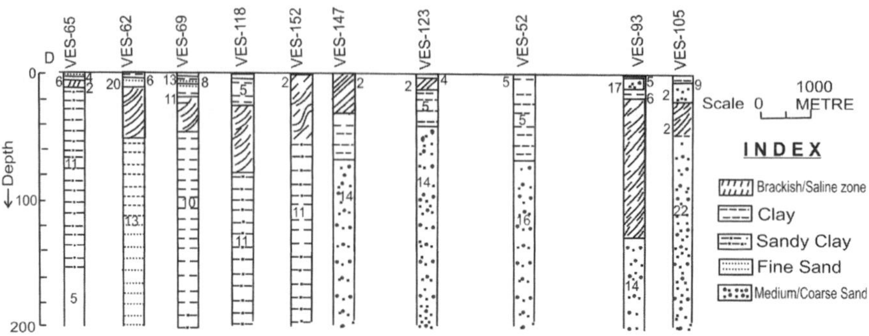

Fig. 3 Saline/brackish water-affected shallow zones of Sonarpur area at the outskirts of Kolkata, West Bengal (taken from Choudhury et al. 1997); an example of encroachment of saline water in an urban area (based on vertical electrical sounding (*VES*))

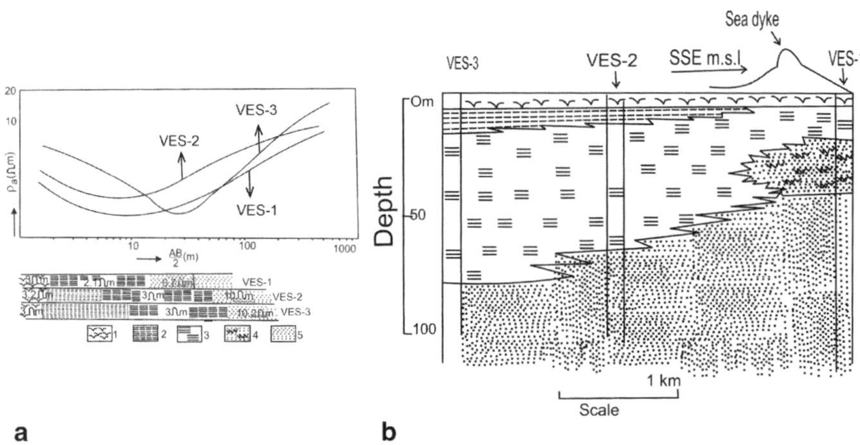

Fig. 4 An example of saline water encroachment in coastal areas near Jaldha, Digha, West Bengal (taken from Patra 1967): vertical electrical sounding (*VES*) curves and the interpreted geoelectric section

(urban pollution encroachment due to over-exploitation (Choudhury et al. 1997)), and the second section (Fig. 4a, b) is presented from the coastal area of Digha, West Bengal, India (Patra 1967). It is a common problem for all the countries having long shore lines.

As early as 1927, Conrad and Marcel Schlumberger from Schlumberger Well Surveying Corporation first observed some measurable voltage in a freshly drilled borehole and confirmed its repeatability. It took more than 15 years to understand wherefrom some potential difference and the measurable voltage came. It was later realised that salinity contrast between the borehole mud and the formation water was responsible for the origin of this potential. The mechanisms for the origin of these potentials were known to the electrochemists much earlier (Glasstone 1962, 1969; Bochris and Reddy 1970). Later, redox potential came in earth sciences in many different fields, including groundwater pollution. Probably this was the starting point of salinity contrast between the two electrolytes and its geophysical applications. Since then, it is a long root march between saline water–fresh water interface and geophysical tools to detect it.

The moment it was known that electrical conductivity of water has strong dependence on salinity and electrical conductivity of water increases very rapidly with the concentration of sodium chloride (NaCl), almost all the branches of electrical and electromagnetic methods that have some say on electrical conductivity of earth formation got entry into the saline water–fresh water problems. Because electrical currents are mostly conducted by the pore fluids; higher the electrical conductivity of the pore fluids, higher will be the electrical conductivity of the rocks. Besides geophysical tools under electrical and electromagnetic methods, some more geophysical methods joined hands to solve the fresh water–saline water problems. The total scenario about the geophysical methodologies include (i) DC resistivity method (Stefanescu et al. 1930; Keller and Frischknecht 1966; Mooney et al. 1966; Adel Zhody 1965; Kalenov 1957; Bhattacharyya and Patra 1968; Koefoed 1979; Ghosh 1970; Zhdanov and Keller 1994; Roy and Apparao 1971; and many others); (ii) induced polarisation (IP) method (Seigel 1959; Keller and Frischknecht 1966; Sumner 1976; Nabighian and Elliott 1976; Roy and Elliott 1980a, b; Roy et al. 1995; and others); (iii) (a) electromagnetic method (artificial sources; Hohmann 1971, 1975, 1983, 1988; Ward 1967; Ward and Hohmann 1988; Wait 1951, 1970, 1982; Keller and Frischknecht 1966; Nabighian 1994; Parasnis 1966; Patra 1967, 1976, Patra and Mallick 1980, Patra and Shastri 1982 Zhdanov and Keller 1994; and others), (b) natural-source electromagnetics (Cagniard 1953; Tikhnov 1950; Berdichevsky and Dmitriev 1960; Vozoff 1989; Maddan et al. 1989; Mackie et al. 1993; Oldenburg 1979; Constable et al. 1998; Hoversten et al. 1998; and others), (c) audio frequency magnetotellurics (AMT; Strangway et al. 1973; Hoover and Long 1976; Hoover et al. 1976, 1978), (d) controlled-source AMT (CSAMT; Zonge and Hughes 1994; Zonge et al. 1989), (e) electromagnic transients (time domain EM; Morrison et al. 1969; Lee and Lewis 1974; Wait 1982; Kaufman and Keller 1983; Fitterman 1987; Fitterman and Stewart 1986; Stewart 1982), (f) GPR (Bares and Haeni 1991; Fisher et al. 1992; Davis and Annan 1989; Annan and Davis 1976; Arcone 1984; Benson 1995), (g) remote sensing (Swain and Davis 1978; Seigel 1980; Richards and Xiuping 1998; Schowengerdt 1963; Rossitier et al. 1975; Jensen 1986; Joseph 2003), (h) high-resolution shallow seismics (Cardimona et al. 1998; Ghose and Slob 2006) and (i) borehole geophysics; (iv) SP and resistivity logging (Doll 1948; Anon 1958, 1972; Pirson 1963; Dakhnov 1962); (v) natural gamma ray and neutron logging (Anon 1958, 1972; Pirson 1963; (vi) sonic logging and density logging (Anon 1958, 1972; Pirson 1963).

Soon it was realised that clay or shale can be highly conducting layers and their resistivity may be of the same order as saline water sands. Firstly, clay or shale may be impermeable, but they can be highly porous and the porosity can be as high as 45 %. So current gets a wider path to flow. Secondly, clay has an alumina (Al_2O_3) layer giving a parallel path for the flow of current. To resolve these problems, many more geophysical tools came forward. First in this list is the membrane polarisation in IP method. Firstly, clay, shale or shaly sands will have some IP effect, whereas saline water sand will have zero IP effect. Secondly, clay or shale is much better absorber of the radionuclides; therefore, gamma ray counts against a shale bed will be much higher than that against saline water beds. GPR came in this field with a new physical parameter, viz electrical permittivity or dielectric constants. Dielectric constant of fresh water and saline water is same, but clay beds and saline water-bearing beds will have very different dielectric constants. Demarcation on the basis of electrical conductivity only will invite ambiguity. SP and resistivity logging can differentiate fresh water and saline water beds. Combination of thermal neutron-capture gamma ray and thermal neutron–thermal neutron logging can easily distinguish saline water and freshwater beds. Although reflectivity of the fresh water and brackish water will be different in remote sensing, its area of major application lies in mapping of fractures, fissures, faults, thrusts, lineaments, etc. This is how remote sensing with geographical information system (GIS) and global positioning system (GPS) entered in this problem especially for hard-rock areas. Firstly, India has a huge pretertiary flood basalt-covered area known as Deccan traps. Secondly, India has Archaean cratons and Proterozoic mobile belt in the east and Dharwar craton and Peninsular gneisses in the south India. Therefore, remote sensing will be a very useful tool for reconnaissance survey related to delineation of fractures, fissures, etc. Since India has a large coastal area, saline water can enter deep inside the onshore land mass. Elastic wave propagation cannot differentiate saline water–fresh water zones, but since high-resolution shallow seismics can resolve the subsurface as accurately as GPR and all advanced seismic processing softwares can be used for multichannel GPR data processing, GPR and high-resolution seismics (HRS) is a good combination of geophysical tools for very shallow investigation. Seismic reflection and refraction join hands with other major geoelectrical methods for better delineation of saline water zones. Sonic log gives primary porosity. If resistivity, density or neutron porosity logs are available, secondary porosity of a formation can be determined. It is an important parameter for hard-rock areas. Sonic and density log together will give more accurate estimate of the contrasts in acoustic impedances of the different stratified layers for seismic reflection data interpretation. This is how porosity logging tools entered into this area. Very briefly, some of these points are discussed in this short chapter.

2 Basic Physics and Electrochemistry Behind These Geophysical Tools

2.1 Electrolytic or Ionic Conduction

Electrolytic conduction takes place in a liquid due to the movement of cations and anions in the opposite direction. In a bath of electrolyte, the flow of current is controlled by the equation

$$I = AF\ (C_1 v_1 + C_2 v_2 + C_3 v_3 + \ldots\ldots\ldots C_n v_n) \qquad (2)$$

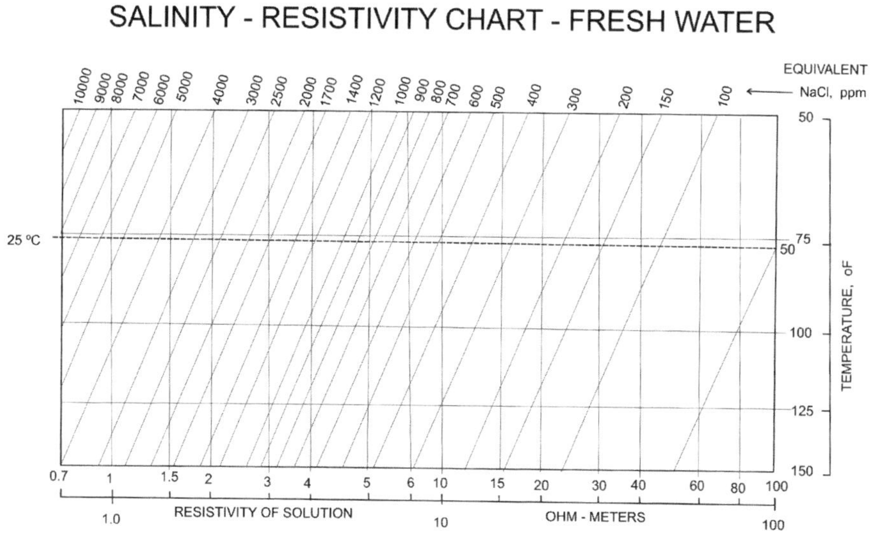

Fig. 5 Variation of electrical conductivity of an electrolyte with salinity. (Taken from Scott Keys and Mac Cary 1971, USGS book)

where I is the current flowing between the anode and cathode in an electrolytic bath, and C_n and v_n represent the concentration and mobility of nth species of the ions. Mobility of an ion (anion or cation) is the velocity under unit potential gradient between the positive and negative electrodes in the bath. A is the cross-sectional area through which the current flows. F is the Faraday number, i.e. 96,500 C. Therefore, higher the concentration of anions and cations, higher will be the flow of current I through the electrolytic bath. Since J (= I/A), the current density $= \sigma E$, from the theory of DC flow field, electric field remaining constant, electrical conductivity σ will increase with the increase in saline water. Electrical resistivity of the fluid will decrease since electrical resistivity ρ and σ has inverse relation. All the geoelectrical tools, viz. DC resistivity sounding and traversing, electromagnetic sounding and traversing, AMT and CSAMT, electromagnetic transients (TEM) and GPR, will search for this property. Electrolytic current conduction through geological medium takes place mostly through pore or void spaces filled with electrolytes. Fig. 5 shows the variation of electrical conductivity with salinity of water.

2.2 *Electrochemical Cells*

Liquid Junction Cell

Whenever two electrolytes of different NaCl concentrations are in direct contact, electrochemical cells are developed there. These electrochemical cells that can generate electric current are known as liquid junction cells or diffusion potential cells. They follow the Handerson's equation

$$E_l = \left\{ \frac{(U_- - U_+)}{(U_- + U_+)} \right\} \left(\frac{RT}{nF} \right) \ln \left(\frac{a_2}{a_1} \right) \tag{3}$$

Where U_- and U_+ are the mobilities of the anions and cations, respectively, *R* is the universal gas constant, *T* is the absolute temperature, *n* is the valency, *F* is the Faraday number, a_2 and a_1 are respectively the chemical activities of the two electrolytes in contacts and E_l is the liquid junction potential. The values of these parameters are respectively given as

$U_- = 67.6 \times 10^{-5}$ cm/sec/volt
$U_+ = 45.6 \times 10^{-5}$ cm/sec/volt
R = universal gas constant = 8.314 J
F = Faraday number = 96,500 C
T = absolute temperature, i.e. 298° at 25 °C

$$E_l = -11.6 \ln \left(\frac{a_2}{a_1} \right) \text{ millivolts}$$

Membrane Cell

Whenever two electrolytes of two different concentrations are in contact through a cationic membrane, the mobility of the anions becomes zero ($U_- = 0$). Potentials generated through these cells are termed as membrane potentials or diffusion–adsorption potential. Under normal circumstances, the diffusion–adsorption potential will have opposite polarity to that of diffusion potential. Since $U_- = 0$, Handerson's equation changes to the form

$$E_M = -\left(\frac{RT}{nF} \right) ln \frac{a_2}{a_1}. \tag{4}$$

$$E_M = -59.1 \ln \frac{a_2}{a_1} \text{ millivolts.}$$

This equation is known as Nernst equation. These two cells can generate measurable potential across a saline water–fresh water contact in the coastal areas. In principle, these potentials can be measured on the surface with the help of a very sensitive potentiometer. These two electrochemical cells will show opposite polarity for a particular concentration ratio of the two electrolytes in contact. Inside a borehole on the other hand, these two electrochemical cells combine together to form the SP electrochemical cells. The flow of current through the earth formation adjacent to the borehole will change direction depending on whether the NaCl concentration of the formation water is more or less than that of the drilling mud. Fig. 6 shows the series connection of E_j and E_M such that the two potentials get added up. SP current will be anticlockwise when $C_2 > C_1$ and clockwise when $C_2 < C_1$. This is termed as SP reversal, a qualitative diagnostics for saline water–fresh water problems.

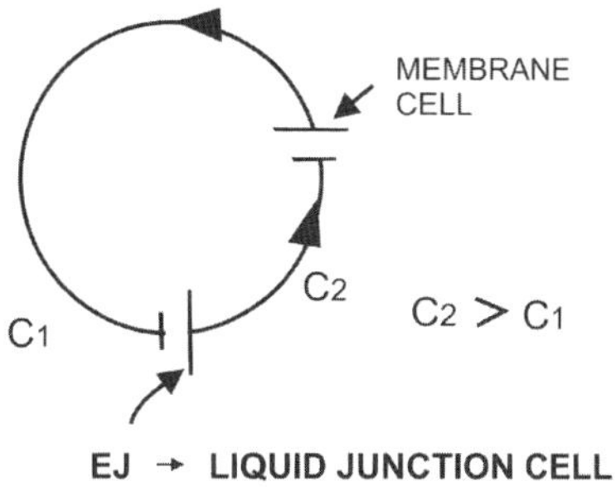

Fig. 6 Liquid junction and membrane potential cells are in series connection in borehole sand shale sequence condition; C_1 and C_2 are the NaCl concentrations of two electrolytes in contact, respectively

Redox Cell

Oxidation–reduction potential is a measure of the tendency for chemical species in ionic solution to change their oxidation or reduction state. It is also a quantitative measure of the tendency of the redox system to evolve towards oxidised or reduced state. From chemical thermodynamics point of view, the measured redox potential of a solution containing various ionic species in various oxidation states is given by the Nernst equation:

$$E_h = E_0 + (RT/nF)\ln(a_{ox}/a_{red}) \quad (5)$$

if the potential determining reaction is

$$Ox + ne \rightarrow Red \quad (6)$$

where E_h is the redox potential of the system with reference to the standard hydrogen electrode; E_0 is the redox potential of the system under standard condition, i.e. when $a_{ox} = a_{red} = 1$ at 25 °C; R, T, n and F have already been defined; and a_{ox} and a_{red} are, respectively, the chemical activities of the oxidised and reduced species. Redox potential of a solution depends on the ratio of the oxidants and the reductants, and it is not affected by the neutral ionised salts such as NaCl. Therefore, redox potential measurement is not an important parameter for salinity detection, but it is an important indicator of other types of pollutants, viz. arsenic, fluoride, phosphate, chromium, nitrate, etc (Roymahashay and Sen 2010).

2.3 *Direct Current Resistivity*

DC resistivity sounding or profiling is the most well-known topic to the geophysicist, and it is the most widely used geophysical tool in groundwater exploration as well as for demarcation of saline water–fresh water zones. The basic physics is based on Ohm's law and one of the three basic electromagnetic equations, i.e. $J = \sigma E$, coming from DC field theory, where J is the current density, σ is the electrical conductivity and E is the electric field created by sending DC through the

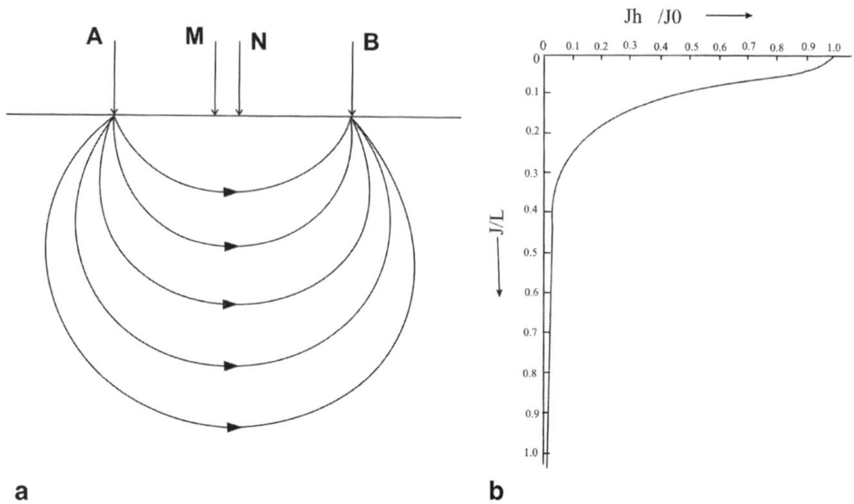

Fig. 7 **a** For a homogeneous and isotropic medium, the current can take a very long route to reach the negative electrode, but the path is equal-resistance path, so current can enter deep inside a medium. **b** Variation of current density with depth

ground. Since, earth is an inhomogeneous and anisotropic medium, the scalar σ changes to a tensor σ_{ij}. The basic physics demands that current will always flow through the least resistance path, and in a homogeneous and isotropic medium, the current will flow through the equal-resistance path. The modified version of Ohm's law gives R = ρL/A, where R is the resistance offered by the ground in ohms, ρ is the specific resistivity, i.e. the resistance offered by the two opposite faces of an unit cube and it is in ohm metre, L is the length of the path the current travels and A is the cross-sectional area for the flow of current. The most important point is as long as L/A remains constant, the resistance offered by the ground also is same.

If the cross-sectional area for the flow of current is more, current flow path at greater depth is possible (Fig. 7a). Deeper the flow path of the current, greater will be the cross-sectional area for the flow of current. Still, the resistance offered remains the same. Thus, current can flow through deep inside the medium even though the current density is less (Fig. 7b). The current does not flow through the least-length path but through least-resistance path or equal-resistance path. This is the starting point of the electrical method of prospecting. Hence, higher the current electrode separation, deeper will be the flow of current, greater will be the depth of exploration (Frolich 1967) and the depth of investigation (Roy and Apparao 1971). This made resistivity sounding possible. Potentials are measured using two non-polarisable electrodes at two independent points. For an inhomogeneous ground, the measured apparent resistivity ($\rho_a = \text{K}\Delta\text{V/I}$) is based on measured current flow through the ground and potential difference between the two potential electrodes multiplied by the geometric factor K. Mailet (1947) first proposed that in DC methods, the ratio of the layer thickness to the layer resistivity in some cases and

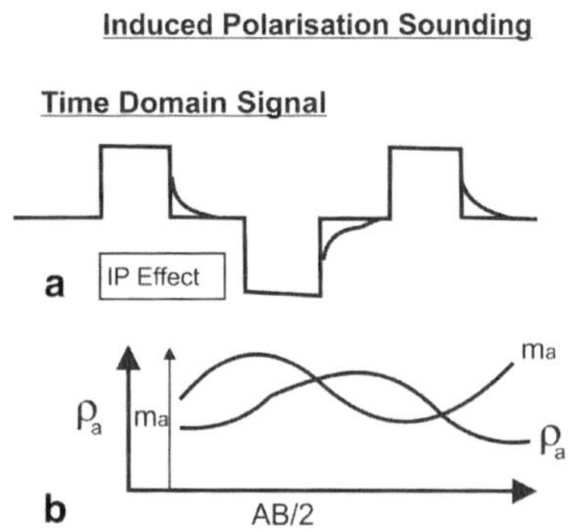

Fig. 8 **a** Time domain pulsating DC for IP measurement. **b** Illustration of joint resistivity and IP sounding curves

the product of the layer resistivities and thicknesses in some other cases are better determined parameters than the individual layer thicknesses and resistivities in 1D problems. Based on this, Adel Zhody (1965) published Dar Zarrough parameters. The basic physics behind scalar potential theory for DC resistivity method is based on the solution of Laplace equation in cylindrical and spherical polar coordinates for 1D/2D boundary value problems. Poisson's equation was used for solution of 2D/3D boundary value problems. This topic is a well-developed subject taking into account that inversion softwares for 1D/2D/3D problems are also available. Therefore, it is an important widely used geophysical tool used in groundwater industries. VES stands for vertical electrical sounding. Figures 3 and 4 are the field examples of applications of VES for solving saline water fresh–water problems.

2.4 *Induced Polarisation*

Membrane Polarisation

Membrane polarisation is one of the important branches of IP method. The other branches are electrode polarisation and electrokinetic polarisation. Here also the concept of 'Overvoltage' and Faradaic double layer from electrochemistry entered in the area of geophysics.

Figure 8a shows the nature of pulsating DC used in joint resistivity and IP sounding, and Fig. 8b shows the field records in terms of apparent resistivity and apparent chargeability versus electrode separation. Figure 9 shows the nature of flow of depolarisation current in time domain IP when a polarisable bed is sandwiched between the two nonpolarisable beds. For time domain IP, the forward problem is solved following the theories of Seigel (1959), Nabighian and Elliott (1976) and Sumner (1976). On-time voltage is recorded for DC resistivity, and off-time voltage is recorded for IP. Polarisation in electrical methods in geophysics means separation of charges and formation of double layers. Membrane polarisation is a much weaker signal than electrode polarisation, and it generally originates in shale or clay sands. Electrokinectic potential is negligible in comparison with other two

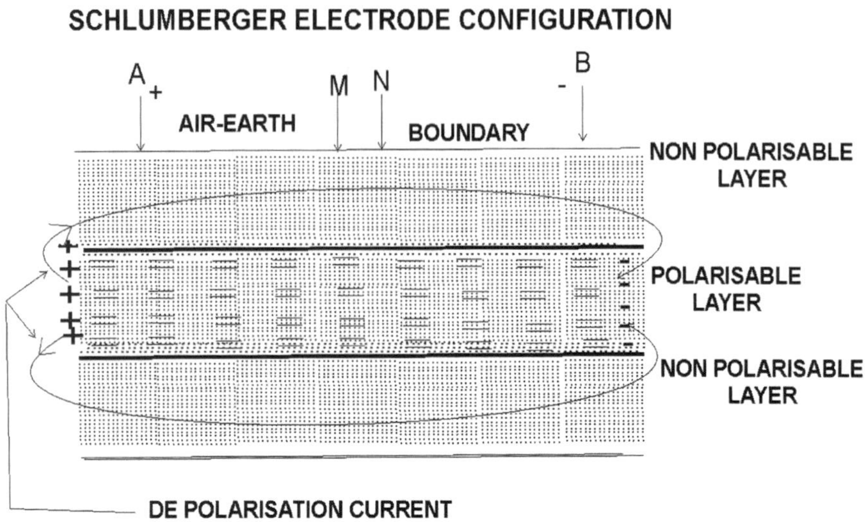

Fig. 9 The origin of membrane polarisation: polarisable layer confined between two nonpolarisable layers, and flowpath for depolarisation current

Fig. 10 Variation of polarisability of shaly sand with percent of clay content by volume; clay type also controls the polarisabolity. (Taken from Keller and Frischknecht 1966)

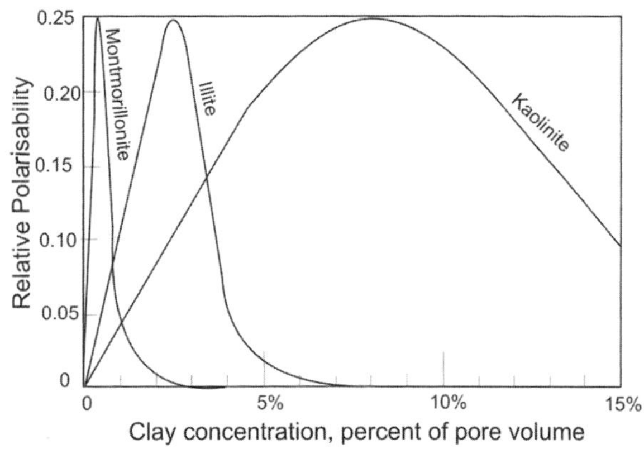

signals and is ignored in geophysical exploration. Membrane polarisation is the most prominent signal in shale or clay sand when the percentage of shale or clay will be about 0.5 % for montmorillonite, 2.5 % for illite and 8 % for kaolinite for peak IP effect. With gradual increase in the percentage of shale, IP effect gradually decreases (Keller and Frischknecht 1966; Fig. 10). Membrane polarisation effect depends on the type of clay. In time domain IP survey in a sedimentary area, the charging–discharging time is increased to enhance the signal (Roy and Elliott 1980). Clean sand/sandstone has negligible chargeability. Saline water sand has zero chargeability. The unit of IP can be either millivolt second/volt (= millisecond) or it can be a dimensionless quantity. If the ratio of primary to secondary voltage at a certain instant of time after the current is switched off is taken, it will be dimensionless. In time domain IP, if the chargeability is measured by measuring the area under the decay voltage curve after the current is swiched off and applying a certain delay time, the unit will be millivolt second. This area under the curve depends on the delay time, charging–discharging time and the integration time, and it also depends on the time the depolarisation current takes for the total discharge. Higher the salinity or higher the electrical conductivity of the medium, faster will be the decay of the depolarisation current, and the medium will get discharged quickly. The recorded chargeability for saline water sand will be zero. Electrical conductivity of the depolarisation current path, absence of double layer and decay rate of the depolarisation current in time domain IP are the physics behind this tool, and this is how it entered into this subject.

2.5 Electromagnetics

Magnetic field originates when a charge moves. Electromagnetic field originates and electromagnetic waves start propagating when the flow of charge becomes time varying. These electromagnetic waves are used in many branches of science and engineering. Modern civilisation is totally dependent on electromagnetic waves. Electromagnetics entered in geophysics in a very big way. Far too many branches of geophysics are based on propagation of electromagnetic waves. Details are beyond the scope of this short communication. Geophysicists use the widest range of the electromagnetic spectrum with gamma ray on one end and sun spot cycle with 11-year periodicity on the other. Coils carrying time-varying alternating currents (AC) for generation of electromagnetic waves are called oscillating magnetic dipoles. These coils are called the transmitters or transmitting antenna. These electromagnetic waves are transverse in nature, with electric and magnetic fields orthogonal to each other in a plane wavefront and are at right angles to the direction of propagation. Electromagnetic waves can travel through vacuum, air, insulators and earth materials. When these waves, because of their time-varying nature, encounter a good conductor of electricity, eddy currents are generated having different amplitude and phase but same frequency. These eddy currents generate secondary fields. A coil that receives both primary waves originated from the transmitter and the secondary fields originated from a conductor is called a receiver or a receiving antenna (Fig. 11). The signals received by the receiving coil will always be a complex quantity because the two signals having different amplitude and phase are interacting. The signal will be available in $\mathrm{A+iB}$ or $\mathrm{A}e^{i\varphi}$ form. In-phase part with the primary signals originated from the transmitting antenna is a real component. Ninety degree out-of-phase part with the primary signal is an imaginary or quadrature component. Phase is a very important part in any electromagnetic measurement. Secondary field through eddy currents brings information about the subsurface. Two basic laws of physics, viz. Faraday's law of electromagnetic induction and Lenz's law, are behind this process. Since electromagnetic waves travel through vacuum, air, insulators,

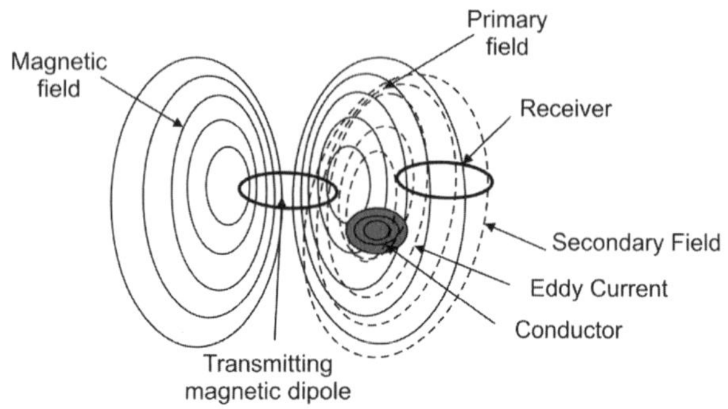

Fig. 11 Basics of man-made source electromagnetics: formation of eddy currents and secondary field in a conducting zone

etc., galvanic contact with the ground is not necessary. This is how airborne and satellite-borne geophysical tools originated.

Basic physical principle for electromagnetic sounding or profiling is as follows: higher the frequency of the electromagnetic signal and higher the electrical conductivity of the medium, stronger will be the eddy currents, stronger will be the secondary field and EM signal, greater will be the attenuation of the EM signals because of the stronger force of opposition as per Lenz's law of electromagnetic induction and lesser will be the depth of penetration of the EM signals. This is the basis for electromagnetic method to become a powerful tool to handle saline water–fresh water problem. One more basic point of electromagnetics is higher the frequency, shorter will be the wavelength and greater will be the resolving power of the EM tool. Because frequency and wavelength have inverse relation, GPR (frequency range: 200 MHz to 10 GHz) has very low depth of penetration and very high resolving power. The other end of the spectrum is geomagnetic depth sounding, where EM signals have very long wavelength and have very poor resolving power. The entire theoretical development for solution of geophysical forward problems is based on the Maxwell's equation and Helmholtz's electromagnetic wave equations. Electromagnetics also is a very developed subject having 1D/2D/3D/forward and inverse problem software. Electrical conductivity and electrical resistivity have inverse relation. In desert, hard-rock and permafrost areas, electromagnetic method will appear to be a much better geophysical tool where DC resistivity method will fail. Dipole frequency sounding (DFS) and central frequency sounding (CFS) are important electromagnetic tools for 1D earth problems. For shallow 2D groundwater EM problems, very low frequency electromagnetics (VLF-EM) can be used. For deep-seated aquifers, TEM, AMT and controlled-source AMT (CSAMT) can be used, provided the survey spot is reasonably away from the coastline to minimise the coast effect.

In Audio and Subaudio Frequency Range, Conduction Current Dominates over the Displacement Current

Starting point of the electromagbetic theory for oscillating vertical and horizontal magnetic dipole placed on the surface of the earth or in the air is based on Helmholtz's electromagnetic wave equation derived from Maxwell's electromagnetic equations. The Helmholtz's wave equations are

$$\begin{aligned} \nabla^2 H &= \gamma^2 H, \\ \nabla^2 E &= \gamma^2 E \\ \nabla^2 \Pi &= \gamma^2 \Pi \\ \nabla^2 F &= \gamma^2 F, \end{aligned} \tag{7}$$

where ∇ is the differential operator; E and H, respectively, are the electric and magnetic field vectors; and Π and F, respectively, are Hertz and Fitzerald vector potentials. Here γ is the propagation constant $\gamma = \sqrt{i\omega\mu(\sigma + i\omega\varepsilon)}$, where ω = angular frequency, μ = magnetic permeability in Henry/metre, σ = electrical conductivity in mho/metre and $\in$ = electrical permittivity in farad/metre. For audio and subaudio range, electrical permittivity is negligible and EM response is conductivity dependent. Since saline water is much more conducting than fresh water, electromagnetic depth sounding using oscillating magnetic dipole is a powerful tool for detecting saline water–fresh water boundaries in the areas inaccessible to DC resistivity sounding. CFS (Patra 1976; Patra and Shastri 1982) and DFS (Wait 1951, 1970, 1982; Ryu et al. 1970; Nabighian 1994) are the proper geophysical tools.

Earth's Natural Electromagnetic Field in Groundwater Exploration: Audio Frequency Magnetotellurics

The basic physics is that the ratio of the electric field to the transverse magnetic field gives the impedance of the ground. The unit is ohm. This fact was known to the physicists long before Cagniard (1953) and Tikhnov (1950) proposed their theory of magnetotellurics (MT). MT is based on plane wave electromagnetics. Since electrical conductivity of an earth formation increases due to an increase in both temperature as well as salinity, any geophysical tool prescribed for geothermal exploration can also be used for the study of saline water-invaded zones. Thus AMT and CSAMT entered in this subject for deep-seated groundwater exploration.

Within the frequency range of interest in AMT, only the EM signals of atmospheric origin are important. The AMT signals range from 1.0 Hz to 16 KHz and are created by worldwide thunderstorm activity in the lower atmosphere. These signals from the thunderstorm and lightning activity are concentrated in the tropics and are referred to as spherics or atmospherics. Cloud to ground thunder is the principal component. It is surrounded by several cloud to cloud thunders. The EM wave propagates around the globe along the earth-ionosphere wave guide of about 50–60-km thickness. There are three storm centres, located in Brazil, Central Africa, and Malaysia, which have on an average 100 storm days per year. The geographical distribution of these storm centres is such that there is a storm in progress in one of the centres during any hour of the day. Therefore, detectable high-frequency EM signals are always available around the globe. These signals propagate around the

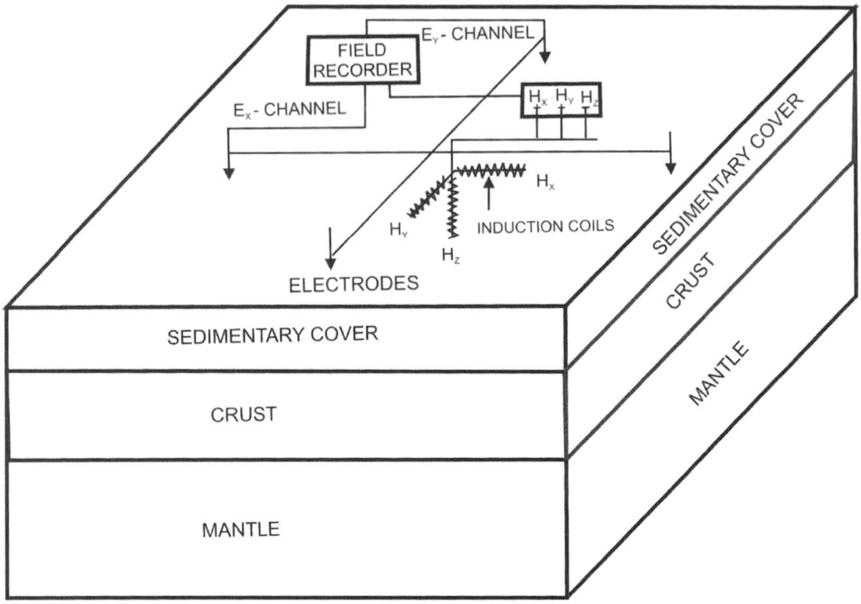

Fig. 12 Field set-up for magnetotellurics and audio frequency magnetotellurics; two pairs of current electrodes for electric field and three induction coils for magnetic field measurement

world, trapped in the wave guide. The energy is reflected back and forth between the lowermost layer of the ionosphere and the ground surface and the spherics attain their peaks in the early afternoon, local time. Some of the high-frequency man-made signals from broadcasting stations or defense installations or powerlines get mixed with the thunderstorm signals. These high-frequency signals have altogether different origin. The plane wave electromagnetic theory applicable for MT is applicable for AMT also. The difference lies in frequency, signal origin, instrumentation and duration of the field work.

Both electric and magnetic components of the electromagnetic field are measured on the surface, the way it is done in MT (Fig. 12). Two pairs of current electrodes measure east–west component Ex and north–south component Ey. Three induction coils measure three magnetic components, i.e. Hx—east–west component, Hy—north–south components and Hz—the vertical component. The ratio of electric components to transverse magnetic components is used to determine the apparent resistivity of a multilayered earth. Both scalar and tensor MT measurements are possible. It is a tool suitable for studying the first 10 km of the earth. It is an effective tool for mineral exploration/shear zone/contact zone/collision zone/deep-seated groundwater exploration and contamination studies. Deep-seated connate water pockets created due to crustal movements or formation of major lineament and faulted traps are the probable targets to be explored by AMT.

Controlled-Source Audio frequency Magnetotellurics

Controlled-source AMT is one of the artificial-source frequency soundings in electromagnetic methods where the nature of measurements is similar to that of MT and AMT and the nature of qualitative and quantitative interpretation is similar to that of AMT. The man-made source here is either a grounded dipole or a horizontal loop. The length of the grounded dipole can be within 1–4 km. Syed et al. (1985) did some work on groundwater contamination by brine leakage from pipelines using CSAMT. For further details, see Zonge and Hughes (1994). Inverse problem of 1D CSAMT data has been solved by Routh and Oldenburg (1999) and Xiaobo et al. (2000). With little modification, Mackey (1993) approach can be used for forward modelling and inversion of 2D/3D CSAMT data. Both AMT and CSAMT pseudosections are highly informative qualitatively (Fig. 13).

Both AMT and CSAMT can map the fracture/fissure/fault zones very effectively over and above detecting deep-seated conductors. Therefore, these tools can join hands with remote sensing for solving the problems of hard-rock groundwater geophysics. Depth of exploration is about 1–2 km, and the frequency range is about 0.1 Hz–10 KHz. The sources are kept at least four skin depth $(\delta = \sqrt{2 / \omega\mu\sigma})$ away from the site of measurements. Similar to MT and AMT, in total five components (two electrical (Ex and Ey) and three magnetic components (Hx, Hy, Hz)) are measured. Data processing and computation of apparent resistivity and phase are similar to those of MT. So it can be an ideal tool for detection of deep-seated saline connate water pockets.

In Radar Frequency Range, Displacement Current Dominates over the Conduction Current

Ground-Penetrating Radar In megahertz (10^6 Hz) and gigahertz (10^9 Hz) range, displacement current becomes significant, and therefore electrical permittivity or relative dielectric constant becomes an important physical parameter. The propagation constant γ becomes approximately equal to $i\omega\sqrt{\mu \in}$. This part is much larger than γ $(= \sqrt{i\omega\mu\sigma})$ in the audio frequency range. Electrical conductivities (σ) of the fresh water and saline water are significantly different, but dielectric constant (ϵ) is more or less of the same order.

In contrast, dielectric constant of saline water sand and clay will be significantly different, although their electrical conductivities may be of the same order. This is how GPR entered into the arena of saline water–fresh water problems with electrical permittivity $(\in)$ as an additional physical parameter not available in the audio frequency range. GPR is a high-resolution shallow geophysical mapping tool. Table 1 shows the electrical conductivity and relative electrical permittivity of the common earth materials of our interest for the present discussion (Bares and Haeni 1991). Resolution of the subsurface is wavelength or frequency dependent, and the resolving power varies within $\lambda/4$ to $\lambda/2$, where λ is the wavelength of

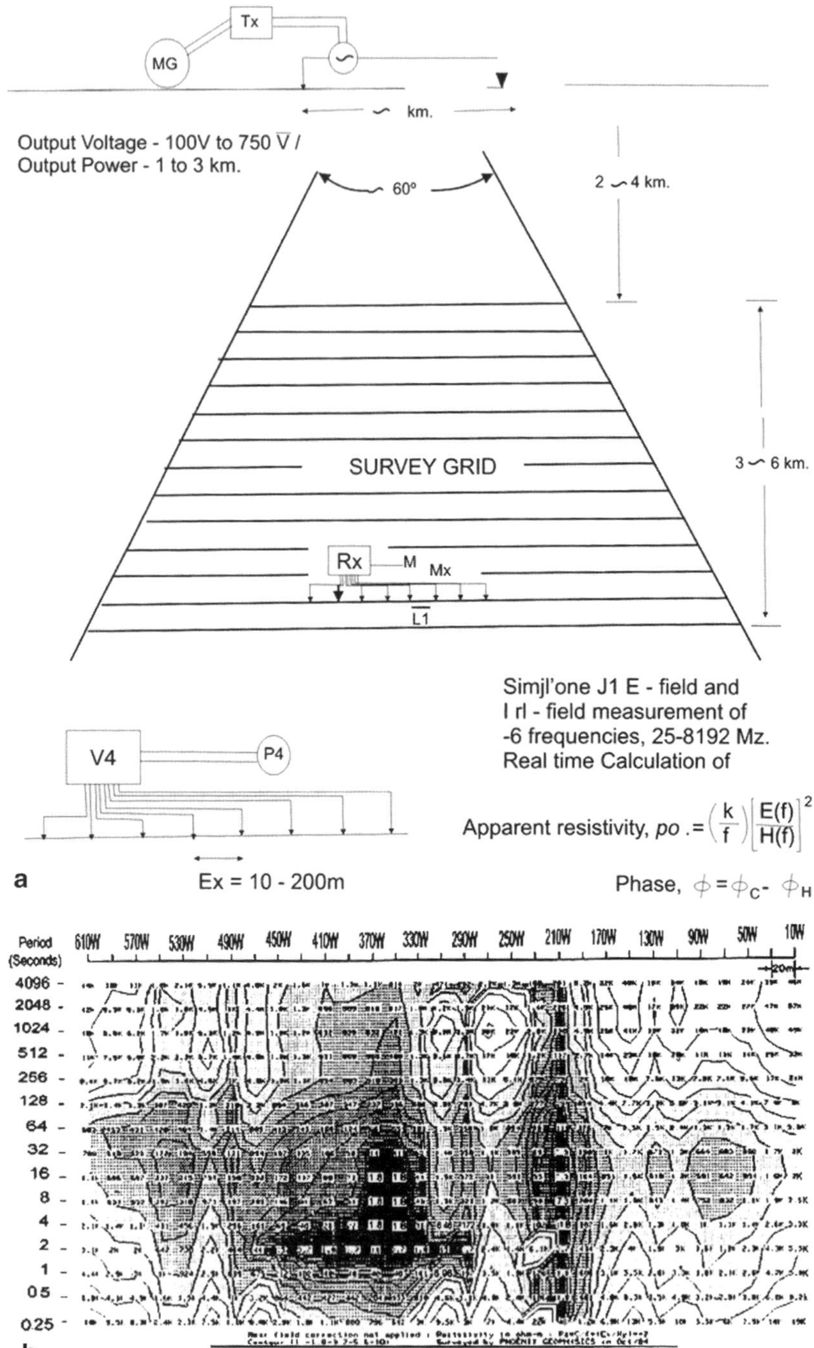

Fig. 13 **a** Field set-up for scalar CSAMT survey; Phoenix configuration and grid for scalar CSAMT; single transmitter and many receivers (taken from Mitsuru Yamashita 1984, Phoenix advertisement pamphlet). **b** CSAMT pseudosection. (Taken from Mitsuru yamashita 1984 Phoenix advertisement pamphlet)

Table 1 Electrical conductivity and dielectric constants of common earth materials in a sedimentary sequence

Common earth materials of present interest	Electrical conductivity σ (mho/metre)	Relative dielectric permittivity ϵ
Fresh water	10^{-4} to 3×10^{-2}	81
Salt water	4	81
Sand (dry)	10^{-7} to 10^{-3}	4–6
Sand (saturated with water)	10^{-4} to 10^{-2}	30
Silt (saturated with water)	10^{-3} to 10^{-2}	10
Clay (saturated with water)	10^{-1} to 1	8–12
Shale	10^{-1}	7

the radar signals. The principal limiting factor in the depth of penetration of GPR signal is attenuation of electromagnetic waves in the earth materials. The attenuation results from spherical spreading losses, exponential losses through materials with contributions from conversion of electromagnetic energy to thermal energy and through electrical conduction losses and dielectric relaxation in water, clay, silt, etc. The penetration depth through good conducting clay with higher porosity or saline water sands can be as low as 1 m. In areas having low electrical conductivity, and with little low GPR frequency, the penetration depth can be as high as 30 m. For same order of resolution, seismic signal frequency is of the order of about a few hundred hertz, whereas radar frequency is in the order of a few megahertz, and seismic travel time is in milliseconds, whereas GPR travel time is in nanoseconds (10^{-9} s). Elastic wave velocities are in a few thousand metres per second, whereas radar signal velocities are in a few million metres per seconds. In earth formation, velocity of electromagnetic waves goes down considerably. Despite so much of difference between seismics and GPR, they follow many of the basic laws of optics in physics.

Transmission, Reflection, Emission and Absorption Electromagnetic Waves

Remote Sensing Geophysical observations are taken almost always from a certain distance from the target. However, by remote sensing, we mean that the observations are collected from space satellites. We get high flight imageries from the sensors housed in these space satellites. From 1960 onwards, man-made artificial orbiting satellites are in the upper atmosphere. The sensors in these satellites measure the electromagnetic waves, which are reflected solar rays coming from the earth's surface as well as EM waves that are directly emitted by the earth's surface. Both the emissions from the sun with a surface temperature of 6,000 °C and from the earth with an average surface temperature of 300 °K are detected by the sensors housed in the orbiting satellites. The distribution of energy follows Planck's law of blackbody radiation. It relates the energy radiated by a black body at a given temperature with a given wavelength of emission. These electromagnetic sensors are passive sensors that record electromagnetic spectrum restricted to visible and infrared band of the solar spectrum. The band of interest in the EM spectrometry ranges from 0.3 μm to about 3 cm or in terms of frequencies from 10^{10}–10^{13} Hz. Visible range of the spectrum is from 0.4 to 0.7 μm. High-frequency side of VIBGYOR has near- and far-ultraviolet rays. The low-frequency side has infrared rays of the spectrum. Solar radiation-generated EM spectrum from the sun is in this frequency range. Higher-frequency radiations such as X-rays, gamma rays, radio waves, etc. are also present. Different objects, viz. ocean water, vegetation, rocks, dry and wet land, etc., on the earth surface are either sources of EM energy or reflectors of energy received from the sun. This is the approach for the study of the earth by remote sensing within this narrow band of 0.3–10-μm wavelengths. Earth's surface receives energy in this prominent wavelength and reacts either by absorbing it, reflecting it or transmitting it. When a signal is transmitted at a particular frequency band for a specific geological problem, the receiving sensor will be in active mode. Since the nature of the interaction of the electromagnetic radiation with an object depends on its properties, the study of these interactions can lead to an understanding of the objects under observation. The basic property that allows identification of objects is called signature. This concept of signature is the identification mark. Remote sensing system (Joseph George 2003) comprises the following: (i) origin of electromagnetic energy (sun, transmitter carried by the sensor), (ii) transmission of energy from the source to the surface of the earth and its interaction with the intervening atmosphere, (iii) interaction of the energy with the earth's surface (reflection/absorption/emission) or self-emission, (iv) transmission of the reflected or emitted energy to the remote sensor placed in a suitable platform through the intervening atmosphere, (v) detection of energy by the sensor, converting it to photographic images or electrical output, (vi) transmission and recording of the sensor output, (vii) preprocessing of data and generation of data products (collection of ground truths and other collateral information), (viii) data analysis and interpretation and (ix) integration of interpreted images with other data. The nature of the solar energy spectrum band is shown in Fig. 14a and b. Fig.14a shows the energy distribution spectrum in terms of the wavelengths of the electromagnetic waves. Fig. 14b shows the spectrum band for (i) photography, (ii) thermal scanner and (iii) radar and passive microwaves.

2.6 Elastic Waves

Elastic waves generated within the earth due to sudden man-made disturbances of any form (explosives, weight dropping, hammer throw, vibroseis, etc.), mostly needed in seismic exploration, do not have any direct relation with saline water–fresh water problems in geophysics. How did this subject then enter into the arena of the current topic of discussion is a question to be answered first. Green's theorem of equivalent layers in potential theory (Stratton 1941; Kellog 1953; Blakely 1996; Roy 2007) and ambiguity in geophysical interpretation (Roy 1962) made in roads for entry of these geophysical tools based on the propagation of elastic waves through earth. Green's theorem of equivalent layers states that infinitely many earth

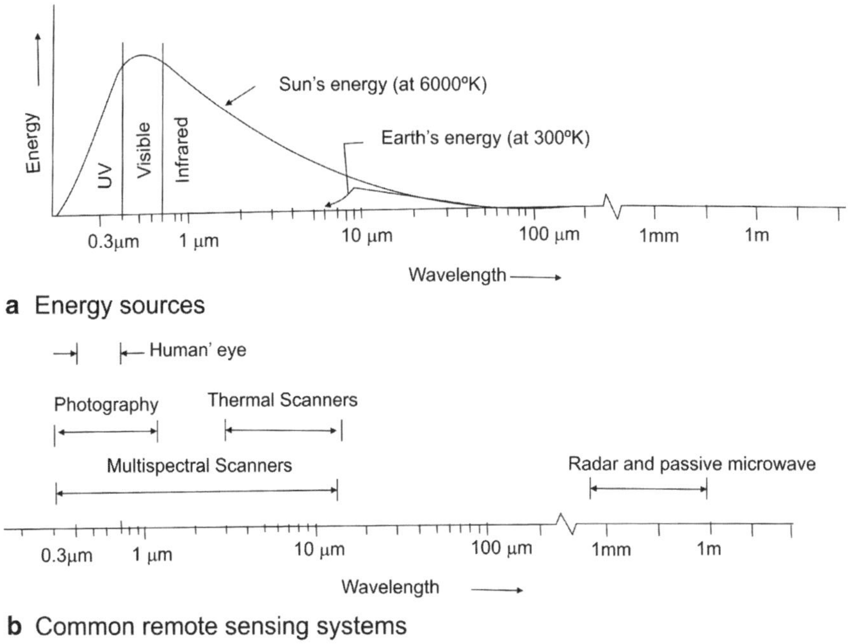

Fig. 14 **a** Energy spectrum of the sun and earth for black-body radiation. **b** Frequency spectrum of the common remote sensing system. (Taken from Sabnabis and Patangay 1998)

models can generate the same response on the surface of the earth. Besides gravity and magnetics, the principle of equivalence does exist in DC and electromagnetic methods. In DC electrical method principles of equivalence ($S = \frac{h}{\rho}$ = longitudinal conductance; $T = h\rho$ = transverse resistance) do exist. That is, for a certain type of subsurface setting (K-, A-type or H-, Q-type models; Keller and Frischknecht 1966; Bhattacharyya and Patra 1968), S or T is the better determined parameter and not the individual layer resistivities or thicknesses. For many combinations of resistivities and thicknesses, one may get the same apparent resistivity sounding curves. Similarly in electromagnetic sounding, σh product is a better determined parameter and not the σ or h separately. Since both seismic reflection and refraction used for groundwater exploration can determine the bed thicknesses with better accuracy, electrical conductivity or resistivity becomes a better determined parameter when geoelectrical tools join hands with seismics. This is how seismic reflection and refraction have entered in the arena of saline water–fresh water problem. These waves generated within the earth or on the surface and propagates through the solid earth are basically dependent on the elastic properties of the rocks, i.e. bulk modulus, shear modulus of elasticity or modulus of rigidity of the rocks and their bulk densities. These waves are both longitudinal and transverse. The primary waves or P waves are longitudinal waves, and their nature of propagation is similar to that of sound waves. This is why these elastic waves are termed as acoustic waves. The secondary waves or S waves are transverse waves, and their nature

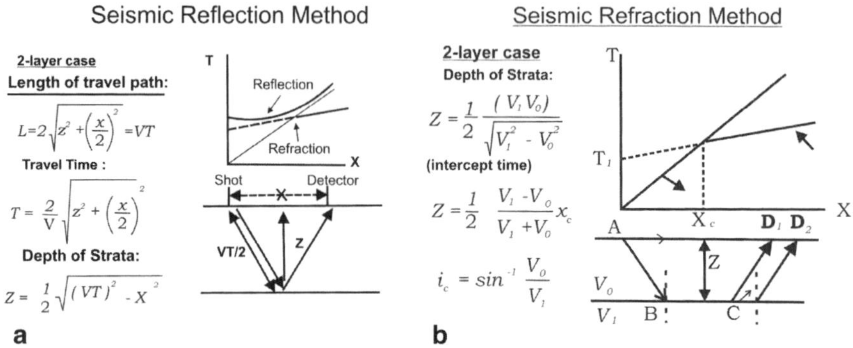

Fig. 15 **a** Basics of seismic reflection and the formulae for depth estimation. **b** Basics of seismic refraction and the formulae for depth estimation

of propagation is similar to that of electromagnetic waves. Electromagnetic waves travel through the air or vacuum with a velocity of light (3×10^8 m/s), whereas elastic waves travel with a velocity of sound in the air (330 m/s). Electromagnetic waves attenuate down significantly while propagating through a good conductor, whereas elastic waves attenuate down significantly while propagating through highly fractured, loose, unconsolidated rocks/sands. Basic physical properties of light waves, viz. reflection, refraction, diffraction, attenuation, dispersion, etc., do exist both in elastic wave theory and electromagnetic theory. For elastic waves, reflection is controlled by the contrasts in acoustic impedance. Acoustic impedance is the product of the velocity and density of a medium. The reflection factor for elastic wave is $(\rho_2 V_2 - \rho_1 V_1)/(\rho_2 V_2 + \rho_1 V_1)$, where $\rho_1 \rho_2$ and $V_1 V_2$, respectively, are the densities and velocities of the first and second medium. Refraction of elastic waves is controlled by the contrasts in velocity (V_2/V_1) of the two media. Guiding equations of elastic wave theory (Aki and Richard 1980) and electromagnetic wave propagation theory (Stratton 1941; Wait 1951) are different.

Many of the basic physical properties of optics, viz. Huygen's principle, Snell's law, Sine law, etc., are valid for elastic wave propagation. Elastic waves proved their superiority over other branches of geophysics on the issue of resolving power. This is why seismic reflection and seismic refraction are the two most dominant geophysical tools. The basics of seismic reflection and refraction are shown in Fig. 15a and b. Further details are beyond the scope of this short chapter. So far as mapping the subsurface with greater details is concerned, they are the best. Phenomenal development in seismic data processing during the past four decades (Yilmaz 1987) made seismic reflection the most dominating geophysical tools for oil exploration. Later, both seismic reflection and refraction entered into the groundwater exploration programmes (Zohdy et al. 1974; Eaton and Watkins 1970; Freeze and Cherry 1979; Geissler 1989; Kelly and Mares 1993; Karanth 1987). The only geophysical tool that could throw challenge to seismic methods on the resolving power issue is the GPR. Resolution of the subsurface is frequency dependent for both electromagnetic and elastic waves, and the frequency of the radar signals are much higher than the seismic signals; therefore, resolving power of GPR is much higher but with very little depth of investigation. Depending on the maximum depth to be seen, frequency of the radar signal can be adjusted. Recent researches on seismic method generated a subject named HRS. Its resolving power could be matched with that of GPR with proper adjustment of frequencies. Thus, HRS and GPR combination appeared as a powerful tool to map the subsurface geology of the saline water–fresh water zones. With the gradual increase in depth of the target, regular seismic reflection and refraction can be used along with electromagnetic tools, viz. CFS, DFS, TEM, VLF-EM, etc., for this problem.

Sonic logging is a powerful borehole geophysical tool that entered in groundwater geophysics for four important reasons. (i) It can map the finer sedimentary stratigraphic boundaries in great detail needed as an important information for interpretation of HRS-GPR data. (ii) It responds to primary porosity (ϕ_s) of a formation. In the presence of neutron porosity (ϕ_N), resistivity porosity (ϕ_R; from combined SP and resistivity log) or density porosity (ϕ_D), one can estimate the secondary porosity of a formation needed in groundwater exploration because neutron, resistivity and density porosities respond to total porosity (primary and secondary). (iii) It can detect faults, fractures and fissures and is needed to a greater extent in hard-rock groundwater geology. It can join hands with remote sensing for delineation of fractures if proper borehole facilities are available in the study area. (iv) Sonic logging, joining hands with gamma–gamma density logging, can estimate the bed thickness, contrasts in acoustic impedance and velocities of the different strata in detail and can help in seismic reflection and refraction data interpretation. Thus, elastic waves-generated geophysical tools are needed in differentiating saline water–fresh water zones.

2.7 *Radiation Logging*

Nuclear Capture Cross-section of Chlorine

Nuclear capture cross-section is a probabilistic cross-section. It is a probability with which thermal neutrons (energy level <0.025 keV) will be captured by earth formations. It is expressed in the units of barns (10^{-24} sq cm). Cadmium, boron and chlorine have much higher nuclear capture cross-sections in comparison with those of other common earth elements, viz. hydrogen, oxygen, silicon, etc. Table 2 shows the order of the nuclear capture cross-sections of the elements of our present interest. This particular property is used to differentiate the fresh and saline water. For neutron logging, in capsule source domain, plutonium–beryllium (Pu-Be) or americium–beryllium (Am-Be) combinations generate high-energy neutrons (about 2.5 MeV). These neutrons in the borehole collide with the earth formations. Neutrons lose energy after each collision. Particle dynamics states that degree of dissipation of energy will be maximum when two particles of the same size collide. In this case, neutrons and hydrogens have almost the same size and mass. This is why neutron logs are hydrogen sensitive. After a few collisions, the neutrons reach

Table 2 Nuclear capture cross-section of common earth elements of present interest

Elements	Σ
Hydrogen	0.33
Oxygen	0.0002
Sodium	0.53
Silicon	0.16
Carbon	0.003
Sulphur	0.52
Potassium	1.0
Calcium	1.0
Chlorine	34.0
Boron	755.0
Cadmium	2,500.0

Σ in Barns (10^{-24} sq cm)

the thermal level and lose capability to collide with the earth formations further. This process is known as thermalisation of neutrons. Higher the concentration of hydrogen, lesser will be the number of collisions before thermalisation and lesser will be the distance of the thermal neutrons from the neutron source in the borehole. At this stage, these thermal neutrons diffuse through the earth formations, and the velocity of the neutrons becomes comparable with the thermal vibration velocities of the earth formations. During the movement of thermal neutrons through the earth formations, some of the thermal neutrons get captured, while the rest of them continue to move. Now how many of them will be captured and how many will diffuse through formation is dictated by the nuclear capture cross-section of the common earth elements. In the detector part of the neutron logging instrument housing sonde, we have to keep both neutron–thermal neutron and neutron-capture gamma detectors. For normal fresh water zone, the log response for both neutron–thermal neutron and neutron-capture gamma response will be more or less the same. In saline water-contaminated zones, the capture gamma ray counts will be much higher than the neutron–thermal neutron counts. Because chlorine has much larger nuclear capture cross-section than common earth elements, joint detection of thermal neutrons and capture gamma ray in two separate detectors can solve delineation of saline water pockets or saline water invasions. The nuclear reactions in neutron logging are as follows:

Energy level

1. Generation of neutrons from capsule sources in neutron logging:

$$\underset{\text{beryllium}}{\underset{\downarrow}{{}_4\mathbf{Be}^9}} + \underset{\alpha\text{ particle}}{\underset{\downarrow}{{}_2\mathbf{He}^4}} \rightarrow \underset{\text{excited carbon}}{\underset{\downarrow}{{}_6\mathbf{C}^{13}}} \rightarrow \underset{\text{stable carbon}}{\underset{\downarrow}{{}_6\mathbf{C}^{12}}} + \underset{\text{source neutrons}}{\underset{\downarrow}{{}_0\mathrm{n}^1}} + \underset{\text{source-emitted }\gamma\text{ rays}}{\underset{\downarrow}{\gamma\text{ rays}}} \tag{8}$$

Table 3 Average radioactive level of rocks of present interest

Rock type	Uranium (ppm)	Thorium (ppm)	Potassium (ppm)
Granite	5	18	3.86×10^4
Shale	3.7	12	1.7×10^4
Sandstone	0.45	1.7	0.64×10^4
Limestone	1.3	1.1	0.27×10^4
Basalt	0.5	2	0.5×10^4

2. Neutron-capture gamma ray logging

$$\underset{\text{atoms of earth formation}}{{}_{Z}X^{A}} + \underset{\text{thermal neutrons}}{{}_{0}n^{1}} \rightarrow \underset{\text{excited atom of earth formation}}{{}_{Z}X^{A+1}} + \underset{\text{capture } \gamma \text{ rays}}{\gamma \text{ rays}} \quad (9)$$

3. Neutron–thermal neutron logging

$$\underset{\text{boron}}{{}_{5}B^{10}} + \underset{\text{thermal neutron}}{{}_{0}n^{1}} \rightarrow \underset{\text{lithium}}{{}_{3}Li^{7}} + \underset{\alpha \text{ particle}}{{}_{2}He^{4}} \quad (10)$$

4. Accelarator source high-energy neutron

$$\underset{\text{deuterium}}{{}_{1}H^{2}} + \underset{\text{tritium}}{{}_{1}H^{3}} \rightarrow \underset{\alpha \text{ particle}}{{}_{2}He^{4}} + \underset{\text{high-energy neutron}}{{}_{0}n^{1}} \quad (11)$$

Natural Gamma Ray Logging

Uranium (${}_{92}U^{238}$ and ${}_{92}U^{235}$), thorium (${}_{90}Th^{232}$) and potassium (${}_{19}K^{40}$) and many of their daughter products are the sources of radionuclides for natural gamma rays in the earth formations. Granites and granodiorites are having the most abundant sources of alpha and beta particles and gamma rays. Shale and clay have the property to absorb the radionuclides much more than sands/sandstones/limestones/dolomites. Potassium in the form of sylvites and carnallites is absorbed by the shale, and as a result, the shale has the second highest level of radioactivity among the common earth rocks and minerals of our present interest. Average radioactive content of these rocks is as shown in Table 3.

Therefore, natural gamma ray counts of shale or clay will be more than sands/sandstones/limestones. Hence, natural gamma ray logging can distinguish between sand and clay. Since, in general, the resistivity of the fresh water sandstone is more than that of shale or clay, there will be no problem to distinguish between fresh water sand and shale/clay beds. However, in the saline water-invaded areas, the order of resistivity of shale and saline water sand may be more or less of the same. Here, natural gamma ray logging will be able to distinguish between shale and saline water sand. This is how natural gamma ray logging has entered into this geophysical problem.

Gamma–Gamma Density Logging

Gamma–gamma density logging or scattered gamma ray logging is an auxiliary tool in groundwater industries. It joins hand with sonic logging for accurately estimating the acoustic impedance (velocity multiplied by density) of the earth formation surrounding the borehole with considerable details. This is needed for improving the accuracy of seismic reflection data interpretation. As a result, the depth estimation of the different reflecting boundaries will be more accurate. Moreover, sonic–density duo can give estimate of the secondary porosity of a formation.

The basic physics behind this tool is when gamma ray passes through the formation, it can interact with the earth formation in three different ways, viz. (i) pair production, (ii) Compton scattering and (iii) photoelectric effect, depending on the energy level of the gamma rays. When energy of the incident gamma ray is more than 2.0 MeV, it generates electrons and positrons; in other words, energy is converted into mass according to Einstein's energy to mass conversion relation $E = mC^2$. Positrons are positively charged electrons having same mass. In contrast, when the gamma ray energy is less than 1 MeV, the most prominent effect is the photoelectric emission, which follows Einstein's equation $1/2\,mv^2 = h(\nu - \nu_0)$, where h is the Planck's constant, ν is the frequency of incident radiation and ν_0 is the threshold frequency of the medium through which gamma ray passes and $1/2\,mv^2$ is the kinetic energy of the photoelectrons. In between approximately 1–2-MeV energy level, gamma ray gets a band where Compton scattering is prominent. Here, gamma rays are scattered after each collision with the earth formation and lose a part of their energy. These scattered gamma rays with reduced energy are detected in the borehole counter. cobalt-60 (Co^{60}) or cesium-137 (Cs^{137}) are chosen as sources of gamma rays, which emits gamma rays of the required energy level where Compton scattering will be effective. Here the energy loss per collision will be proportional to the electron density of the formation, which in turn will be proportional to the bulk density of the formation.

3 Geophysical Tools for Saline Water–Fresh Water Problems

For studying saline water–fresh water problems, one geophysical technique for field survey may not always be sufficient for proper diagnosis. Several combinations of geophysical tools may prove to be better for unambiguous interpretation. Proper combinations of these tools are to be chosen depending on the geology of the area, depth of subsurface to be examined with a specific problem in mind and the expertise available. Almost all quantitative interpretations of geophysical data these days are based on inverse theory.

The tools are as follows:

1. Joint resistivity and IP sounding
2. CFS or DFS
3. Transient electromagnetic sounding
4. Ground-penetrating radar
5. Remote sensing, with GIS and GPS
6. High-resolution seismic reflection and GPR
7. Seismic reflection and refraction
8. Natural gamma ray and neutron logging
9. Combined SP and resistivity logging
   - SP reversal
   - Single-point resistance logging
   - Normal and lateral resistivity logging

3.1 *Joint Resistivity and Induced Polarisation Sounding*

Since resistivity of water decreases sharply with salinity as shown in the Figs. 2–5, saline water-contaminated aquifers can be distinguished from fresh water aquifers by resistivity sounding alone if the geology of the area is known. However, both saline water-saturated aquifers and clay or shale layers in some areas may have the same order of resistivity. Combined resistivity and IP sounding and joint inversion of resistivity and IP data can resolve this ambiguity because clay and shale have some IP effect due to membrane polarisation across the cationic membrane. Saline water-saturated clean sand will have negligible to zero chargeability. The charging and discharging times have to be increased because membrane polarisation is a weak signal compared with electrode polarisation. On-time voltage and current are measured for computation of apparent resistivity for DC sounding, and off-time area under the decay curve is measured for IP measurement. A delay time of 200–300 ms is given such that electromagnetic coupling and capacitative coupling effect due to sudden switching off of the current gets time to subside. Introduction of delay time in electronic circuit and removal of electromagnetic coupling are important steps in time domain IP. Stefanescu's integral (1930; Keller and Frischknecht 1966; Bhattacharyya and Patra 1968; Koefoed 1979; Ghosh 1970) is used for solution of the forward problem for a 1D n-layered earth. It is given by

$$\rho_a = \rho_1 \int_0^{\infty} \theta_n(\lambda) J_1(\lambda r) \lambda \, d\lambda \qquad (12)$$

where ρ_a is the apparent resistivity for an n-layered earth, ρ_1 is the resistivity of the first layer, $\theta_n(\lambda)$ is the kernel function for an n-layered earth, $J_1(\lambda r)$ is the Bessel's function of the first order and first kind and λ is the integration variable. For a two-layered earth,

$$\theta n(\lambda) = \frac{k_1 u^{h1}}{1 - k_1 u^{h1}} \tag{13}$$

For a three-layered earth, the kernel function is

$$\theta n(\lambda) = \frac{k_1 u^{h1} + k_1 u^{h2}}{1 - k_1 u^{h1} - k_1 u^{h2} + k_1 k_2 u^{(h2-h1)}} \tag{14}$$

where h_1 is the thickness of the first layer, $k_1 = (\rho_2 - \rho_1)/(\rho_2 + \rho_1)$, ρ_1 and ρ_2, respectively, are the resistivities of the first and second layer and $U = e^{-2\lambda}$. Further details about the kernel function for a multilayered earth is available in Bhattacharyya and Patra (1968), Ghosh (1970), Koefoed (1979), Zhdanov and Keller (1994) and Roy (2007). Inversion format is available in (Inman 1975).

For computation of apparent chargeability curves for a 1D n-layered earth, the formula for computation is

$$m_a = m_1 B_1 + m_2 B_2 + m_3 B_3 \text{-------------------------} m_n B_n \tag{15}$$

where m_a is the apparent chargeability for an n-layered earth, m_1-----------------m_n, respectively, are the layer chargeabilities and B_1-----------------B_n, respectively, are the weighting functions for an n-layered earth. Seigel (1959) has defined the values of these weighting functions, and Roy et al. (1995) used them for computation of n-layered apparent chargeability sounding curves. Here

$$\text{(a)}\quad B_1 = \partial log\rho_a / \partial log\rho_1 = 1 + 2L^2 \frac{\rho_1^2}{\rho_a} \int_0^\infty \lambda \frac{\partial \theta_n(\lambda)}{\partial \rho_1} J_1(L, \lambda) d\lambda \tag{16}$$

$$\text{(b)}\quad B_2 = \partial log\rho_a / \partial log\rho_2 = 2L^2 \frac{\rho_1 \rho_2}{\rho_a} \int_\theta^\infty \lambda \frac{\partial \theta_n(\lambda)}{\partial p_2} J_1(L\lambda) d\lambda \tag{17}$$

$$\text{(c)}\quad B_n = \partial log\rho_a / \partial log_n = 2L^2 \frac{\rho_1 \rho_n}{\rho_a} \int_\theta^\infty \lambda \frac{\partial \theta_n(\lambda)}{\partial p_n} J_1(L\lambda) d\lambda \tag{18}$$

are the respective weighting functions of the first to nth layers.

Here sum of the weighting functions is equal to unity, i.e. $\sum_{i=1}^{n} Bi = 1$. It also suggests that for quantitative interpretation of IP sounding data, simultaneous resistivity and IP sounding are a must. Generally interactive joint inversion of resistivity and IP sounding data has to be done in each iteration. Weighting functions are to be calculated with the modified apparent resistivity and layer true resistivity values. The two field examples of joint resistivity and IP sounding data and their joint inversion interpreted models are shown in Fig. 16a and b. Low resistivity and low chargeability are signals for saline water sand. Relatively high resistivity and low chargeability are signals for fresh water sand (Roy et al. 1980, 1995). IP forward

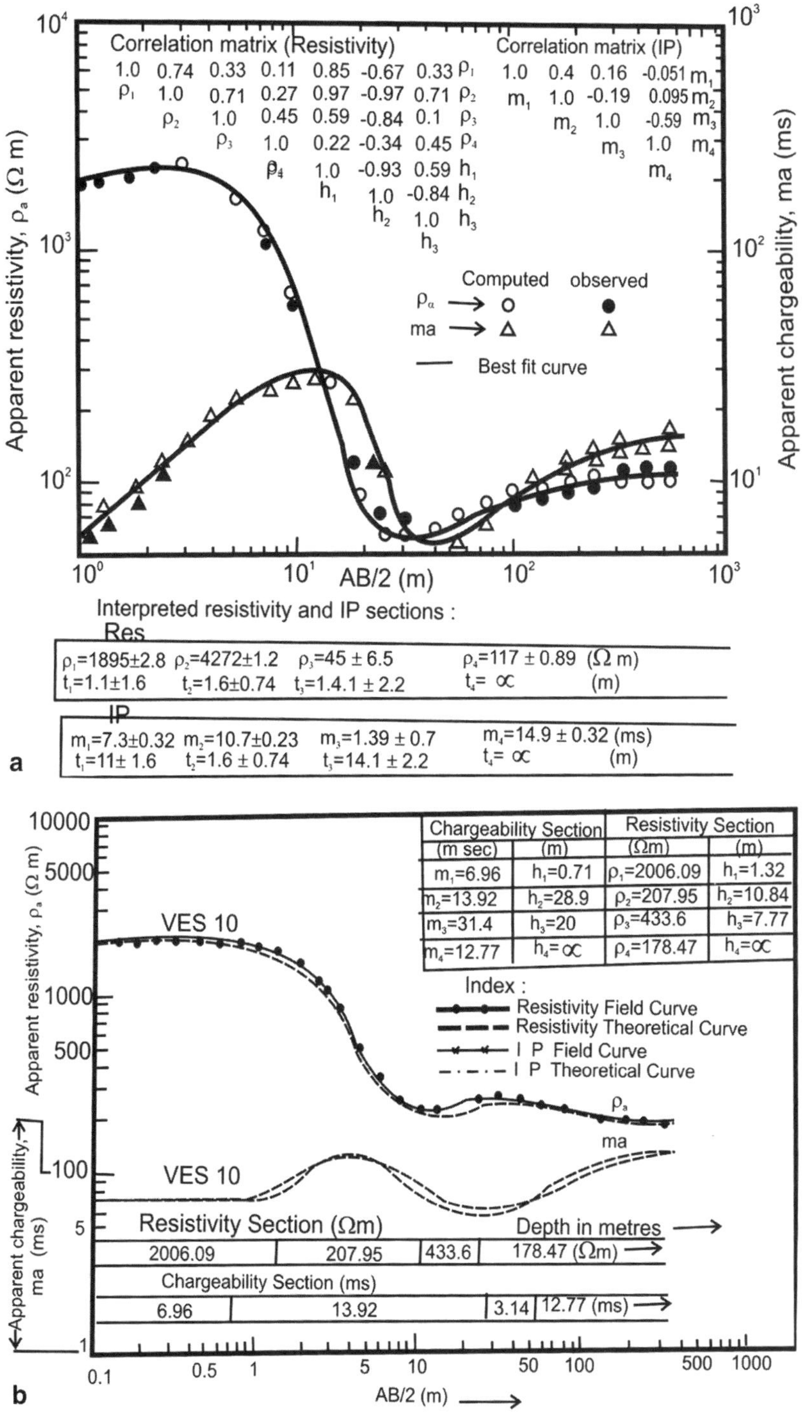

Fig. 16 **a** Joint resistivity and IP sounding field curves over a saline water pocket and joint 1D inversion results; simultaneous decrease of resistivity and IP is an indication of saline water zone. **b** Joint resistivity and IP sounding curves over a shaly sand and the interactive joint 1D inversion results; increase in resistivity and decrease in IP are indicators of fresh water zone

model curves are resistivity model dependent. Resistivity and IP sounding together can differentiate saline water and clay zones more effectively. To standardise the quantitative values of chargeabilities, same charging discharging time, delay time and integration time is to be used. Joint inversion of resistivity and IP sounding data is to be done for field data interpretation.

3.2 Central Frequency Sounding or Dipole Frequency Sounding

An insulated circular loop, placed on the ground, and through which AC of audio and subaudio frequency range (20/30 Hz to 20/30 KHz) are sent, is a transmitter or a transmitting antenna.

It is also called a vertical oscillating magnetic dipole. The vertical component of the magnetic field at the centre of the loop is measured by another small coil (receiver) placed concentrically at the centre (Patra 1970). This is called CFS. The field technique involves measurement of vertical component of the magnetic field, induced at the centre of the loop, and at variable frequency. DFS (Ryu et al. 1970), in contrast, is an electromagnetic depth sounding technique where two horizontal coplanar dipoles are placed at a certain distance. One is the transmitting dipole through which AC in audio and subaudio frequency range is circulated, and the other is the receiving coil. Both the transmitting and receiving coils are vertical oscillating magnetic dipoles because the direction of magnetic vector is vertical. In this tool, also electromagnetic depth sounding is done by changing frequency of the transmitting dipole. Lower the frequency, greater will be the depth of investigation. It is meant for studying shallow (10–90 m) hydrogeological and engineering geophysical problems in the areas where a large horizontal ground, suitable for DC geoelectric sounding, is not available.

Otherwise, also CFS and DFS have field logistic advanges. Moreover, for dry loose sand dessert areas, permafrost region and hard-rock (hard and compact granites and basalts) areas where DC find it difficult to penetrate, CFS and DFS are good geophysical tools. Skin depth of the signal will approximately tell the depth of exploration. Frequency sounding will have logistic advantage in the field over geometric sounding. Figure 17 shows the field layout of CFS. The basic Helmholtz electromagnetic wave equation (Eq. 19) is solved using Π, the Hertz vector potential,

$$\nabla^2 \Pi = \gamma^2 \Pi \tag{19}$$

in cylindrical polar coordinates by the method of separation of variables. One gets the expressions for vertical component of the magnetic field for CFS and DFS as follows (Patra and Shastri 1982):

$$(\mathrm{h}_z)_{\mathrm{CFS}} = 2a^2 \int_\theta^\infty F_2(m) m^2 J_1(ma) dm \tag{20}$$

$$(\mathrm{h}_z)_{\mathrm{DFS}} = 2r^2 \int_\theta^\infty F_2(m) m^2 J_0(mr) d(m) \tag{21}$$

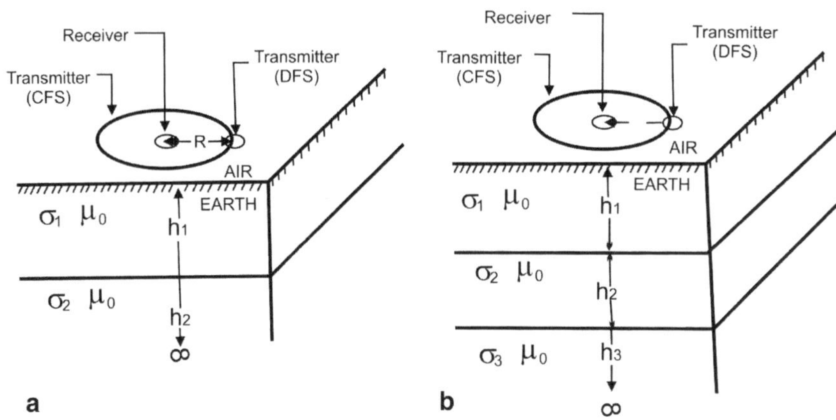

Fig. 17 Field layout for central frequency sounding (*CFS*) and dipole frequency sounding (*DFS*) for **a** a two-layered earth and **b** a three-layered earth

Here $F_2(m)$ is the kernel function and is a function of layer conductivity and layer thickness, assuming magnetic permeability to be free-space magnetic permeability and negligible electrical permittivity in audio and subaudio frequency range. Here

$$\mathrm{F}_2(m) = \frac{(m_1 + m_2) + (m_1 - m_2)e^{-2h1m1}}{(m + m_1)(m_1 + m_2) + (m - m_1)(m_1 - m_2)e^{-2h1m1}} \tag{22}$$

R is the distance between the centres of the transmitting and receiving dipole. Here

$$\begin{aligned} m_1 &= \sqrt{m^2 - \gamma_1^2} \text{ and } \quad m_2 = \sqrt{m^2 - \gamma_2^2} \\ \gamma_1 &= \sqrt{i\omega\mu\sigma_1} \; and \quad \gamma_2 = \sqrt{i\omega\mu\sigma_2}\,. \end{aligned} \tag{23}$$

Same instruments can be used for both CFS and DFS.

In case of an insulated circular loop of radius 'a' carrying an AC $Ie^{-i\omega t}$, the expressions for the vertical magnetic fields for CFS and DFS are given in Eqs. 20 and 21. Here J_1 is the Bessel's function of first kind and order one, r is the radial distance of the point of observation from the centre of the loop, a is the CFS loop radius, z is the depth of the point of observation from the plane of the loop and m is the integration variable.

The expressions in Eqs. (20) and (21) become zero as $J_l\,(mr) \to 0$ as $r \to 0$.

Therefore, radial magnetic component and the Azimuthal electric component are zero at the centre of the loop in CFS. Only the vertical magnetic component will exist. This is in nut shell the central theme of the CFS tool. In DFS, the vertical component is measured at the centre of the receiving dipole. The secondary field that picks up information of the subsurface combine with the primary field from the transmitting field, and the real part of the complex quantity is analysed. The sensitivities of the CFS and DFS three-layered electromagnetic depth sounding curves for different values of the conductivity of the target bed are shown in Fig. 18a and b. It is a powerful tool to study the saline water–fresh water problems.

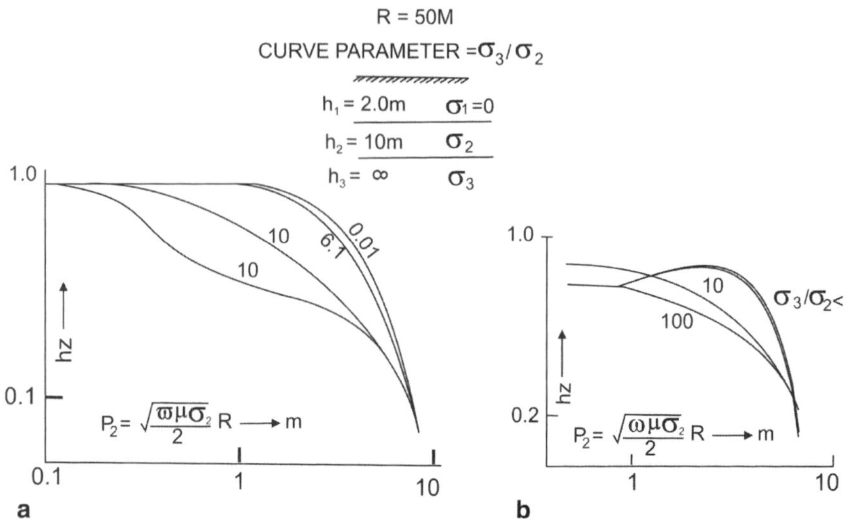

Fig. 18 **a** CFS and **b** DFS forward model sounding curves for a three-layered earth

3.3 Transient Electromagnetic Sounding

TEM or time domain EM sounding is done with a square or a circular loop carrying steady current, with a receiver loop at the centre of the coil. The steady current generates vertical magnetic field at the centre of the magnetic dipole. When the transmitter current is abruptly switched off, eddy current is induced into the ground due to abrupt changes of the primary magnetic field lines of forces. Since the primary magnetic field dies down with time in microseconds, eddy currents and secondary magnetic field also die down in microseconds. The induced currents flow in horizontal circles under the transmitter loop. Initially, the current is concentrated near the surface, but as time passes, the current density maximum moves downward and outward (Nabighian 1979). Diffusion and dissipation of current density are controlled by the resistivity of the ground (Fitterman 1987). The more resistive the ground is, faster the current maximum diffuses downward and the more rapid is the dissipation of current. For a layered earth, the current will reside longer in the conductive zones than in resistive zones of similar thickness. There are several features of the transient sounding methods. The high sensitivity of the technique to conductivity changes results in very good vertical resolution. To sound deeper, it is only necessary to record the transients to later time. In principle, the transmitter loop need not be expanded. In practice, noise considerations require the loop size to be larger for deeper targets. Exploration depths of one to three loop diameters are typical. The sounding tends to be minimally affected by the lateral variation in resistivity. Because the induced current flows on all the sides of the receiver loops, the recorded transient itself does not give an easily understood representation of

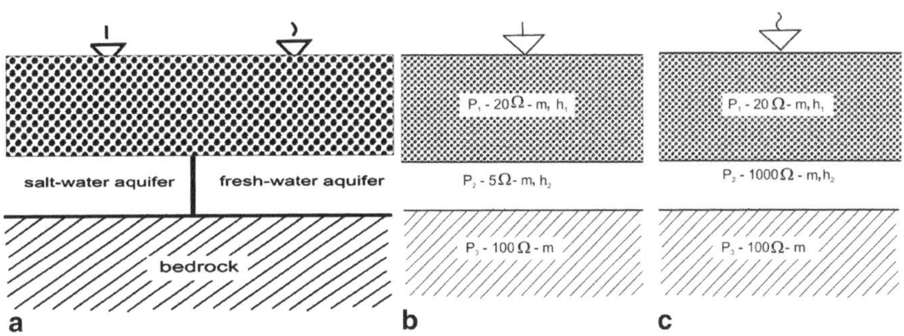

Fig. 19 **a** Three-layered earth model for saline water and fresh water aquifers. **b** Resistivity model for saline water. **c** Resistivity model for fresh water. (Taken from Fitterman and stewart 1986)

the subsurface conditions. The transient must be normalised for the transmitter, and receiver parameters such as current intensity, loop size and received gains are compared with the voltage measured over a half space of uniform resistivity to obtain an apparent resistivity as it is done in DC resistivity and MT. The use of apparent resistivity allows comparison of data from different locations (Fitterman and Stewart 1986). There are several methods of computing apparent resistivity for transient soundings. The so-called late-stage apparent resistivity is used here (Fig. 19).

The expressions for the apparent resistivities are as follows:

After measuring the voltage V_{obs} *(t)*, V_{un}, a voltage measured over a uniform homogeneous and isotropic half space of resistivity ρ_1, after a time *t*, is measured. The apparent resistivity is given by

$$\frac{\rho_a}{\rho_1} = \left| \frac{V_{un}(\rho_1 t)}{V_{obs}(t)} \right|^{\lambda} \tag{24}$$

where λ is a real number and V_{obs} is the observed voltage at time *t*. In this case, ρ_1 must be known to compute apparent resistivity ρa. This problem is avoided by assuming that the field V_{un} follows the late-stage behavior and by choosing $\lambda = 2/3$ (Kaufman and Keller 1983). Then apparent resistivity becomes

$$\rho_2 = \frac{\mu}{4\pi t} \left| \frac{2\pi \mu r^2 M_r I}{5tv} \right|^{2/3} \tag{25}$$

where μ is the magnetic permeability, *r* is the transmitter loop radius, M_r is the receiver coil moment, *I* is the transmitter current, *t* is the time after the current is switched off and *V* is the voltage induced in the receiver coil (Fig. 20).

In practice, rectangular loops are used and the data are analysed assuming equal-area circular loop radius. Fig. 21 shows the delineation of saline water-bearing layer below the fresh water zone above the granitic basement.

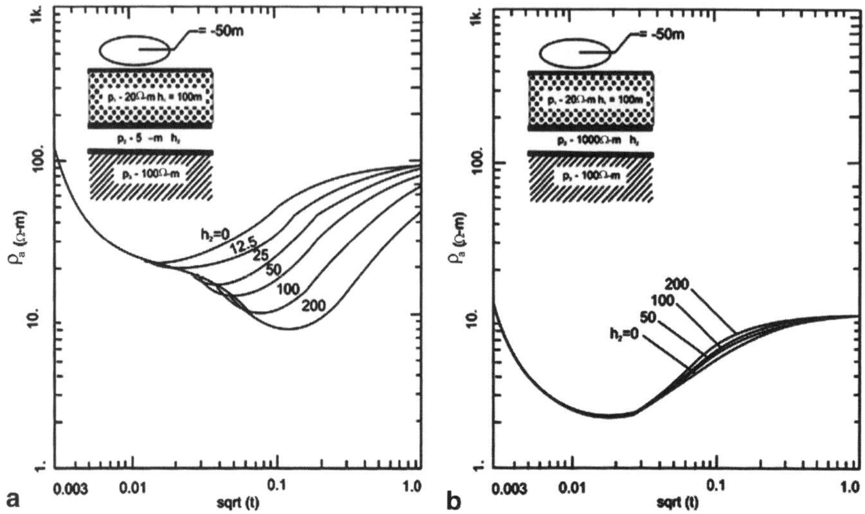

Fig. 20 **a** Transient sounding curve for saline water aquifer. **b** Transient sounding curve for fresh water aquifer. (Taken from Fitterman and Stewart 1986)

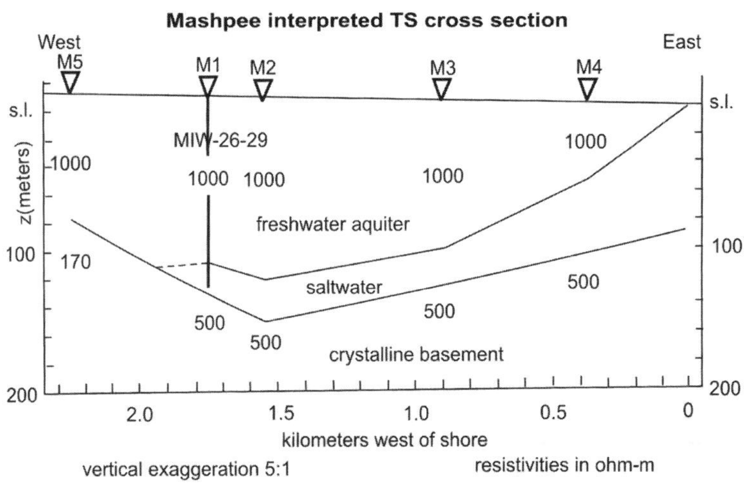

Fig. 21 Delineation of saline water-bearing zone by transient electromagnetics. Location: Mashpee, Massachusetts. (Taken from Fitterman 1987)

3.4 Ground-Penetrating Radar

GPR is a surface geophysical tool having a transmitter and a receiver like many other EM tools, but transmitted frequency is in the range of a few hundred megahertz to a few gigahertz for shallow investigation and better resolution. Since higher the electrical conductivity, lesser will be the depth of penetration as mentioned, GPR can map the shallow saline/brackish water pockets using the relative electrical permittivity ϵ because ϵ will be the more dominant parameter than σ at that frequency. In GPR, the transmitter emits signals in radiofrequency range towards the subsurface, and when it reaches an interface having contrast in electrical permittivity, a part of the signal is reflected back to the receiving antenna and the rest is transmitted down. Because of very high frequency, the transmitted signals attenuate down very fast to negligible value in this type of geological settings we are talking about. The receiver converts the radio frequency signals of only a few nanoseconds in duration to audio frequency signals of tens of milliseconds in duration (Bares and Haeni 1991) by repetitively sampling many waveforms. This signal is then sent to the control unit. The travel time for a signal to pass through the subsurface, get reflected back from the inhomogeneity and return back to the receiver is only a few nanoseconds (10^{-9} s). Base to the reflector is determined using the two basic relations (i) $d = tv/2$ and (ii) $v = c/\epsilon^{0.5}$, where d = depth to the reflector, t = two-way travel time, c = velocity of light in free space and ϵ = relative dielectric permittivity. This is the basics of single-channel GPR technology used for groundwater hydrology and can be used for detection of groundwater tables and saline water and fresh water interface. Much more sophisticated multichannel (48-channel) GPR tool has come in the market that uses the entire seismic processing softwares (Yilmaz 1987) that reduces noise more efficiently, and more accurate value of electrical permittivity is obtained (Fisher et al. 1992). Multichannel GPR improves (i) noise reduction level, (ii) velocity estimation, (iii) effective depth of penetration and (iv) accuracy in spatial positioning of features in the resultant depth section. Sophistication in GPR has gone to the extent that with the help of HRS, GPR can go to the extent of determining porosity and water saturation of a formation (Ghose et al. 1998) for groundwater geophysics problem. GPR has one transmitter and one receiver quite close to each other with a frequency of about 100 MHz for single-channel survey. They are towed by a truck kept far ahead of the equipment housing to avoid any interference of the truck in GPR signals from the earth formations, and the ground coverage can be as high as 3–5 km/h. It is a low-cost survey strategy (Bares and Haeni 1991). Data can be used with processing (Bares and Haeni 1991) for quantitative and without processing for qualitative interpretation. The GPR technique is similar in principle to seismic reflection technique. Pulse mode GPR systems radiate short pulses of high-frequency (10–1,000 MHz) electromagnetic energy into the ground from a transmitting antenna. The propagation of the radar signal depends on the frequency-dependent electrical properties of the ground. Electrical conductivity of the soil or rock materials along the propagation paths introduces significant absorptive losses that limit the depth of penetration into the earth formations and is primarily dependent on the moisture content and mineralisation present. When the radiated energy encounters an inhomogeneity of the subsurface, part of the incident energy is reflected back to the radar antenna and a part is transmitted through the inhomogeneity. The electrical properties of the geological materials are governed primarily by the water content, dissolved minerals, clay and heavy minerals. Reflected signals are amplified and transformed to the audio frequency range, recorded, processed and displayed. Subsurface features can be identified from these records. The GPR data are presented as a two-way travel time measured in nanoseconds. If the propagation velocity of the electromagnetic pulse is known, the depth of the reflector can be known from

$$d_r = vt_r/2 \tag{26}$$

where d_r is the depth of the reflector, t_r is the two way travel time and v is the velocity through the subsurface materials. For low-loss medium, the propagation velocity is related to the relative dielectric constant by the relationship

$$v = c/\sqrt{\epsilon_r} \tag{27}$$

where c is the velocity of electromagnetic waves in free space (30 cm/ns) and ϵ_r is the relative dielectric constant. The relative dielectric constant is a measure of the capacity of a material to store a charge when an electric field is applied to it, relative to the same capacity in a vacuum. Thus, in an otherwise uniform low-loss medium of known or estimated relative dielectric constant ϵ_r, the depth of the reflector (Bensen 1995) is

$$d_r = ct_r / 2\sqrt{\epsilon_r} \tag{28}$$

The penetration capability of GPR is site specific and depends on the frequency spectrum of the source of excitation signal, the antenna radiation efficiency and electrical properties of the subsurface materials. Attenuation loss is caused by (i) conversion of radiated energy to heat through electrical conduction losses, (ii) dielectric relaxation losses in water and (iii) chemical diffusion in clay mineral. The effect of signal scattering by small-scale heterogeneities can also increase attenuation with increasing frequency; materials with high conductivity such as clayey soils will rapidly reduce the depth of penetration. The radar frequency selected for a particular study is chosen to provide an acceptable compromise between deeper penetration and higher resolution. GPR can locate faults, lineaments and groundwater contamination, including saline water invasion.

3.5 Remote Sensing with GIS and GPS

Swain and Davis (1978), Seigel (1980), Schowengerdt (1963), Rossitier et al. (1975), Richards and Xiuping (1998), Joseph George (2003), and Jensen (1986) are some of the important references of this vast topic. This subject is a fairly new subject, and numerous references can be cited on this topic. Satellites commonly used in groundwater application are Landsat TM (ThematicMap), SPOT, Soyouz, ERS and others. The data are collected over a large area (34,000 km^2) in one image depending on the type of the satellite. The data can be acquired two to three times a month, and therefore best seasonal data can be chosen. The data can be obtained at different wavelengths (visible and invisible). Remote sensing data are generally collected from the government or nongovernment organisations, e.g. National Remote Sensing Agency of India. Depending on the nature of the problem, different types of data are collected. For groundwater exploration Earth Resource Satellites ERS-1,2,3 and Linear Imaging self-Scanning LISS-I, II, III, IV, V, Land Sat Thematic Map, SPOT, along with GIS, GPS data are taken together for joint interpretation. Often analogue thematic or land-use maps are digitised manually in the absence of any suitable digitiser. GIS includes local topography, local geology, watersheds, drainage patterns, vegetation, crop patterns, previous geological and geophysical survey records, bore hole information from local office on earth sciences and detailed maps. GPS is a satellite-based navigation system. Its purpose is to know the exact location (latitude, longitude and height from the mean sea level) of any ground observation. This satellite-based navigation is based on the principle of trilateration. Trilateration is a basic geometric principle that allows one to find out the exact location of a point if its distances from the three well-established stations are known or can be determined. The principle is exactly similar to the way the epicentre of an earthquake is determined on the surface of the earth.

Remote sensors are instruments that measures the electromagnetic radiation, which sense the surface or medium due to scattering or emission. Radiance is a property measured as a function of wavelength or other parameters. The technical design of the instrument is frequency dependent. Therefore, the design of an equipment in the microwave range is different from that designed to sense in the infrared, visible light or ultraviolet region. The information collected by remote sensors is meant to identify and map various earth surface objects. The performance of the sensor is evaluated based on its classification as well as mapping accuracy. It is reasonable to assume that this will depend on the instrument's ability to detect small differences in the emittance/reflectance of the earth's surface in a number of spectral bands for an object as small as possible.

Remote sensing is a very powerful tool for detection of faults, fractures, fissures and lineaments in the flood basalt-covered hard-rock areas. Since western part of India is extensively covered by pretertiary flood basalts and since saline water has encroached upto 13–14 km deep inside the Indian subcontinent in western India, especially in Gujarat and Maharashtra, digital remote sensing will be an appropriate geophysical tool. The entire east–west stretch of Narmada–Son lineament and its adjacent faults and fractures can be made the potential targets. Detailed analysis of fractures will be necessary to differentiate between the tensional and shear fractures because tensional fractures are expected to be the suitable zones for accumulation of groundwaters, vis a vis the suitable location for encroachment of coastal saline water.

Since groundwater exploration in the hard-rock areas demands complex decisions in most of the cases, Land Sat imagery with spatial, spectral and temporal characteristics needs to be analysed for proper choice of borehole sites. Lineaments suitable for hydrogeological studies are to be identified by ground geological and geophysical survey taking remote sensing data. It will be necessary to club them with GIS data to identify proper target areas for drilling and borehole geophysics.

3.6 High-Resolution Seismic Reflection and GPR

Once it was established that the entire package of seismic softwares (filtering, static correction, common midpoint gathering, velocity analysis, normal and dip moveout corrections, stacking, depth migratiom, inversion) can be used for GPR data processing and no separate softwares are necessary, multichannel GPR has started. The results show significant improvement over single-channel data in noise reduction, depth of penetration (by stacking), spatial positioning and reduction of diffraction artefacts. These characteristics increase the potential for reliable interpretation of structural and stratigraphic details. This concept of using GPR and high-resolution shallow seismics is being implemented (Fisher et al. 1992). Cardimona et al. (1998) have shown first that shallow high-resolution seismic data and GPR data can be collected from the same depth and can be used for joint interpretation. For shallow high-resolution seismic reflection, the source can be a portable vibrator (Ghose et al. 1998) or 22-calibre rifles. The geophones can be placed as close as 10 cm. The reflection event with zero offset can be 14 ms for shallow zone and 24 ms for slightly deeper zone. High-frequency signals are needed for HRS. The most dominant frequency will be of the order of 600 Hz. Small portable vibrators can generate several hundred hertz. Cardimona et al. (1998) took shallow seismic and GPR data from the same depth and along the same line for credible images from both HRS and GPR. The water saturation in the soil increases the seismic velocity. In contrast, GPR resolution improves as soil saturation increases because the radar velocity decreases. Low-frequency GPR data are compared with the high-frequency seismics because they have similar horizontal and vertical resolution.

A seismic velocity of 475 m/s for a 100-Hz seismic signal in the unsaturated zone and a radar velocity of 0.1 m/ns for a 20-MHz radar signal yield data with a common wavelength of 4–5 m. Comparison of reflection features between high-frequency seismics and low-frequency radar shows many common features. Ghose and Slob (2006) quantitatively integrated the HRS and GPR data to estimate porosity and water saturation of the shallow unsaturated zone by proper mixing of λ, μ, ρ and $\in$, where the parameters, respectively, are Lame's constant, rigidity modulus, bulk density and electrical permittivity.

3.7 Seismic Reflection and Refraction

Seismic method has altogether five phases. First phase is the earthquake seismology where the energy is released from an earth shaking rock rapture and is used to study the internal constitution of the earth. The second phase is deep seismic sounding typically meant for crustal studies and to delineate the Mohorovicic discontinuity and upper and lower crust boundary. The third phase is seismic reflection and refraction for oil exploration mostly from sedimentary rocks. The fourth phase deals with seismic reflection and refraction for both groundwater and engineering geophysics. The fifth phase, i.e. high-frequency and HRS, join hands with GPR for finer detailing of near-surface (a few metres) features. Phase 1 and 3 are the two most important branches of the subject and are responsible for the development of the subject (both hardware and software).

Seismic reflection and refraction are also important subjects in groundwater industries. For saline water–fresh water problems, these are auxiliary tools and lend helping hands to geoelectrical tools as mentioned earlier. The operational part of these tools in shown in Fig. 15a and b. Both these tools are more accurate in depth estimations of the subsurface reflectors and refractors and mapping subsurface structures (fault, dip, fractures and fissures). The estimated velocities give an idea about the subsurface lithologies. Here we search the discontinuities in elastic properties that create reflecting or refracting zones. The data are interpreted in terms of their nature, depth location and configuration. In most of the cases, they are significant geological boundaries. An important contribution of seismics is their capabilities to delineate the continuity of the aquifers. The differences between the reflection and refraction method lie in (i) ray path geometry, (ii) geophone array, (iii) accuracy, (iv) size, (v) shape of the target, (vi) number of discontinuities to be mapped and (vii) vertical succession of the velocities. In general, seismic reflection is used mostly in oil exploration, whereas seismic refraction has more application in groundwater industries. Seismic refraction can determine the depth of the water table sharply because of a sharp increase in velocity in the water-saturated zone. Seismic refraction maps a sedimentary groundwater basin more accurately.

3.8 Combined Natural Gamma Ray and Neutron Logging Tool

Combined natural gamma ray and neutron logging is a powerful tool for solving saline water–fresh water problems. Natural gamma ray logging can differentiate shale/clay beds from saline water beds because the gamma ray count will be much higher for shale bed than saline sand bed, although the resistivity may be of the same order. Natural gamma ray response of the fresh water and saline water sand will be of the same order (Fig. 22).

Neutron-capture gamma ray count will be much higher than neutron–thermal neutron counts against a saline water bed. For fresh water sand, both the counts will be more or less of the same order. Fig. 23 shows an illustration of simultaneous recording of neutron–thermal neutron and neutron-capture gamma ray log against a saline water bed. Neutron-capture gamma counts will be significantly more than that of neutron–thermal neutron counts.

Sophisticated pulsed neutron logging tools that are used for neutron activation technique and chemical analysis of the earth formation as well as for oil exploration using accelerator source neutron (Eqn. 11) can in principle be used for saline water–fresh water problems also. However, such expensive logging tools, viz. thermal neutron decay 'time logging', are not used in groundwater industries.

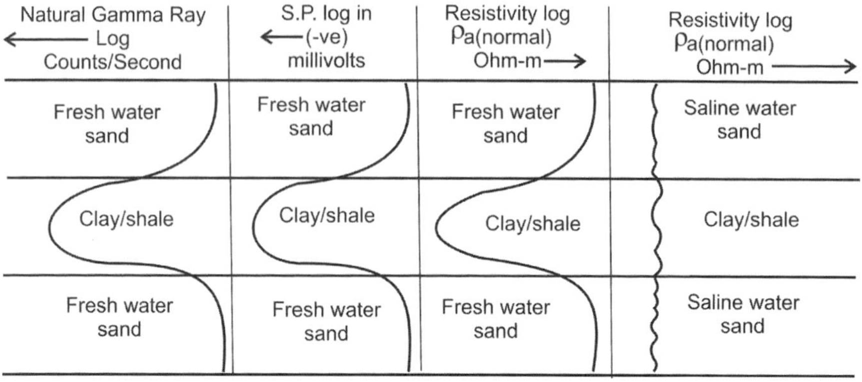

Fig. 22 An illustration of the (i) natural gamma ray, (ii) SP and (iii) DC resistivity two-electrode normal log response across a sand (fresh and saline water filled) and clay section

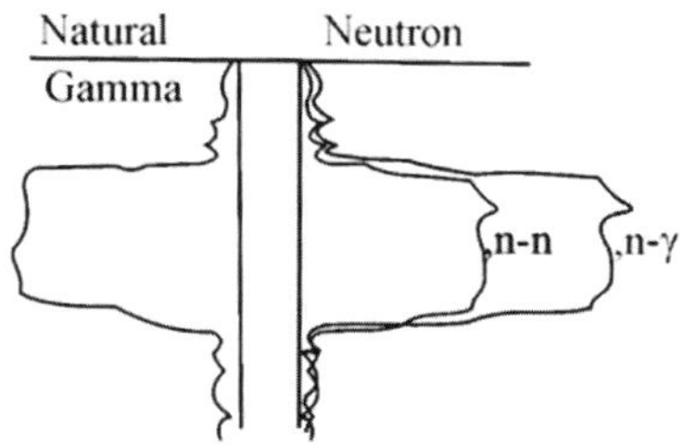

Fig. 23 Illustration of joint neutron–thermal neutron and neutron-capture gamma log response in a saline water bed

3.9 Combined SP and Resistivity Logging

SP Reversal

Combined SP and resistivity logging is also a powerful tool to differentiate between the saline water and fresh water aquifers. SP reversal is the most important qualitative diagnostics. It has already been pointed out that in a sand shale sequence, both liquid junction potential or diffusion potential and membrane potential or diffusion–adsorption potential cells get added having the two electrochemical cells in series, and current flows in the anticlockwise direction in the upper boundary and clockwise direction in the lower boundary when the NaCl concentration in the formation water is more than that in the borehole mud (Fig. 24).

For a fresh water aquifer, however, polarity of the L-J cell and the membrane cell will be in the opposite direction. Therefore, the current flow across the sand–shale boundaries will also be in the opposite direction, i.e. clockwise in the upper bed and anticlockwise in the lower bed. For saline water sand, SP anomaly will be negative, whereas for fresh water sand, SP anomaly will be positive. This is known as the SP reversal, i.e. SP anomaly changes sign. One gets positive SP supported by relatively higher resistivity for fresh water sand and negative SP supported by low resistivity for saline water sand in joint SP and resistivity logging.

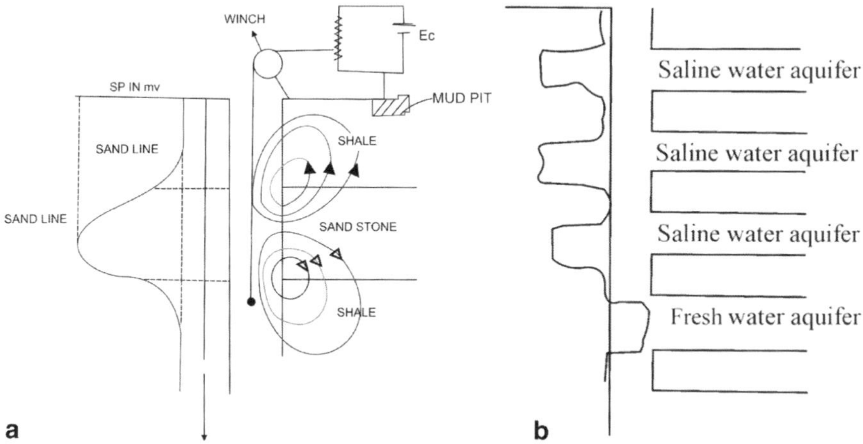

Fig. 24 **a** Illustration of SP current across the sand–shale boundary. **b** Illustration of negative SP for saline water aquifers and positive SP for fresh water aquifers

Single-Point Resistance Logging

In single-point resistance logging, positive current electrode is inserted in the borehole and negative current electrode is inserted in a mud pit on the surface adjacent to the borehole. Low-frequency AC is sent through the electrodes such that AC and DC currents can be isolated with proper circuitry; DC currents are SP currents inside a borehole. AC currents are sent through the ground, and ground resistance is measured. Ground resistance can delineate the bed boundaries and can differentiate saline water from fresh water zones qualitatively. This is the cheapest, simplest and the most widely used borehole geophysical tool used in saline water–fresh water problems (Fig. 25).

Normal and Lateral Resistivity Logging

In two-electrode system, one current electrode and one potential electrode are placed in the borehole.

Return current electrode and one potential electrode are kept on the surface as shown in Fig. 26a. In three-electrode system, one current electrode and two potential electrodes are kept inside the borehole and the return current electrode is on the surface (Fig. 27a; three electrode, pole–dipole).

Resistivity logs are widely used in groundwater industries and especially in saline water–fresh water problems. In Figs. 26b and 27b, the symmetrical and asymmetrical apparent resistivity curves for normal and lateral logs are shown across a sand–shale sequence; in both normal and lateral logs, one gets quantitative estimates of resistivities and thicknesses of both the layers. SP short-normal, long-normal and lateral resistivity logging tools are very popular in detecting saline water and fresh water zones. These tools are widely used in groundwater industries.

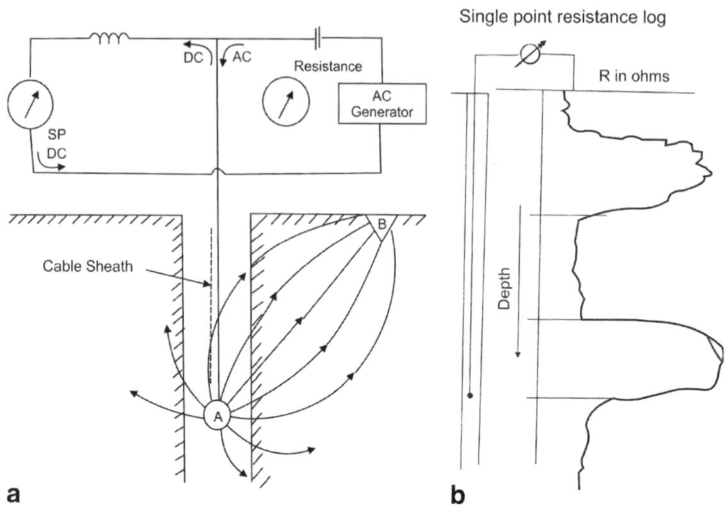

Fig. 25 **a** Field set-up for single-point resistance logging. **b** An illustration of a point resistance log

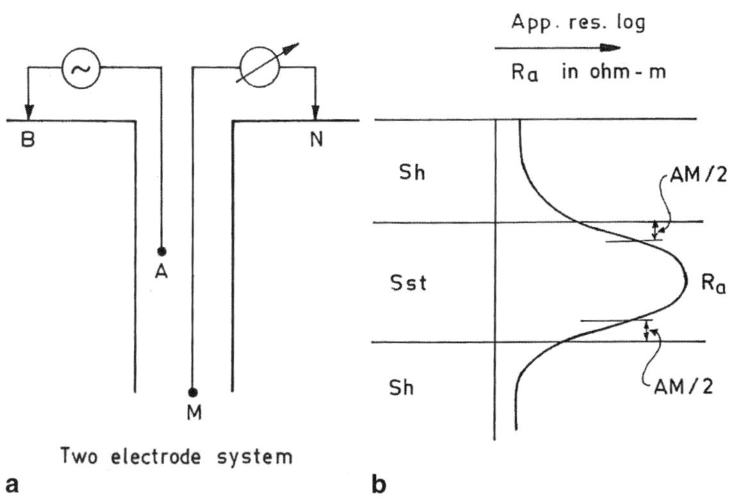

Fig. 26 **a** Normal electrode logging set-up in a borehole. **b** Apparent resistivity curve across a sand–shale sequence

4 Concluding Remarks

An attempt has been made to present almost the total perspective of the geophysical tools used for handling saline water–fresh water problems in actual practice. Contrasts in physical and electrochemical properties of fresh water, saline water,

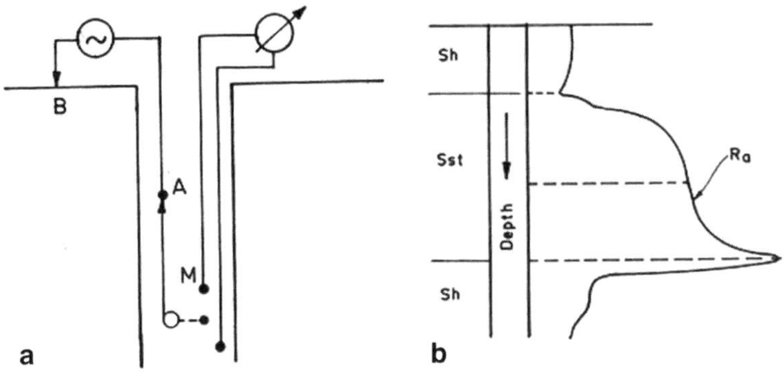

Fig. 27 **a** Lateral electrode logging set-up in a borehole. **b** Apparent resistivity curve for a lateral electrode configuration across a sand–shale sequence

clay and shale and the geophysical tools needed for their detection, delineation and differentiation on one hand and different geological scenario-dependent matching geophysical tools needed to handle the problems on the other are highlighted in this chapter. Always a proper combination of geophysical tools will be needed for removing ambiguity and reconfirming the interpretation, within the depth range of 1–2,000 m or beyond. Since electrical conductivity of water is very strongly dependent on salinity, most of the electrical methods had direct entry in this issue. GPR came with an additional physical parameter, viz. electrical permittivity. All the geophysical tools based on elastic wave propagation do not have any direct say on salinity issue, but they are excellent auxiliary tools to estimate the depth of the reflectors or refractors, and therefore, better estimate of the electrical conductivity is obtained from longitudinal conductance 'S' and transverse resistance 'T'. Moreover, this combination of tools can map the subsurface with better precision. Sonic log in collaboration with other porosity logs can determine the secondary porosity and can map the finer stratigraphic boundaries with precision. Neutron logging is a very powerful tool to locate saline water–fresh water zones. Natural gamma ray logging can differentiate between clay/shale and saline water beds because clay and shale are very good absorbers of radionuclides. Availability of borehole information will give additional advantage in interpretation. Since electrical conductivity of the earth formation increases with temperature as well as with salinity, any geoelectrical tool used for the exploration of geothermal reservoir can also be used for deep-seated saline water pockets. Thus, AMT and CSAMT became tools for the detection of deep-seated saline water pockets (Connate water). For hard and compact rocks, permafrost areas and loose and desert sand areas, electromagnetic tools are to be used because DC resistivity method will fail. Remote sensing is a very powerful tool for delineation of fractures, fissures, faults, lineaments, etc. over a large area. In the soft-rock areas, remote sensing can also map rock moistures. Appropriate geophysical tools best suited for different depth ranges are as follows: (i) GPR and HRS are for 1–10 m, (ii) DC resistivity, IP, CFS and DFS along with shallow seismic reflection or refraction are for 5–100-m sand or a little beyond, (iii) transient electromagnetic sounding, DC resistivity sounding/traversing using collinear dipole–dipole array and high-power low-frequency electromagnetic sounding jointly with seismic reflection sounding can go from 50 to 500 m or a little beyond and (iv) deep resistivity sounding, AMT and CSAMT are for 100–2,000-m depths or a little beyond. Times are in (i) nanoseconds for GPR, (ii) microseconds to a few hundreds milliseconds for TEM, (iii) milliseconds for time domain-induced polarisation and HRS and (iv) fraction of a millisecond to 1 s for AMT, CSAMT, CFS, DFS and seismics. Longer-period signals go outside the periphery of groundwater geophysics. It is an important problem for the countries having a long coastline.

Acknowledgements The author is grateful to Prof. Debashis Sengupta, Professor of Geophysics, IIT, Kharagpur, for his constant encouragement towards completion of this work. The author is immensely grateful to Prof. R. K. Majumdar, retired professor of geophysics, Department of Geological Sciences, Jadavpur University, Kolkata, and Mr. Arkoprovo Biswas, Senior Research Fellow, Department of Geology and Geophysics, IIT, Kharagpur, for providing valuable literature for consultation.

References

Aki K, Richards PG (1980) Quantitative seismology: theory and methods, vol. 1. W. H. Freeman and Co., New York

Anon (1958) Schlumberger well logging document no. 8. Schlumberger Well Surveying Corporation, Houston

Anon (1972) Schlumberger log interpretation principles. Schlumberger Well Surveying Corporation, Houston

Annan AP, Davis JL (1976) Impulse radar sounding in permafrost. Radio Sci 11:383–393

Archie GE (1942) The electric resistivity log as an aid in determining some reservoir characteristics. Trans AIME 146:54–62

Arcone SA (1984) Field observations of electromagnetic pulse propagation in dielectric slabs. Geophysics 49:1763–1773

Baker GS, Steeples D, Schmeissner C, Plumb R (2001) Near surface imaging using coincident seismic and GPR data. Geophys Res Lett 28(4):627–630

Bares M Jr, Haeni FP (1991) Application of GPR methods in hydrogeological studies. Groundwater 29(3):375–386

Benson AK (1995) Applications of ground penetrating radar in assessing some geological hazards: examples of groundwater contamination, faults, cavities. J Appl Geophys 33:177–193

Baker GS, Steeples D, Schmeissner C, Plumb R (2001) Near surface imaging using coincident seismic and GPR data. Geophys Res Lett 28(4):627–630

Bares M Jr, Haeni FP (1991) Application of GPR methods in hydrogeological studies. Groundwater 29(3):375–386

Benson AK (1995) Applications of ground penetrating radar in assessing some geological hazards: examples of groundwater contamination, faults, cavities. J Appl Geophys 33:177–193

Berdichevsky MN, Dmitriev VI (1960) Electrical prospecting with the telluric current method. Quarterly of the Colorado school of mines (trans: GV Keller)

Bhattacharyya PK, Patra HP (1968) Direct current geoelectric sounding, principles and interpretation. Elsevier, Amsterdam

Blakely RJ (1996) Potential theory in gravity and magnetic applications. Cambridge University Press, Cambridge

Bockris J, Reddy AKN (1970) Modern electrochemistry, vol. 1. Plenum, New York

Cagniard L (1953) Basic theory of the magentotelluric method of geophysical prospecting. Geophysics 18:605–635

Cardimona SJ, Clement WP, Kadinsky C (1998) Seismic reflection and ground penetrating radar imaging of a shallow aquifer. Geophysics 63(4):1310–1317

Choudhury K, Saha DK, Rai MK, Naskar DC (1997) Geophysical surveys for hydrogeological and environmental studies in the eastern Calcutta metropolis. Indian Miner 51(1 and 2):41–56

Constable SC, Orange A, Hoversten GM, Morrison HF (1998) Marine magnetotellurics for petroleum exploration. Part I. A sea floor equipment system. Geophysics 63(3):816–825

Dakhnov VI (1962) Theory of well logging, Moscow (in Russian) (trans: GV Keller). Bulletin, Colorado School of Mines

Davis JL, Annan AP (1989) Ground penetrating radar for high resolution mapping of soil and rock stratigraphy. Geophys Prospect 37:531–551

Doll HG (1948) The S. P. log, theoretical analysis and principles of interpretation. Am Inst Min Metall Eng Tech Pub 2463:1–40

Eaton GP, Watkins JS (1970) The use of seismic refraction and gravity methods in hydrologic investigations. In: Morey LW (ed) Mining and groundwater geophysics, geological survey of Canada, Economic Geology Report #26, pp 554–568

Fisher EA, McMechan GA, Annan AP (1992) Acquisition and processing of wide aperture ground penetrating radar data. Geophysics 57:495–504

Fitterman DV (1987) Examples of transient sounding for groundwater exploration in sedimentary aquifers. Groundwater 25(6):685–692

Fitterman DV, Stewart MT (1986) Transient electromagnetic sounding for groundwater. Geophysics 51(4):995–1005

Freeze RA, Cherry JA (1979) Groundwater. Prentice Hall, New Jersey

Frohlich RK (1967) The depth of penetration of dipole arrays compared to the Schlumberger arrangement. Geoexploration 5:195–203

Geissler PE (1989) Seismic reflection profiling for groundwater studies in Victoria, Australia. Geophysics 54(1):31–37

Joseph G (2003) Fundamentals of remote sensing. Universities Press, India

Ghose R, Slob EC (2006) Quantitative integration of seismic and GPR reflections to derive unique estimates for water saturation and porosity of subsoil. Geophys Res Lett 33(L05404):1–4

Ghose R, Nijhof V, Brouwer J, Matsubara Y, Kaida Y (1998) Shallow to very shallow high resolution reflection seismic using a portable vibrator system. Geophysics 63(4):1295

Ghosh DP (1970) The application of linear filter theory, to the direct interpretation of geoelectrical resistivity measurements, monograph. Drukker J. H. Pasmans Gravenhage, The Netherlands

Glasstone S (1962) An introduction to electrochemistry, 10th Printing. D. Van Nostrand and Co., Princeton

Glasstone S (1969) Thermodynamics for chemists, 11th Printing. D. Van Nostrand and Co., New York

Guyod H (1966) Interpretation of electric and gamma ray logs in water wells. The well log analysis, Gearhart and Owen Industries, Inc., 1–16

Haeni FP (1986) Application of continuous seismic reflection methods to hydrologic studies. Groundwater 24:23–31

Hohmann GW (1971) Electromagnetic scattering by conductors in the earth near a line source of current. Geophysics 36(1):101–131

Hohmann GW (1975) Three-dimensional polarization and electromagnetic modelling. Geophysics 40(2):309–324

Hohmann GW (1983) Three-dimensional EM modeling. Geophys Survey 6:27–53

Hohmann GW (1988) Numerical modeling for electromagnetic methods in applied geophysics, vol 1. Theory, Nabhighian MN (ed). SEG Publication, Tulsa, Oklahoma, USA

Hoover DB, Long CL (1976) Audiomagnetotelluric methods in reconnaissance geothermal exploration. Proc 2nd UN Sympos Devel. Geothermal Resources 1059–1064

Hoover DB, Frischknecht FC, Tippens (1976) Audiomagnetotelluric sounding as a reconnaissance exploration technique in Long Valley California. Journal of Geophysical Research 81(5):801–809

Hoover DB, Long CL, Senterfit RM (1978) Some results from audiomagnetotellurics investigation. In geothermal areas, Geophysics 43(7):1501–1514
Hoover DB, Long CL, Senterfit RM (1978) Some results from audiomagnetotellurics investigations in geothermal areas. Geophysics 43(7):1507, 1514
Hoover DB, Frischknecht FC, Tippens CL (1976) Audiomagnetotelluric sounding as a reconnaissance exploration technique in long valley California. J Geophys Res 81(5):801–809
Hoversten GM, Morrison HF, Constable SC (1998) Marine magnetotellurics for petroleum exploration, part II, numerical analysis of subsalt resolution. Geophysics 63(3):826–840
Inman JR (1975) Resistivity inversion with ridge regression. Geophysics 40(5):798–817
Jensen JR (1986) Introductory digital image processing approach–a remote sensing perspective. Prentice Hall, Englewood Cliffs
Joseph G (2003) Fundamentals of remote sensing. Universities Press, Hyderabad
Kalenov EN (1957) Interpretatsia Krivikh Verlikalnogo Electricheskogo, Zondirovanya, Gostoptekhizdat
Kaufman AA, Keller GV (1983) Frequency and transient sounding. Elsevier, Amsterdam
Kelly WE, Mares S (1993) Applied geophysics in hydrogeological and engineering practice. Elsevier, Amsterdam, p 289
Keller GV, Frischknecht FC (1966) Electrical methods in geophysical prospecting. Pergamon Press, Oxford
Kellog OD (1953) Foundations of potential theory. Dover Publications, New York
Koefoed O (1979) Geosounding principles, vol 1. Elsevier, Amsterdam
Lee T, Lewis R (1974) Transient EM response of a large loop on a layered ground. Geophys Prospect 22:430–444
Mackie RL, Madden TR, Wannamaker PE (1993) Three dimensional magnetotelluric modeling using difference equations-theory and comparisons to integral equation solutions. Geophysics 58:215–226
Madden TR, Mackie RL (1989) Three dimensional magnetotelluric modelling and inversion. Proc IEEE 77(2):318–333
Maillet R (1947) The fundamental equations of electrical prospecting. Geophysics 12(4):529–536
Mooney HM, Orellana E, Pickett N, Tornheim L (1966) A resistivity computation method for layered earth models. Geophysics 31(1):192–203
Morrison HF, Phillips RJ, O'Brien DP (1969) Quantitative interpretation of transient electromagnetic fields over a layered half space. Geophys Prospect 17:82–101
Moskva KK (1987) Groundwater assessment, development and management. Tata McGraw Hill, New Delhi, p 720
Murali S, Pangay NS (1998) Principles and applications of groundwater geophysics. Association of Exploration Geophysicists, Hyderabad
Nabighian MN (1994) Electromagnetic methods in applied geophysics, vol 1 and 2, Investigations in Geophysics. SEG Publication, Tulsa
Nabighian MN, Elliott CL (1976) Negative induced polarisation effect in layered media. Geophysics 41:1236–1253
Nabighian MN (1979) Quasi static transient response of a conducting half space an approximate representation. Geophysics 44:1700–1706
Nabighian MN (1984) Time domain electromagnetic method of exploration, (Special Issue). Geophysics 49(7):849–1027
Oldenburg DW (1979) One-dimensional inversion of natural source magnetotelluric observations. Geophysics 44(6):1218–1244
Parasnis DS (1966) The mining geophysics. Elsevier, Amsterdam
Patra HP (1976) Electromagnetic depth sounding for groundwater with particular reference to central frequency sounding: principles, interpretation and application. Geoexploration 14:254–258
Patra HP, Shastri NL (1982) Relative performances of central and dipole frequencies over a layered Earth. Pure Appl Geophys 120:527–537
Patra HP, Mallick K (1980) Geosounding principles, vol 2. Elsevier, Amsterdam
Patra HP (1967) A note on the possibility of saline water invasion around the Jaldha coast, West Bengal India. Geoexploration 5:95–101

Patra HP (1970) Central frequency sounding in shallow engineering and hydrogeological problems. Geophys Prospect 18(2):236–254
Pirson SJ (1963) Handbook of well log analysis on oil and gas formation evaluation. Prentice Hall Inc., Englewood Cliffs
Richards JA, Jia X (1998) Remote sensing, digital image analysis, an introduction. Springer, Germany
Rossitier JR, Strangway DW, Annan AP, Watt RD, Redman JD (1975) Detection of thin layer by radio interferometry. Geophysics 40:299–308
Routh PS, Oldenburg DW (1999) Inversion of controlled source magnetotellurics data for a horizontally layered earth. Geophysics 64(6):1689–1697
Roy A (1962) Ambiguity in geophysical interpretation. Geophysics 27(1):90–99
Roy A, Apparao A (1971) Depth of investigations in direct current methods. Geophysics 36:943–959
Roy KK, Elliott HM (1980) Model studies on some aspects of resistivity and membrane polarisation behaviours over a layered earth. Geophys Prospect 28(5):759–775
Roy KK, Elliott HM (1980) Resistivity and induced polarisation survey for delineating saline water and fresh water zone. Geoexploration 18:145–162
Roy KK, Bhattacharyya J, Mukherjee KK (1995) An interactive inversion of resistivity and induced polarisation soundings for location of saline water pockets. Explor Geophys 25:207–211
Roy KK (2007) Potential theory in applied geophysics. Springer, Heidelberg, 650 p
Roymahashay B, Sen SK (2010) Jal O Jaladushan (in Bengali) water and water pollution, scientific explanation for remedy. Pragatishil Publisher, Kolkata
Ryu J, Morrison HF, Ward SH (1970) Electromagnetic field over a loop source of current. Geophysics 35(5):862–896
Sabnavis M, Patangay NS (1998) Principles and application of ground water geophysics. Association of Exploration Geophysics, Hyderabad
Sandberg SK, Hohmann GW (1982) Controlled source audiofrequency magnetotellurics in geothermal exploration. Geophysics 47:100–116
Schowengerdt RA (1963) Techniques for image processing and classification in remote sensing. Academic press, New York
Scott Keys W, Mac Cary LM (1971) Application of borehole geophysics to water resources investigation, techniques of water resources investigation of USGS, Book No.2, Collection of Environmental data, United States Government Printing Office, Washington
Seigel BS (ed) (1980) Remote sensing in geology. Wiley, New York
Seigel HO (1959) A theory for induced polarization effects, for step function excitation. In: Wait JR (ed) Overvoltage research in geophysical applications. Pergamon, New York, pp 4–21
Stefanescu S, Schlumberger C, Schlumberger M (1930) Sur la distribution Electrique potentielle autour d'une prise de terre ponctuelle dans un terrain a couches horizontales homogenes et isotropes. Journ De Phys et du Radium VII:132–140
Stewart MT (1982) Evaluation of electromagnetic methods for rapid mapping of salt water interfaces in coastal aquifers. Ground Water 20(5):538–545 ((ed, JH Lehr), September–October)
Stratton JA (1941) Electromagnetic field theory. McGraw Hill, New York
Sumner JS (1976) Principles of induced polarisation in geophysical exploration. Elsevier, Amsterdam, p 277
Strangway DW, Swift CM, Holmer RC (1973) The applications of audiofrequency magnetotellurics (AMT) in mineral exploration. Geophysics 38:1159–1175
Swain PH, Davis SM (1978) Remote sensing: the quantitative approach. McGraw Hill, New York
Syed T, Zonge KL, Figgins SJ, Anzzolin AR (1985) Application of the controlled source audiomagnetotellurics (CSAMT) survey to delineate zones of groundwater contamination—a case history: presented at the surface and borehole geophysical methods in groundwater investigations: national conference and expo. Fort worth
Tikhnov AN (1950) On determining electrical characteristics of the deep layers of the Earth crust. Doklady 72(2):295–297
Vozoff K (1989) Magnetotelluric methods, geophysics reprint series no. 5. Society of Exploration Geophysics, Tulsa

Ward SH (1967) Electromagnetic theory for geophysical application in mining geophysics. Society of Exploration Geophysicists 2:13–196
Ward SH, Hohmann GW (1988) Electromagnetic theory for geophysical applications. In: M Nabighian (ed) Electromagnetic methods in applied geophysics, vol 2. pp 131–311
Wait JR (1951) The magnetic dipole over the horizontally stratified Earth. Can J Phys 29:577–692
Wait JR (1970) Electromagnetic waves in stratified media, 2nd edn. The McMillan Company, New York
Wait JR (1982) Geoelectromagnetism. Academic Press, New York
Widess MB (1978) How thin is the thin bed? Geophysics 38:1176–1180
Xiaobo Li, Behrooz O, Laust P (2000) Inversion of controlled source tensor magnetotelluric data for a layered earth with azimuthal anisotropy. Geophysics 65(2):452–464
Yamashita M (1984) Controlled source audiofrequency magnetotellurics. Phoenix Geophysics Limited, Ontario
Yang CH,Tong LT, Huang CF (1999) Combined application of DC and TEM to sea water intrusion mapping. Geophysics 64(2):417–425
Yilmaz O (1987) Seismic data analysis, processing, inversion and interpretation of seismic data, vol 1, 2. (ed, SM Doherty). SEG Publications, Tulsa
Zhdanov MS, Keller GV (1994) The geoelectrical methods in geophycical exploration Elsevier, Amsterdam
Zhody AAR (1965) The auxiliary point method of electrical sounding data interpretation and its relation to the Dar Zarrouk parameters. Geophysics 30(2):644–660
Zohdy AAR, Eaton GP, Maybey DR (1974) Application of surface geophysics to groundwater investigation. USGS publications on techniques of water resources Investigation, pp 1–116
Zonge KL, Hughes LI (1994) Controlled source audiofrequency magnetotellurics in electromagnetic methods in applied geophysics, vol 2, (Ed.Misac Nabighian) Investigations in Geophysics, no. 3, Society of Exploration Geophysics, Tulsa, p 713
Zonge KL,Ostrander AG, Emer DF (1989) Controlled source audiofrequency magnetotelluric measurements in magneto telluric method (ed, K Vozoff), Geophysics Reprint Series No. 5. Society of Exploration, Geophysics, Tulsa, p 749

Prevention and Treatment of Acid Mine Drainage: An Overview

D. P. Tripathy

Abstract Acid mine drainage (AMD) is recognized as one of the serious environmental problems in the mining industry. It is formed by a series of complex geochemical and microbial reactions that occur when water comes in contact with pyrite (iron disulphide minerals) in coal, refuse or the overburden of a mine operation. It has low pH, high TDS, high levels of trace elements, sulphate and total hardness. It can severely contaminate surface and ground water making it harmful to human and aquatic life, concrete structures and mining machineries. Hence, attention must be paid to reduce the generation of AMD as well as to treat it. This paper succinctly describes the genesis of AMD and its associated technical issues. It discusses in detail the factors affecting AMD and characteristics of AMD water in Indian mines vis-à-vis abroad. It also illustrates the various strategies that may be adopted for prevention, control and active and passive methods of treatment of AMD.

Keywords Acid mine drainage · Wetlands · Lime neutralization · Pyrite · Anoxic limestone drains (ALD) · Pyrolusite process

1 Introduction

Acid mine drainage (AMD) is one of the most significant environmental challenges faced by the mining industry worldwide. AMD refers to water drainage flowing from or caused by surface mining, deep mining or coal refuse piles that is typically highly acidic with elevated levels of dissolved metals. The formation of AMD is primarily a function of the geology, hydrology and mining technology employed for the mine site. It is formed by a series of complex geo-chemical and microbial reactions that occur when water comes in contact with pyrite (iron disulphide minerals) in coal, refuse or the overburden of a mine operation. It has low pH, high TDS, high levels of trace elements, sulphate and total hardness. It can severely contaminate surface and ground water making it harmful to human and aquatic life, concrete structures and mining machineries. Hence, attention must be paid to reduce the

D. P. Tripathy (✉)
Department of Mining Engineering, National Institute of Technology,
Rourkela 769008, India
e-mail: debi_tripathy@yahoo.co.in

D. Sengupta (ed.), *Recent Trends in Modelling of Environmental Contaminants*,
DOI 10.1007/978-81-322-1783-1_4,

Table 1 Sources of AMD

Primary sources	Secondary sources
Mine rock dumps	Treatment sludge pounds
Tailings impoundment	Rock cuts
Underground and open pit mine workings	Concentrated load-out
Pumped/nature discharged underground water	Stockpiles
Diffuse seeps from replaced overburden in rehabilitated areas	Concentrate spills along roads
Construction rock used in roads, dams, etc	Emergency ponds

generation of AMD as well as to treat it. Every mine is unique in terms of its AMD potential; thus, the nature and size of the associated risk and feasibility of mitigation options will vary from site to site. Mine planners and managers must therefore be equipped with the knowledge and tools to control AMD specifically, to identify techniques that will minimize AMD impacts on life forms and their support systems (Akcil and Koldas 2006; Sharma 2010).

There are many types of sulphide minerals. Iron sulphides are most common but other metal sulphide minerals may also produce AMD. Upon exposure to oxidizing conditions and in the absence of alkaline materials, some sulphide minerals are oxidized in the presence of water and oxygen to form highly acidic, sulphate-rich drainage. Acidity levels, and metal composition and concentration depend on the type and amount of sulphide mineral and the presence or absence of alkaline materials. In the majority of cases, bacteria play a major role in accelerating the rate of acid generation; the inhibition of bacterial activity can therefore impede the rate of acid generation. Alkaline drainage is not as environmentally damaging as AMD. Some alkaline waters have high concentrations of ferrous iron, which dramatically lowers the pH upon oxidation and hydrolysis, rendering the effluent acidic. Alkaline drainage is more common in underground mines than in surface mines. The primary and secondary sources of AMD are summarized in Table 1.

Pyrite is the mineral of most relevance from an acid-generation perspective, because its concentration, grain size and distribution may be the most important factors affecting the production of acidic mine waters (Nordstrom and Alpers 1999). Other sulphides commonly found in ore deposits are listed in Table 2 (Plumlee 1999). These sulphides may produce acid rock drainage(ARD), neutral mine drainage(NMD) or saline drainage(SD). Secondary minerals resulting from sulphide oxidation include a complex array of soluble sulphates, hydrous sulphates, hydroxysulphates, metal oxides and hydroxides, clays, carbonates and supergene, and digenetic sulphides. Some of these secondary minerals may have deleterious effects on water quality because of the release of additional acidity during their formation (e.g. metal (hydr) oxides) or release of stored acidity, sulphate or metals (or both sulphate and metals) during their dissolution (e.g. iron and aluminium hydroxysulphates).

The type and distribution of sulphide minerals can vary widely according to the type of ore deposit, nature of the mine waste and mine stage. Certain types of ore deposits can be devoid of sulphides (e.g. oxide-facies banded iron formation (BIF) deposits) while others contain very substantial amounts of sulphide (e.g. Vol-

Table 2 Common sulphides known or inferred to generate acid when oxidized (Plumlee 1999)

Mineral	Formula
Common sulphides known (*inferred*) to generate acid with oxygen as the oxidant:	
Pyrite, marcasite	FeS_2
Pyrrhotite	$Fe_{1-x}S$
Bornite	Cu_5FeS_4
Arsenopyrite	FeAsS
Enargite/famatinite	Cu_3AsS_4/Cu_3SbS_4
Tennantite/tetrahedrite	$(Cu, Fe,Zn)_{12}As_4S_{13}/(Cu, Fe,Zn)_{12}Sb_4S_{13}$
Realgar	AsS
Orpiment	As_2S_3
Stibnite	Sb_2S_3
Common sulphides that may generate acid with ferric iron as the oxidant:	
All of the above plus:	
Sphalerite	ZnS
Galena	PbS
Chalcopyrite	$CuFeS_2$
Covellite	CuS
Cinnabar	HgS
Millerite	NiS
Pentlandite	$(Fe, Ni)_9S_8$
Greenockite	CdS

canogenic massive sulphide (VMS) deposits). Similarly, within an individual ore deposit, the distribution of sulphide minerals can range from disseminated evenly throughout the deposit to the sulphides being confined to specific zones. The distribution depends on the nature of the original ore-forming processes or subsequent alteration. The techniques used to extract and process ores can also significantly affect the type and distribution of sulphide minerals, which in turn has ramifications regarding the nature of corresponding mine discharges. The large volume of waste rock resulting from an open pit mine may have lower overall sulphide content than the smaller volume originating from an underground mine. This lower overall sulphide content would be due to a greater dilution with rock not directly associated with the mineralization.

2 Basic AMD Chemistry

Oxygen, water and pyrites are the requisites for generation of AMD. The chemistry of oxidation of pyrites, the production of ferrous ions and subsequently ferric ions, is very complex, and this complexity has considerably inhibited the design of effective treatment options. The commonly accepted chemical reactions that represent the chemistry of pyrite weathering to form AMD are as mentioned below:

$$4\,FeS_2 + 15\,O_2 + 14\,H_2O = 4\,Fe(OH)_3 + 8\,H_2SO_4 \quad (1)$$

The first reaction in the weathering of pyrite includes the oxidation of pyrite by oxygen. Sulphur is oxidized to sulphate and ferrous iron is released. This reaction generates two moles of acidity for each mole of pyrite oxidized.

$$2\,FeS_2 + 7\,O_2 + 2\,H_2O = 2\,Fe^{2+} + 4\,SO_4^{\;2-} + 4\,H^+ \qquad (2)$$

The second reaction involves the conversion of ferrous iron to ferric iron. The conversion of ferrous iron to ferric iron consumes one mole of acidity. This reaction is referred to as the 'rate determining step' in the overall acid-generating sequence.

$$4\,Fe^{2+} + O_2 + 4\,H^+ = 4\,Fe^{3+} + 2\,H_2O \qquad (3)$$

The third reaction which may occur is the hydrolysis of iron. Three moles of acidity are generated as a byproduct. It is observed that the formation of ferric hydroxide precipitate (solid) is pH dependant. Solids form if the pH is above about 3.5 but below pH 3.5 little or no solids will precipitate.

$$4\,Fe^{3+} + 12\,H_2O = 4\,Fe(OH)_3 + 12\,H^+ \qquad (4)$$

The fourth reaction is the oxidation of additional pyrite by ferric iron. It takes place very rapidly and continues until either ferric iron or pyrite is depleted.

$$FeS_2 + 14\,Fe^{3+} + 8\,H_2O = 15\,Fe^{2+} + 2\,SO_4^{\;2-} + 16\,H^+ \qquad (5)$$

A ton of coal containing 1 % pyrite sulphur can generate 33 lb of yellow boy and over 60 lb of sulphuric acid (Sengupta 1996). Inhibition of bacterial activity can lessen the rate of acid generation. Total exclusion of moisture or oxygen will stop acid formation.

3 Factors Influencing the Quality of Acid Mine Drainage

Primary, secondary, tertiary and downstream factors influence the quality and quantity of mine drainage.

- **Primary factors** influencing the amount and quality of acidic water are the relative amount of water and oxygen in the environment. In order for pyrite to oxidize, both oxygen and water must be present. Water serves not only as a reactant, but also as a reaction medium and a product-transport solvent. The primary factors that determine the rate of acid generation are: pH, temperature, oxygen content of the gas phase/water phase, degree of saturation with water, chemical activity of Fe^{3+}, surface area of exposed metal sulfide, chemical activation energy required to initiate acid generation and bacterial activity.

Table 3 Factors affecting sulphide oxidation

Physical factors	Chemical factors	Biological factors
Climate conditions	pH	Microbial ecology
Precipitation events	Redox conditions	Microbial growth kinetics
Water movement	Chemical composition of drainage	–
Temperature	Secondary mineral formation	–
Climate conditions	Sorption	–
–	Neutralization reactions	–
–	Photochemistry	–

- A **secondary factor** is the neutralization of acids by the alkalinity released from the carbonate minerals in the mine waste and surrounding stratum.
- **Tertiary factors** include the physical characteristics of mining waste, the spatial relationship between wastes and the hydrologic regime.
- **Downstream factors** may impact the quality and quantity of acid drainage. Physical processes such as dilution and precipitation and chemical processes such as neutralization will permit a stream to assimilate acid drainage, but not without incurring a great deal of acid damage to the preceding stream area (Forstner and Salomons 1988).

4 Factors Affecting the Rate of Sulphide Oxidation

A large number of factors control the rate of sulphide oxidation. These factors are classified as chemical factors, physical factors and biological factors and are summarized in Table 3.

5 Characteristics of AMD

Characteristics of AMD from five mines in M.P. and Assam as per Gandhi (1990) are shown in Table 4.The characteristics of water of Amjhore pyrite mine (PPCL) as per Prasad (1994) are presented in Table 5. The mine produces about 288 m^3/day of acidic mine water. The most serious problems with AMD are confined to Western Maryland, Northern West Virginia, Pennsylvania, Western Kentucky and along the Illinois-Indiana border (USEPA (1995). The following is a Table (Table 6) of the concentrations of constituents in the coal mine drainage for the USA, Illinois, Kentucky, Maryland and West Virginia.

Table 4 Characteristics of AMD from coal mines of India (Bhole 1994)

Characteristics	Assam		M.P.		
	Ledo	Bargolia	Ravanwara	Chandmeta	Ekalehara
pH	3.9	4.3	2.96	6.45	6.64
S.Conductivity (micromhos/cm)	0.003	0.002	0.38	0.48	0.22
Total hardness($CaCO_3$) mg/l	720	840	750	1670	700
Chloride (mg/l)	1488	1235	9.93	4.26	11.35
Calcium (mg/l)	168	212	248	519	180
Magnesium (mg/l)	72	72	130	370	250
Sulphate (mg/l)	–	–	2875	1818	843
Iron (mg/l)	100	120	5.6	–	–

Table 5 Characteristics of mine water of Amjhore pyrite mine (Prasad 1994)

Parameter	Unit mg/l except pH
pH	2.6
TDS	17000
Fe	1000
Chloride	10
Calcium as CaO	526
Magnesium as MgO	118
Sulphate	5576
Total acidity as H_2SO_4	5918

6 Prevention, Disposal and Treatment of AMD

There are a number of methods by which AMD can be prevented or reduced or to a great extent disposed off or treated as given in Table 7.

6.1 Prevention/Mitigation of AMD

In preventing acid drainage, water and air contact with the acidic material must be eliminated. Prevention effectiveness depends on the nature of the mine and the strata's geological characteristics. Preventing water from reaching underground mines involves the use of diversion ditches and pipes to divert water from acidic areas. Another method to prevent acid drainage is to prevent the material from oxidizing. By burying mine waste, or covering the waste with an impermeable liner, pyrite cannot oxidize and sulfuric acid cannot form. Some mines use a method in which an asphalt emulsion of polyurethane sealant encapsulates the mine waste (Mehrotra and Singhal 1992). Different options for prevention of AMD are summarized in Table 7.

Research on acid prevention and mitigation has focused on three main areas: chemical inhibition of the acid generating reactions; inhibition of the microbes

Table 6 Characteristics of AMD water in USA

Substance	U.S. (1)	Illinois (2)	Kentucky (3)	Maryland (4)	West Virginia (5)
pH	3.2–7.9	3	1.8–3.5	3.6–6.0	2.6 ± 0.1
Acidity CaCO3(mq/L)	–	–	–	90–438	–
SO4- (mg/L)	–	1300	500–12000	620–1600	1527 ± 12
Ca (mg/L)	–	–	–	183–489	191 ± 10
Mg (mg/L)	–	–	–	17–48	50.5 ± 3.9
Fe (mg/L)	0.6–220	57	57–500	28–92	162 ± 23
Mn (mg/L)	0.3–12	6.4	–	2–5.5	203 ± 0.21
Al (mg/L)	–	37	–	4.0–28	80.8 ± 7.4
Cu (mg/L)	0.01–0.17	–	–	ND–0.08	0.08 ± 0.02
Ni(mg/L)	–	–	–	0.57–1	1.01 ± 0.15
Zn (mg/L)	0.03–2.2	–	–	0.5–3	2.72 ± 0.34
Cd (mg/L)	0.01–0.10	–	–	–	–
Pb (mg/L)	0.01–0.40	–	–	–	–
As (mg/L)	0.002–0.20	–	–	–	–

Table 7 Prevention, disposal methods and treatment of AMD

Prevention/mitigation	Method of Disposal	Treatment methods
Reducing contact of air and water with pyrite	Lagooning	Neutralization(chemical treatment)
Coating and Sealant	Deepwell disposal	Aeration
Addition of alkaline agents and phosphates	Using AMD for coal washing	Ion exchange
Bactericides		Activated sludge
Controlled placement of pyrite		Flash distillation
Submergence		Reverse osmosis
Isolation of above water inflow		Electro-dialysis
Encapsulation/physical barrier		Foam separation
Mine spoil hydrology		Biological treatment
		Passive treatment methods

responsible for catalyzing the acid generating reactions; and physical or geotechnical treatments to minimize water contact and leaching.

Chemical Methods

Alkaline Addition

Alkaline placement strategies involve either mixing alkaline agents directly with pyritic material or concentrated placement to create a highly alkaline environment. Direct mixing places alkaline materials in intimate contact with pyritic spoil to inhibit acid formation and neutralize any generated acidity in situ.

The strategy is to charge infiltrating waters with high doses of alkalinity sufficient to overwhelm any acid produced within the backfill. This approach is highly dependent on the placement of the alkaline trenches to provide maximum inflow to the acid producing zones.

Alkaline Agents

The benefits of adding lime (calcium carbonate) and other alkaline agents have long been recognized in mitigating acid drainage. Direct mixing and contact with pyritic materials appears most effective but an optimum lime to pyrite ratio remains unknown. Indirect treatments such as alkaline recharge (Caruccio and Geidel 1989) and borehole injection (Aljoe and Hawkins 1991; Ladwig et al. 1985) have also yielded mixed results.

Phosphate

The application of rock phosphate is another technique under study as a pyrite oxidation inhibitor (Renton et al. 1988; Evangelou et al. 1991). Dissolution of rock phosphate in acid media releases highly reactive phosphate ions, which will combine with iron to form insoluble iron phosphate compounds. The formation of insoluble iron phosphates would halt or inhibit the cyclic reaction of iron and pyrite.

Coatings and Sealants

Other ongoing research activities are focusing on the surface chemistry of pyrite and development of various types of sealers, coatings and inhibitors to halt acid production.

Biological Agents/Bactericides

The catalytic role of bacteria in pyrite oxidation has been well documented (Kleinmann et al. 1981). Many compounds have been screened as selective bactericides and the anionic surfactants sodium lauryl sulfate and alkyl benzene sulphonate are considered to be the most reliable inhibitors.

Physical or Geochemical Treatments

Controlled Placement

Controlled placement (special handling) is a preventative measure involving the placement of pyritic or alkaline material during mining to minimize or neutralize the formation of AMD. According to the generally accepted chemical equations for pyrite oxidation, oxygen and water are necessary to initiate acid formation. Exclusion of either reactant should preclude or inhibit acid production. Placement of pyritic material encompasses either an attempt to exclude oxygen, usually by complete submergence below the water table; or an attempt to isolate the material from water contact to avoid leaching of acid salts. Placement of alkaline materials has a twofold role:

- Inhibition of the acid-forming reactions by maintaining neutral to alkaline pH and
- Neutralization of any acid formed.

Submergence

Submergence relies on several physico-chemical phenomena for success. Oxygen diffuses very slowly and has limited solubility in water. For this approach to succeed, a stagnant or no flow condition and relatively thick saturated zone appears critical. Stagnant flow conditions leading to the development of anoxic (oxygen free) conditions and a saturated thickness on the order of several tens of feet appear to effectively curtail oxygen diffusion. This approach is most successful in large mines in flat terrain where ground water gradients are low, the saturated zone is thick and aquifers are of large areal extent. Submergence or flooding is also applied to prevent AMD from underground mines. Key considerations include:

- Whether the mine is located above or below drainage.
- The ability of mine seals and outcrop barriers to prevent seepage.
- Potential for mine seals and outcrop barriers to fail under hydraulic pressure.

In general, flooding to prevent AMD is believed to be more successful in below drainage mines. Kim et al. (1982) concluded that sealing and flooding above drainage mines does reduce acid loading but is technically more difficult and less effective than other methods in AMD prevention.

Isolation Above the Water Table

Placement of pyritic material above a water table is an attempt to isolate the material from contact with water, and preclude leaching of acid weathering products. Compaction and capping with clay or other materials may be employed to reduce permeability.

Encapsulation/Physical Barriers

Techniques to isolate or encapsulate pyritic material include the use of fly-ash, cements, bentonite and other clays; these are a few of the materials studied as sealants and flow barriers by Skousen et al. (1987) and Bowders and Chiado (1990).

Water Management

Water management strategies, both during and after mining, are another option for reducing acid generation. Water management can include the following:

- Active mining operations can incorporate diversions to route surface drainage away from pyritic material or through alkaline material.
- Spoil material can be placed and rough graded to prevent ponding and subsequent infiltration.
- Prompt removal of pit water can lessen the amount and severity of acid generated.
- Polluted pit water can be isolated from non-contaminated sources (no commingling) to reduce the quantity of water requiring treatment.
- Constructed under-drain systems can be used to route water away from contact with acid forming material.

Special handling (controlled placement), alkaline placement and water management strategies alone or in combination can substantially reduce or mitigate generation of acid drainage. Optimal strategies are site-specific and a function of geology, topography, hydrology, mining method and cost effectiveness.

Mine-Spoil Hydrology

It plays a crucial role in determining drainage quality. Relatively few studies of hydro-geologic processes have been conducted in the context of controlling mine drainage quality. For economic reasons, treatment is the most practical solution to acid drainage pollution.

6.2 Active & Passive Treatment of AMD

Treatment methods to address AMD focus on neutralizing, isolating, stabilizing and/or removing problem pollutants through various chemical, physical and biological processes. There are two basic types of treatment systems. Active treatment involves addition of alkaline chemicals such as lime, soda ash or ammonia, with the contaminated drainage to decrease its acidity and speed up the removal of metals. Passive systems clean contaminants from mine drainage by exposing it to air, limestone, cattails and other vegetation that form carefully designed components of ponds, neutralizing ditches, buried channels and wetlands. Flow chart to guide selection of type of treatment is shown in Fig. 1. The main steps involved in active treatment of AMD are pretreatment, dosing with alkali, oxidation and sedimentation (Younger 2000).

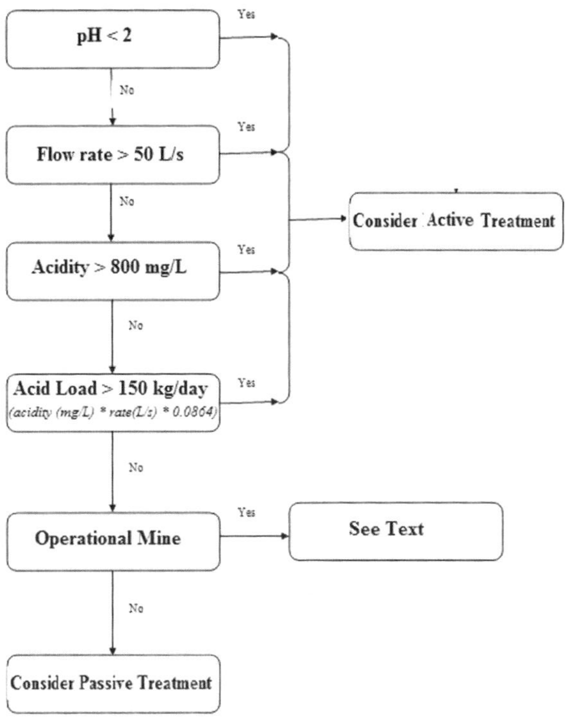

Fig. 1 Flow chart to guide selection between active and passive treatment for AMD (modified from Waters et al. 2003)

Active Treatment Methods for AMD

Neutralization of AMD

Standard AMD treatment methods involve neutralization of the acid by addition of a base. The acidity is buffered by the addition of alkaline chemicals such as calcium carbonate, sodium hydroxide, sodium bicarbonate or anhydrous ammonia etc. These chemicals raise the pH to acceptable levels and decrease the solubility of dissolved metals. Precipitates formed are settled in the solution. But these chemicals are expensive and the treatment system requires additional costs associated with operation and maintenance as well as the disposal of metal-laden sludge.

Treatment, as normally applied to AMD, involves chemical neutralization of the acidity followed by precipitation of iron and other suspended solids. Treatment systems include:

- Equipment for feeding the neutralizing agent to the acid mine drainage
- Means for mixing the two streams (AMD and neutralizing agent)
- Procedures for ensuring iron oxidation
- Settling ponds for removing iron, manganese and other co-precipitates.

Six primary chemicals have been used to treat AMD (Table 8). Each chemical has characteristics that make it more or less appropriate for a specific condition. The

Table 8 Chemical compounds used in AMD treatment (http://www.cewic.fi/cewic/materiaalit/pehkonen_270509.pdf?PHPSESSID=56a70d3f459417e6db8c3a8223bd085c)

Common Name	Chemical Name	Formula	Conversion	Neutralization (%)
Limestone	Calcium carbonate	CaCO3	1	30
Hydrated Lime	Calcium hydroxide	Ca(OH)2	0.74	90
Pebble Quicklime	Calcium oxide	CaO	0.56	90
Soda Ash	Sodium carbonate	Na2CO3	1.06	60
Caustic Soda (solid)	Sodium hydroxide	NaOH	0.8	100
20% Liquid Caustic	Sodium hydroxide	NaOH	784	100
50% Liquid Caustic	Sodium hydroxide	NaOH	256	100
Ammonia	Anhydrous ammonia	NH3	0.34	100

[1]The conversion factor may be multiplied by the estimated tons acid/yr to get tons of chemical needed for neutralization per year

best choice among alternatives depends on both technical and economic factors. The technical factors include acidity levels, flow, the types and concentrations of metals in the water, the rate and degree of chemical treatment needed and the desired final water quality. The economic factors include prices of reagents, labour, machinery and equipment, the number of years that treatment will be needed, the interest rate and risk factors.

Each AMD is unique and the chemical treatment of any particular AMD source is site specific. Each AMD source should be tested with various chemicals by titration tests to evaluate the most effective chemical for precipitation of the metals. The costs of each AMD treatment system based on neutralization (in terms of the reagent cost, capital investment and maintenance of the dispensing system), floc volumes and disposal should be evaluated to determine the most cost effective system.

Under low flow situations, pebble quicklime and ammonia are the most cost effective. Under high flow situations, hydrated lime and pebble quicklime are the most cost effective due to their low reagent cost compared to the other chemicals. Each chemical reacts differently with a specific AMD. Therefore, it is essential that each AMD source be treated and evaluated with each chemical to determine which is most environmentally sound, efficient and cost effective. Coagulants and flocculants may be used in water treatment where retention time in sedimentation ponds is insufficient for metal precipitation. Oxidants can be used to meet more stringent effluent limits and to make chemical treatment more efficient. Floc and the metal hydroxides collected in ponds after chemical treatment, are disposed of in abandoned deep mines, refuse piles or left in collection ponds.

The following discussion highlights some of the characteristics of each of these neutralizing agents:

Limestone (Calcium Carbonate)

Limestone has been used for decades to raise pH and precipitate metals in AMD. Dolomitic limestone is less reactive and generally ineffective in treating AMD It has the lowest material cost and is the safest and easiest to handle, of the AMD chemicals. It has low solubility and tendency to develop an external coating, or armor, of ferric hydroxide when added to AMD. In cases where pH is low and mineral acidity is also relatively low (low metal concentrations), finely ground limestone may be dumped in streams directly or the limestone may be ground by water-powered rotating drums and metered into the stream. These applications have been tried recently in West Virginia in AMD-impacted streams with great success. Limestone has also been used to treat AMD in anaerobic (anoxic limestone drains) and aerobic environments (open limestone channels) (Faulkner 1996).

If limestone is used as the neutralizing agent, the reaction is:

$$H_2SO_4 + CaCO_3 = CaSO_4 + H_2O + CO_2 \tag{6}$$

To remove the ferrous iron, the neutralized water is aerated to produce ferric ions, which reacts with the base to form insoluble ferric hydroxides.

$$Fe_2\left(SO_4\right)_3 + Ca\left(OH\right)_3 = Fe\left(OH\right)_2 + CaSO_4 \tag{7}$$

$$Fe_2\left(S0_4\right)_3 + 3CaCO_3 + 3H_2O = 2Fe\left(OH\right)_3 + 3CaSO_4 + 3O_2 \tag{8}$$

Advantages

- Low cost, ease of use and formation of a dense, easily handled sludge.

Disadvantages

- Slow reaction time, loss in efficiency of the system because of coating of the limestone particles with iron precipitates, difficulty in treating AMD with a high ferrous-ferric ratio and ineffectiveness in removing manganese.
- Limestone treatment is generally not effective for acidities exceeding 50 mg/l.

Hydrated Lime (Calcium Hydroxide)

Hydrated lime is a commonly used chemical for treating AMD. It is sold as a powder that tends to be hydrophobic and extensive mechanical mixing is required to disperse it in water. Hydrated lime is particularly useful and cost effective in large flow, high acidity situations where a lime treatment plant with a mixer/aerator is constructed to help dispense and mix the chemical with the water (Skousen and Ziemkiewicz 1995). Proper storage of hydrated lime is important in order to maintain its flow characteristics and thus ensure efficient use The length of time that the system will be in operation is a critical factor in determining the annual cost of a lime treatment system due to the large initial capital expenditure that can be amortized over time. Hydrated lime is easy and safe to use, effective and relatively inexpensive. It produces the voluminous sludge (when compared to limestone) and has high initial costs.

Pebble Quicklime (CaO)

The amount of chemical applied is dictated by the movement of the water wheel, which causes a screw feeder to dispense the chemical. The hopper and feeder can be installed in less than an hour. This system was initially used for small and/or periodic flows of high acidity because calcium oxide is very reactive. Test results reveal an average of 75 % cost savings over caustic systems and about 20–40 % savings over ammonia systems.

Soda Ash (Sodium Carbonate)

Soda ash briquettes are especially effective for treating small AMD flows in remote areas. Soda ash comes as solid briquettes and is gravity fed into water by the use of hoppers mounted over a basket or barrel. The number of briquettes to be used each day is determined by the flow and quality of the water to be treated. Major disadvantages are higher reagent cost (relative to limestone) and poor settling properties of the sludge.

Caustic Soda (Sodium Hydroxide)

Caustic soda is especially effective for treating low flows in remote locations and for treating AMD having a high manganese content and in low flow, high acidity situations. The system can be gravity fed by dripping liquid caustic directly into the AMD. Caustic is very soluble in water, disperses rapidly and raises the pH of the water quickly. Caustic should be applied at the surface of pond water because the chemical is denser than water and sinks. Major disadvantages are its high cost, the dangers involved with handling the chemical, poor sludge properties and freezing problems in cold weather (USEPA 1983).

Ammonia

Anhydrous ammonia is effective in treating AMD having a high ferrous iron and/or manganese content. It costs less than caustic soda. It is difficult and dangerous to use and can affect biological conditions downstream from the mining operation. The possible off-site impacts are: toxicity to fish and other aquatic life forms, eutrophication and nitrification (Faulkner 1991).

Injection of ammonia into AMD is one of the quickest ways to raise water pH. It should be injected into flowing water at the entrance of the pond to ensure good mixing because ammonia is lighter than water. A cost reduction figure of 50–70 % can be realized when ammonia is substituted for caustic if the target pH for metal precipitation is <9.8 (Skousen et al. 1990).

Major disadvantages of using ammonia include:

- Hazards associated with handling the chemical
- Potential biological implications
- The consequences of excessive application rates (Faulkner 1990).

Passive Treatment Methods for AMD

Passive treatment systems have been implemented on full-scale sites throughout the USA with promising results. The concept behind passive treatment is to allow the naturally occurring chemical and biological reactions that aid in AMD treatment to occur in the controlled environment of the treatment system, and not in the receiving water body. AMD treatment using wetlands are of two types: aerobic and anaerobic. The wetlands act as 'black box' into which AMD water flows and emerges after purification. This approach utilizes small marsh ponds for treatment of acidic water and has been extensively and successfully used in USA. Acid tolerant species plants have been used and are effective in metal removal by adsorption. The success of wetlands depends on depth of water, maximum flow path length and detention time and are more suited for small flows of AMD (Singh 2005). Designing a passive treatment system for AMD requires the understanding of mine water chemistry, available treatment techniques and experience. Analytical sampling of the AMD is extremely important in the selection of appropriate treatment technologies.

It conceptually offers many advantages over conventional active treatment systems. The use of chemical addition and energy consuming treatment processes are virtually eliminated with passive treatment systems. Also, the operation and maintenance requirements of passive systems are considerably less than active treatment systems.

Active chemical treatment of AMD to remove metals and acidity is often an expensive and long term liability. In recent years, a variety of passive treatment systems have been developed that do not require continuous chemical inputs and that take advantage of naturally occurring chemical and biological processes to cleanse contaminated mine waters. The primary passive technologies include constructed wetlands, anoxic limestone drains (ALD), vertical flow reactors (VFR), diversion wells (DO), pyrolusite process and open limestone channels (OLC). Flow chart to select suitable AMD passive treatment system based on water chemistry is given in Fig. 2.

At their present stage of development, passive systems can be reliably implemented as a single permanent solution for many types of AMD and at a much lower cost than active treatment. Relative to chemical treatment, passive systems require longer retention times and greater space, provide less certain treatment efficiency and are subject to failure in the long term. However, many passive systems have realized successful short-term implementation in the field and have substantially reduced water treatment costs at many mine sites (Faulkner and Skousen 1994).

The first passive technology involved the use of natural Sphagnum wetlands that could improve the water quality of AMD without causing other detrimental impacts on the ecosystem. Designing a passive treatment system for AMD requires the understanding of mine water chemistry, available treatment techniques and experience. Analytical sampling of the AMD is extremely important in the selection of appropriate treatment technologies.

Selection of an appropriate passive system is based on water chemistry, flow rate, local topography and site characteristics (Hyman and Watzlaf 1995), and

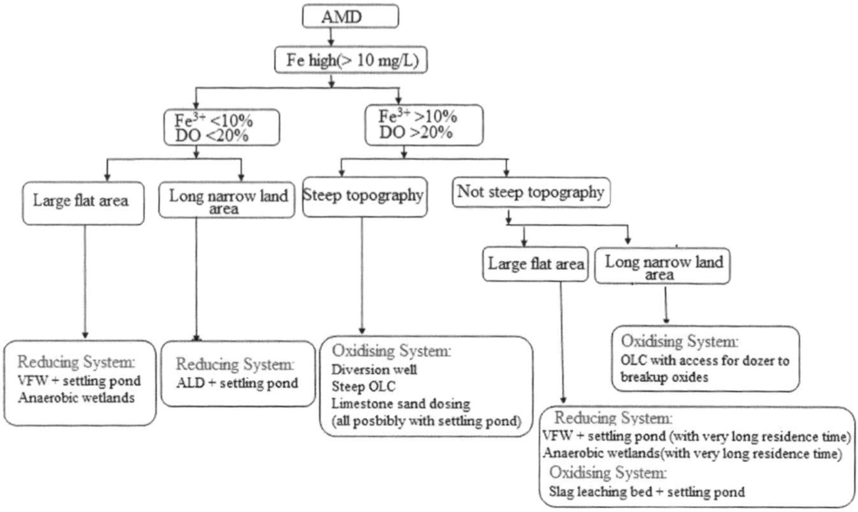

Fig. 2 Flow chart to select among AMD passive treatment systems based on water chemistry (high Fe), topography and available land area (Trumm 2007)

refinements in design are ongoing. In general, aerobic wetlands can treat net alkaline water; ALDs can treat water of low Al, Fe^{3+} and DO; and SAPS, anaerobic wetlands and OLCs can treat net acidic water with higher Al, Fe^{3+}, and DO. As scientists and practitioners improve treatment predictability and longevity of passive systems, they will be able to treat the more difficult waters of high acidity and high Al content.

Passive treatment systems are a valuable option for treating AMD at remote locations. The advantages of passive treatment systems are that they:

- do not require electrical power
- do not require any mechanical equipment, hazardous chemicals, or buildings
- do not require daily operation and maintenance
- are more natural and aesthetic in their appearance and may support plants and wildlife
- are less expensive

The disadvantages of passive treatment systems are that they:

- may require complex discharge permits
- may not meet stringent water-quality-based effluent standards
- may fail because of poor design or severe winter conditions
- are a relatively new technology and an area of active research; as such, there have been failures along with success stories

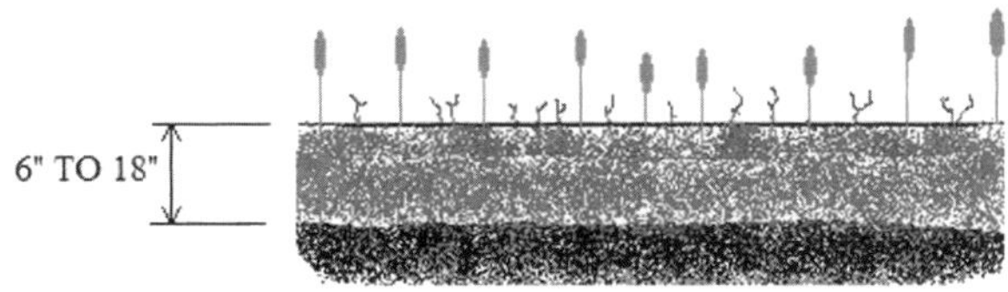

Fig. 3 Typical section of an aerobic wetland

Constructed Wetlands

Constructed wetlands are ecological systems designed to optimize a variety of natural physical, chemical, microbial and plant-mediated processes. In a constructed wetland, influent AMD drains by gravity through the wetland, progressively undergoing metal removal and neutralization. Metals are removed by precipitation, chelation and exchange reactions, while neutralization is primarily achieved by the activity of sulphate reducing bacteria (SRB), or the increase in alkalinity from the chemical and microbial reactions including limestone dissolution.

Passive systems for AMD treatment have commonly used combinations of natural or constructed wetlands, Sphagnum peat and open ponds, supplemented by chemical amendments (mostly limestone) and organic substrate to increase alkalinity and reduce acidity. Sequential treatment of AMD to remove iron by oxidation, hydrolysis and settling in the aerobic stage, followed by SRB activity in an anaerobic stage to raise pH, is an effective combination.

Aerobic Wetlands

An aerobic wetland (Fig. 3) consists of a large surface area pond with horizontal surface flow. It has a water depth of 6–18 inches. They may be lined or unlined. The pond may be planted with cattails and other wetland species. Aerobic wetlands can only effectively treat water that is entirely alkaline. They facilitate natural oxidation of the metals and precipitate iron, manganese, and other metals. In this system, metals are precipitated through oxidation reactions to form oxides and hydroxides. This process is more efficient when the influent pH is >5.5. Aeration prior to the wetland, via riffles and falls, increases the efficiency of the oxidation process and therefore the precipitation process. Iron concentrations are efficiently reduced in this system but the pH is further lowered (Hedin et al. 1994).

Compost/Anaerobic Wetlands

Compost wetlands or anaerobic wetlands (Fig. 4) consist of a large pond with a layer of organic substrate. The flow is horizontal within the substrate layer of the basin. Piling the compost a little higher than the free water surface can encourage the flow within the substrate. Typically, the compost layer is made from spent mushroom compost that contains about 10% calcium carbonate. Other compost materials include peat, moss, wood chips, sawdust or hay. A typical compost wetland will have 12–24 inches of organic substrate and be planted with cattails or

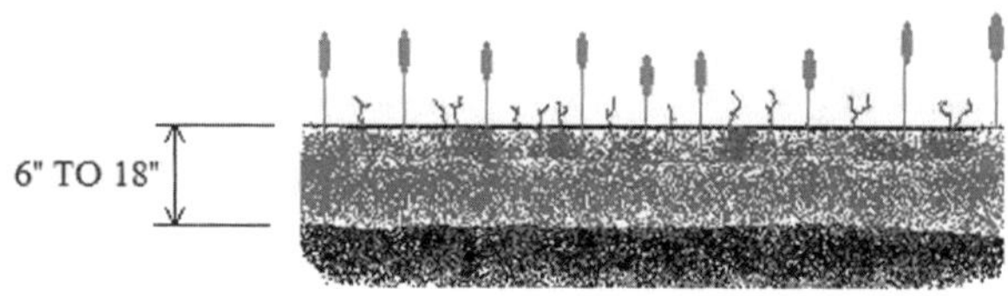

Fig. 4 Typical section of an anaerobic or compost wetland

Fig. 5 Cross section of an open limestone channel

other emergent vegetation. The vegetation helps stabilize the substrate and provides the sulfate reduction reactions.

Compost wetlands can treat discharges that contain dissolved oxygen, Fe^{3+}, Al^{3+} or acidity less than 300 mg/l. The compost wetland acts as a reducing wetland where the organic substrate promotes chemical and microbial processes that generate alkalinity and increase the pH. Microbial respiration within the organic substrate reduces sulphates to water and hydrogen sulphide. The anoxic environment within the substrate also increases the dissolution of limestone (Hedin et al. 1994). Anaerobic wetlands are sized according to U.S. Bureau of Mines criteria for AML sites as follows:

$$\text{Minimum wetland size } (m^2) = \text{acidity loading (g/day)} \div 0.7$$

Open Limestone Channels

OLC (Fig. 5) may be the simplest passive treatment method. OLC are constructed in two ways. In the first method, a drainage ditch is constructed of limestone and AMD-contaminated water is collected by the ditch. The other method consists of placing limestone fragments directly in a contaminated stream. Dissolution of the limestone adds alkalinity to the water and raises the pH. The coating of the limestone by $Fe(CO)_3$ and $Fe(OH)_3$ produced by neutralization reduces the generation of alkalinity, hence large quantities of limestone are needed to ensure long-term success. High flow velocity and turbulence enhance the performance by keeping precipitates in suspension thereby reducing the armoring of the limestone. Impervious liners are sometimes used under the limestone to prevent infiltration of the AMD into the ground water (Ziemkiewicz et al.1994).

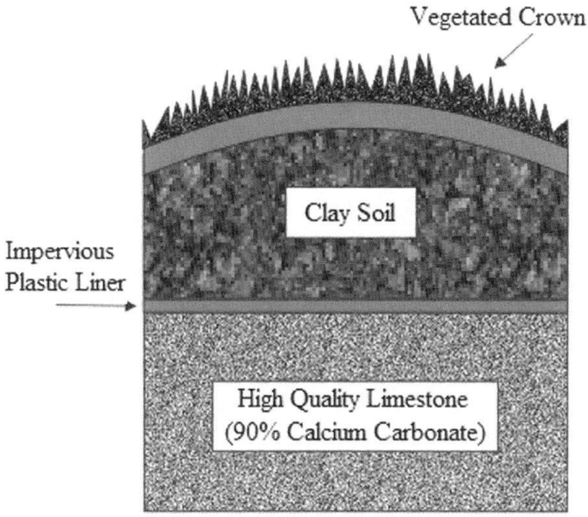

Fig. 6 Cross section of an anoxic limestone drain

Anoxic Limestone Drains (ALD)

ALD (Fig. 6) can be considered a pretreatment step to increase alkalinity and raise pH before the water enters a constructed aerobic wetland. It consists of a buried bed of limestone constructed to intercept subsurface mine water flows and prevent contact with atmospheric oxygen. It is suitable if $Fe^{+3} < 2$ mg/l, net acidity < 300 mg/l of $CaCO_3$ and DO < 1 mg/l (Hedin and Watzlaf 1994). The process of limestone dissolution generates alkalinity. The effectiveness and longevity of an ALD can be substantially reduced if the AMD has high concentrations of ferric iron, dissolved oxygen or aluminium (Brodie 1993). ALD's have been shown to be most effective for influent with dissolved oxygen, ferric iron (Fe3+) and aluminium concentrations of less than 1 mg/L, and sulphate concentrations below 2,000 mg/L. At higher concentrations, the limestone may become armored with oxides or gypsum, reducing the rate of limestone dissolution or plugging the system. In either instance, the ability of the ALD to generate alkalinity may be significantly reduced, and failure of the system may occur.

Vertical Flow Reactors

VFR (Fig. 7) were conceived to overcome the alkalinity producing limitations of ALD and the large area requirements for compost wetlands. The VFR consists of a treatment cell with an under-drained limestone base topped with a layer of organic substrate and standing water. The water flows vertically through the compost and limestone, and is collected and discharged through a system of pipes. The VFR increases alkalinity by limestone dissolution and bacterial sulphate reduction. Highly acidic waters can be treated by running the AMD through a series of VFRs. VFRs

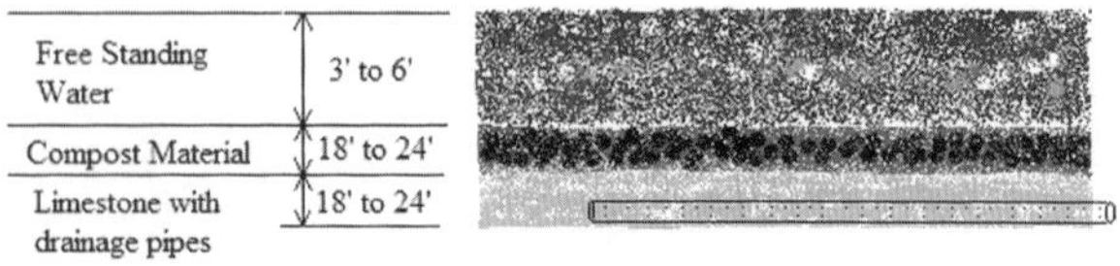

Fig. 7 Typical section of a vertical flow wetland

are sized based on retention times(12–15 h) required to produce the necessary alkalinity (Ziemkiewicz et al. 1997).

Diversion Wells

Diversion wells (Fig. 8) are another simple way of adding alkalinity to contaminated waters. Acidic water is conveyed by a pipe to a downstream 'well' which contains crushed limestone aggregate. The hydraulic force of the pipe flow causes the limestone to turbulently mix. The water flows upward and overflows the 'well' where it is diverted back into the stream. Diversion wells require frequent refilling with clean limestone for continued treatment (Arnold 1991).

Pyrolusite Process

The pyrolusite process (Fig. 9) utilizes site-specific laboratory cultured microbes to remove iron, manganese and aluminium from AMD. The process consists of a shallow bed of limestone aggregate inundated with AMD. The microorganisms are introduced to the limestone bed by inoculation ports located throughout the bed. The microorganisms grow on the surface of the limestone chips and oxidize the metal contaminants, increase the alkalinity and raise the pH of the water (Vail and Riley 1997).

7 Conclusion

AMD is recognized as one of the serious environmental problems in the mining industry. It can severely contaminate surface and ground water making it harmful to human and aquatic life, concrete structures and mining machineries. Hence attention must be paid to reduce the generation of AMD as well as to treat it. Every mine is unique in terms of its AMD potential; thus, the nature and size of the associated risk and feasibility of mitigation options will vary from site to site. Mine planners and managers must therefore be equipped with the knowledge and tools to control AMD specifically, to identify techniques that will minimize AMD impacts on life forms and their support systems.

Prevention of AMD is the most desirable option. The most effective method of control is to minimize penetration of air and water through the waste pile using a

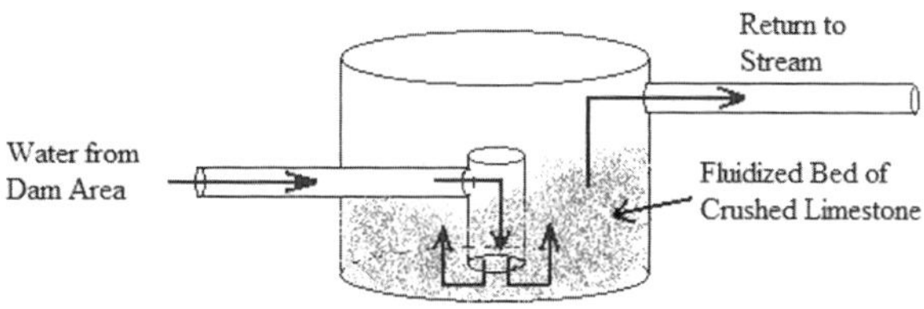

Fig. 8 Schematic view of a diversion well

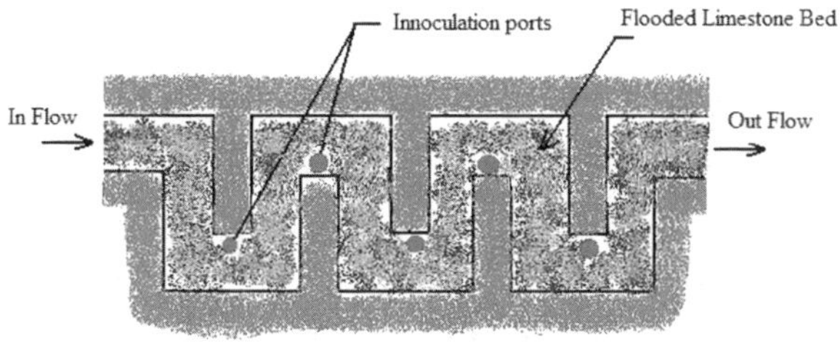

Fig. 9 Plan view of a pyrolusite Bed

cover, either wet (water) or dry (soil), which is placed over the waste pile. Despite their high cost, these covers cannot always completely stop the oxidation process and generation of AMD. Early diagnosis of the AMD problem, identification of appropriate prevention/control measures and effective implementation of the methods to the sites would reduce the potential risk of AMD generation. AMD prevention/control measures broadly include use of covers, control of the source, migration of AMD and treatment. Treatment is not the best solution to most AMD problems. Treatment has the disadvantage of being necessary for as long as the acid discharge continues and thus requires manpower, surface facilities and sludge disposal areas indefinitely. Immediate and timely action must be taken to control the source and the migration of AMD. Suitable cost effective active and passive treatment methods must be chosen to treat the problem so as to minimize its adverse manifested hazards.

References

Akcil A, Koldas S (2006) Acid mine drainage: causes, treatment and case studies. J Clean Prod 14:1139–1145

Aljoe WW, Hawkins JW (1991) Hydrologic characterization and in-situ neutralization of acidic mine pools in abandoned underground coal mines. In: Proceedings second international conference on the abatement of acidic drainage, Montreal, Canada, 16–18 Sept 1991 Vol 1, pp 69–90

Arnold DE (1991) Diversion wells—A low-cost approach to treatment of acid mine drainage. In: Proceedings, twelfth West Virginia surface mine drainage task force symposium, Morgantown, West Virginia, 3–4 April 1991

Bhole AG (1994) Acid mine drainage and its treatment Proc. Int. Symp Paithankar AG (ed). On the impact of mining on the environment: problems and solutions, VNIT, Nagpur, 11–16 Jan, Oxford & IBH, New Delhi, pp 131–141

Bowders J, Chiado E (1990) Engineering evaluation of waste phosphatic clay for producing low permeability barriers. In: Proceedings 1990 mining and reclamation conference and exhibition, West Virginia University, Vol 1, pp 11–18

Brodie GA (1993) Staged, aerobic constructed wetlands to treat acid drainage: case history of Fabius Impoundment 1 and overview of the Tennessee Valley Authority's Program. pp 157–166. In: Moshiri GA (ed) Constructed wetlands for water quality improvement. Lewis, Boca Raton, p 632

Caruccio FT, Gediel G (1989) Water management strategies in abating acid mine drainage—Is water diversion really beneficial? In: Proceedings 1989 multinational conference on mine planning and design, University of Kentucky, Lexington, Kentucky

Evangelou V, Sainju U, Portig E (1991) Some considerations when applying limestone/rock phosphate materials on to acid pyritic spoils. In: Proceedings twelfth annual West Virginia surface mine drainage task force symposium, Morgantown, West Virginia, 3–4 April 1991

Faulkner BB (1990) Handbook for the use of ammonia in treating mine waters. West Virginia Mining and Reclamation Association, Charleston, West Virginia

Faulkner B (ed) (1991) Handbook for use of ammonia in treating mine waters. West Virginia Mining and Reclamation Association, Charleston, West Virginia

Faulkner BB (1996) Acid mine drainage treatment recommendations. West Virginia Mining and Reclamation Association, Charleston, West Virginia

Faulkner BB, Skousen JG (1994) Treatment of acid mine drainage by passive treatment systems. In: Proceedings of the international land reclamation and mine drainage conference and the third international conference on the abatement of acidic drainage, Pittsburgh, 24–29 April 1994, Vol 2, pp 250–257

Forstner U, Salomons W (eds) (1988) Environmental management of solid waste. Springer-Verlag, Berlin

Gandhi PK (1990) Khanan Paryavaran Aur Prabandha. Western Coalfields Limited, Nagpur

Gusek J, Mann C, Wildeman T, Murphy D (2000) Operational results of a 1200 gpm passive bioreactor for metal mine drainage, West Fork, Missouri. In: Proceedings of 5th International Conference on Acid Rock Drainage (ICARD), 21–24 May, Denver, Society for Mining, Metallurgy, and Exploration, Inc., Littleton, Vol 2, pp 1133–1137

Hedin RS (1989) Treatment of acid coal mine drainage with constructed wetlands. In: Majumbar SK et al (eds) Wetlands ecology, productivity and values: emphasis on Pennsylvania. Pennsylvania Academy of Science, Easton, pp 349–362

Hedin RS, Watzlaf GR (1994) The effects of anoxic limestone drains on mine water chemistry. In Vol 1 of Proceedings of the international land reclamation and mine drainage conference and the third international conference on the abatement of acidic drainage, Pittsburgh, 24–19 April, pp 185–194

Hedin RS, Nairn RW, Kleinmann RLP (1994) Passive treatment of coal mine drainage. Bureau of Mines Information Circular 9389. US Bureau of Mines, Pittsburgh, p 35

Hedin RS, Watzlaf GR, Nairn RW (1994) Passive treatment of acid mine drainage with limestone. J Environ Qual 23(6), 1994

http://www.cewic.fi/cewic/materiaalit/pehkonen_270509.pdf?PHPSESSID=56a70d3f459417e6db8c3a8223bd085c

Hyman DM, Watzlaf GR (1995) Mine drainage characterization for the successful design and evaluation of passive treatment systems. In: Proceedings 17th Annual Conference of the Association of Abandoned Mine Lands, French Creek, pp 170–185

Kim AG, Heisey BS, Kleinmann RLE, Deul M (1982) Acid mine drainage: control and abatement research. U.S. Bureau of Mines Information Circular 8905, Pittsburgh

Kleinmann RLP, Crerar DA, Pacelli RR (1981) Biogeochemistry of acid mine drainage and a method to control acid formation. Mining Engineering, March 1981

Ladwig K, Erickson P, Kleinmann R (1985) Alkaline injection: an overview of recent work in control of acid mine drainage, Proceedings of a Technology Transfer Seminar, U.S. Bureau of Mines Information Circular 9027

Mehrotra A, Singhal R (eds) (1992) Environmental issues and waste management in energy and minerals production. Vol 2., A.A. Balkema, Rotterdam

Nordstrom DK, Alpers CN (1999) Geochemistry of acid mine waters. In: Plumlee GS, Logsdon MJ (eds) The environmental geochemistry of mineral deposits, part A: processes, techniques and health issues, Reviews in Economic Geology Vol 6A, Society of Economic Geologists, pp 133–160

Plumlee GS (1999). The environmental geology of mineral deposits. In: Plumlee GS, Logsdon MJ (eds) The environmental geochemistry of mineral deposits, part A: processes, techniques and health issues, Reviews in Economic Geology Vol 6A, Society of Economic Geologists, pp 71–116

Prasad SD (1994) Acid mine drainage in pyrite mines. In: Proc. International Conference On ENVIROMIN'94, (Ed. G.S.Khuntia), New Delhi, pp 364–369

Renton JJ, Rymer TE, Stiller AH (1988) A laboratory procedure to evaluate the acid producing potential of coal associated rocks. Mining science and technology. Vol 7, Elseiver Science Publishers, Netherlands, pp 227–235

Sengupta M (1996) Environmental impacts of mining. Lewis, Boca Raton, pp 167–250

Sharma PD (2010), Acid mine drainage (AMD) and its control. Lambert Academic Publishing, Germany, (ISBN 978–3-8383–5522-1)

Singh G (2005) Chemical, microbiological and geological aspects of acid mine drainage and its control aspects. Indian Min Eng J, 44(2–3):20–30

Skousen J, Ziemkiewicz P (1995) Acid mine drainage control and treatment. National Research Center for Coal and Energy, National Mine Land Reclamation Center, West Virginia University, Morgantown, p 243

Skousen J, Ziemkiewicz P (1996) Acid mine drainage control and treatment, 2nd ed. National Research Center for Coal and Energy, National Mine Land Reclamation Center, West Virginia University, Morgantown

Skousen J, Hilton T (1997) Faulkner B Overview of acid mine drainage treatment with chemicals. http://www.wvu.edu/~agexten/landrec/chemtrt.htm. Accessed 30 June 2012

Skousen JG, Sencindiver JC, Smith RM (1987) A review of procedures for surface mining and reclamation in areas with acid-producing materials in cooperation with The West Virginia Surface Mine drainage Task Force, the West Virginia University Energy and Water Research Center and the West Virginia Mining and Reclamation Association, 39 pp, West Virginia University Energy and Water Research Center

Skousen J, Politan K, Hilton T, Meek A (1990) Acid mine drainage treatment systems: chemicals and costs. Green Lands 20(4):31–37

Trumm D (2007) Acid mine drainage in New Zealand. Reclamation Matters (American Society of Mining and Reclamation (ASMR)) 3(1). http://www.crl.co.nz/publications/ASMR%20 Trumm.pdf. Accessed 9 Sept 2011

US Environmental Protection Agency (EPA) (1983) Neutralization of acid mine drainage, Design Manual. US EPA 600/2–83-001, Cincinnati

USEPA (1995) Streams with fisheries impacted by acid mine drainage in Maryland, Ohio, Pennsylvania, Virginia and West Virginia. USEPA, Philadelphia

Vail WJ, Riley RK (1997) The abatement of acid mine pollution using the pyrolusite process. In: Proceedings, Nineteenth Annual Conference, National Association of Abandoned Mine Lands Program. 17–20 August 1997, Davis

Waters JC, Santomartino S, Cramer M, Murphy N, Taylor JR (2003) Acid rock drainage treatment technologies—Identifying appropriate solutions. Proceedings, Sixth International Conference on Acid Rock Drainage (ICARD), 12–18 July 2003, Cairns, Australia. pp 831–843

Younger PL (2000) The adoption and adaptation of passive treatment technologies for mine waters in the United Kingdom. Mine Water Environ 19:84–97

Ziemkiewicz PF, Skousen J, Lovett RJ (1994) Open limestone channels for treating acid mine drainage: a new look at an old idea. Green Lands 24(4):36–41

Ziemkiewicz PF, Skousen JG, Brant DL, Sterner PL, Lovett RJ (1997). Acid mine drainage treatment with armored limestone in open limestone channels. J Environ Qual 26:560–569, 1017–1024

Very Low-Frequency Electromagnetic Method: A Shallow Subsurface Investigation Technique for Geophysical Applications

S. P. Sharma, Arkoprovo Biswas and V. C. Baranwal

Abstract The very low-frequency (VLF) electromagnetic (EM) method is the simplest EM method to delineate shallow subsurface conducting structures. Since the approach utilizes signals transmitted from worldwide transmitters located in coastal areas in the 5–30 kHz frequency band, it is suitable to depict conducting structures up to 200 m depth in highly resistive terrain. Freely and readily available primary field signals anywhere around the Earth make the VLF method very convenient and efficient for field data collection. Further, VLF data processing using digital linear filtering is quite accurate and very efficient in depicting the qualitative information about subsurface conductors, even though quantitative interpretation of VLF data is as complex as other EM data interpretation. In the present study, various aspects of the VLF method such as basic theory, worldwide VLF transmitters, quantities measured, and interpretation procedures are discussed in detail. Finally, the efficacy of the VLF method for groundwater investigation, mineral investigation, and landslide and subsurface pollution monitoring studies has been demonstrated. Even though the VLF method is a rapid technique for subsurface investigation, use of complementary geophysical methods such as gravity, direct current (DC) resistivity, self-potential, radiometric, etc., reduces the ambiguity in the interpretation and yields reliable subsurface information.

Keywords VLF electromagnetic · Groundwater and mineral investigation · Landslide studies · Subsurface pollution studies

1 Introduction

Electromagnetic (EM) method of geophysical prospecting deals with the propagation of low-frequency time varying EM field (primary field) into the earth. Such fields are generated by passing alternating current in a number of transmitters (a

S. P. Sharma (✉) · A. Biswas
Department of Geology and Geophysics, IIT Kharagpur, Kharagpur, West Bengal 721302, India
e-mail: spsharma@gg.iitkgp.ernet.in

V. C. Baranwal
Geological Survey of Norway (NGU), Trondheim, Norway
e-mail: vikas.baranwal@ngu.no

D. Sengupta (ed.), *Recent Trends in Modelling of Environmental Contaminants*,
DOI 10.1007/978-81-322-1783-1_5, © Springer India 2014

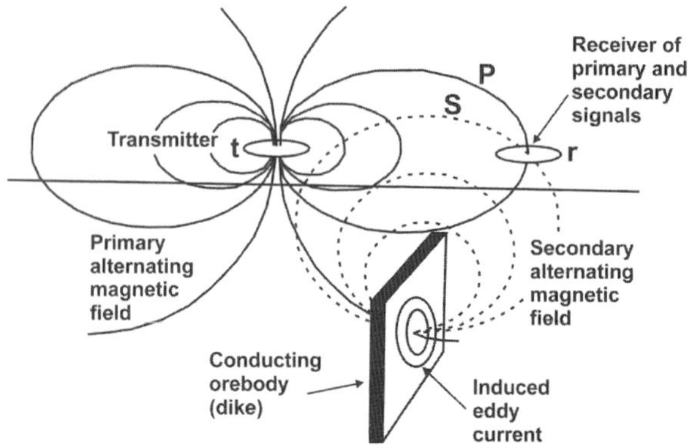

Fig. 1 General principle of electromagnetic (EM) surveying

small circular loop, a large rectangular loop, a grounded wire, a long vertical cable, etc.). Due to low frequency, the EM field penetrates into Earth's interior and interacts with subsurface conductors. Due to this interaction, according to Faraday's law of EM induction, an induced electromotive force (emf) is generated in the subsurface conductor. Further, an induced current is set up in the subsurface conductor. This induced current (secondary current) produces a secondary field (Fig. 1) which propagates to the earth's surface. On the Earth's surface a vector combination of the primary and secondary fields is recorded. The whole process is governed by the well known Maxwell's equations (Faraday's and Ampere's laws) and is known as EM induction phenomenon in the earth.

EM methods utilize a broad frequency range (1 GHz to 10^{-6} Hz). Ground penetrating radar (GPR) uses the highest frequency in the band of 25 MHz to 1 GHz. However, GPR method is based on reflection principle, and EM induction does not take place at such a high frequency. Very low-frequency (VLF) EM method uses the frequency band 5–30 kHz. Audio frequency magnetic field method which is based on thunderstorm (lightening) activity has a frequency of 1–1,000 Hz. Controlled source EM method that uses transmitter and receiver in field used a frequency range of 100–5,000 Hz. Magnetotelluric method which is based on ionosphere current has the lowest frequency of $1–10^{-6}$ Hz or even smaller. Depth of investigation is frequency dependent and increases in the order mentioned above.

Transmitter is a very important component in EM surveying. VLF method has advantage in this regard, as transmitters are freely available for measurement. In principle, the VLF method uses transmitter located in coastal areas worldwide. The primary aim of these transmitters is long distance marine communication or communication with submarines. These signals travel worldwide between the Earth's surface and ionosphere; hence, these signals travel thousands of kilometer distances without much attenuation, as their frequencies are in the lower band (5–30 kHz) in comparison to the normal communication frequencies (thousands of kHz to GHz range). Therefore, they are referred to as VLF methods, even though they use the highest frequency in EM methods based on EM induction phenomenon.

Figure 2 shows a schematic diagram of the working principle of VLF EM method. Primary field from the transmitter travels intercontinental distances, and mea-

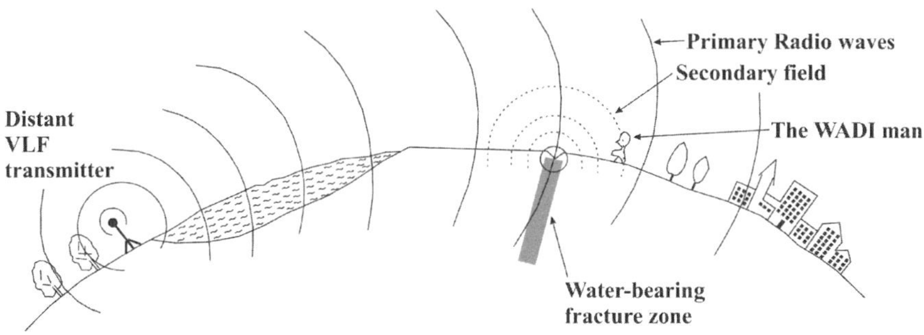

Fig. 2 Basic principle of VLF method. (Source ABEM, Wadi)

surements are performed in uniform primary time varying magnetic field in a small survey area.

2 VLF Transmitters

VLF transmitters are basically long grounded vertical wires of several hundred meters length, carrying alternating current, and operating as vertical electrical dipole with a transmitting power of the order of 1 MW. The monopole transmitting antennas are electrically very small in comparison to the wavelength at VLF frequencies which is of the order of 15 km. An addition of "top-loading," which consists of enormous horizontal wire arrays located at (and connected to) the top of the antenna is required to get high transmission power in the order of 1 MW (McNeill and Labson 1991). Therefore, VLF transmitters, in general, are large, complicated, and expensive structures. These transmitters are set for communication with submarines, and the VLF method takes advantage of the transmitted EM fields by using them as primary field source. Table 1 shows a list of the transmitters working in the VLF range worldwide.

The EM fields transmitted from VLF transmitters at a large distance are a combination of ground and sky waves. Ground wave travels over the earth's surface, whereas the sky wave is refracted and reflected by the ionized layers in the upper atmosphere (~ 50 km and higher). Since the power of a transmitter is very large (~ 1 MW), it is possible to detect these fields over continental distances, nearly half way around the world. The magnetic field lines are horizontal circles concentric about the transmitter. At distances of several hundred kilometers, this field is practically uniform and at right angles to the transmitter direction which leads to assumption of the plane wave (Fig. 3). There is no guarantee to receive enough strong VLF fields everywhere in the world due to inaccessibility of proper transmitter for some regions and also sometimes the VLF stations may be switched off due to maintenance work. Therefore, using portable VLF transmitters may be an alternative for such regions.

Table 1 VLF stations with operating frequency and their locations

Frequency (kHz)	Stations	Frequency (kHz)	Stations	Station code	Place	Country	Latitude	Longitude
15.1	FUO HWUVTI	22.8	NWC	3SA	Changde	China	29N04	111E43
15.3	NHB NPN NPM NLK NEJ	22.9	JJI	3SB	Datong	China	39N56	113E15
15.5	NWC NPM NAA NSS	23.3	JJI	DHO	Burlage	Germany	53N05	007E37
15.6	EWB	23.4	NPM	EWB	Odessa	Ukraine	46N29	030E44
15.7	NPM NSS NAK NPL NPG	24.0	NPM BA NSS NLK	FTA	Assise	France	48N32	002E34
16.0	GBR	24.8	NLK	FUO	Croix	France	44N45	000W48
16.2	UGK JAP	25.3	NAA	GBR	Rugby	UK	52N22	001W11
16.3	VTX	25.5	3SB	GBZ	Rugby	UK	52N22	001W11
16.4	JXN	25.8	NSS NAA	GQD	Rugby	UK	52N22	001W11
16.6	NPM NSS NAK	26.1	NPM NLK NPG NEJ	HWU	Le Blanc	France	46N37	001E05
16.8	FTA	27.0	RCV NAU	ICV	Tavolara	Italy	40N55	009E45
17.0	VTX	27.5	NAU	JAP	Yosami	Japan	34N58	137E01
17.1	UMS	27.7	3SB	JJH	Kure	Japan	34N14	132E34
17.4	NDT	28.0	DHO 3SB	JJI	Ebino	Japan	32N05	131E51
17.6	JXZ	28.5	NAU NPL	JXN	Helgeland	Norway	66N25	013E01
17.8	NPM NAA NSS	28.6	RAM	JXZ	Helgeland	Norway	66N25	013E01
17.9	UBE	29.0	3SA	NAA	Cutler ME	USA	44N39	067W17
18.0	NBA NPL NPG NLK	30.0	UNW	NAK	Annapolis	USA	38N59	076W28
18.1	UPD	21.0	3SA	NAU	Aguada	Puerto Rico	18N23	67W11
18.2	VTX NSS JJH	21.2	JJI	NBA	Balboa	Panama	09N04	079W39
18.3	HWU	21.4	NPM NAA NSS	NDT	Yokosukaichi	Japan	34N58	137E01
18.5	DHO NAA	21.6	3SB	NEJ	Seattle	USA	47N41	122W15
18.5	NHB NPN NPM NAA NLK NPG NEJ	21.8	TBA	NHB	Kodiak	Alaska	57N45	152W30
18.7	JJI	21.9	JJI	NLK	Oso Wash	USA	48N12	121W00
18.9	UMB	22.2	JJI	NPC	Seattle	USA	47N35	122W32
19.0	GQD NPM NSS	22.3	NWC NAA NLK NPC NSS NPM	NPG	S Francisco	USA	38N06	122W16

Table 1 (continued)

Frequency (kHz)	Stations	Frequency (kHz)	Stations	Station code	Place	Country	Latitude	Longitude
19.1	JJI	22.6	GBR	NPL	S Diego	USA	32N44	117W05
19.2	VTX	22.8	NWC	NPM	Pearl Harbor	Hawai	21N25	158W09
19.4	NHB NPN NPM NEJ NLK	22.9	JJI	NPN	Guam	Guatemala	13N34	144E50
19.5	3SA	23.3	JJI	NSS	Washington	USA	38N59	076W27
19.6	GBZ	23.4	NPM	NWC	North West Cape	Australia	21S47	114E09
19.8	NWC NPM NLK NPL NPG TBA	24.0	NPM NBA NSS NLK	RAM	Moscow	Russia	55N49	037E18
19.9	JJI	24.8	NLK	RCV	Rostov	Russia	47N18	039E48
20.2	JJI ICV	25.3	NAA	TBA	Antalya	Turkey	36N53	030E43
20.3	JJI	25.5	3SB	UBE	Petrolovsk	Russia	52N59	158E39
20.5	SA 3SB	25.8	NSS NAA	UGK	Kaliningrad	Russia	54N42	020E30
20.8	ICV	26.1	NPM NLK NPG NEJ	UMB	Rostov	Russia	57N14	039E48
21.0	3SA	27.0	RCV NAU	UMS	Moscow	Russia	55N49	037E18
21.2	JJI	27.5	NAU	UNW	Kaliningrad	Russia	54N45	020E30
21.4	NPM NAA NSS	27.7	3SB	UPD	Murmansk	Russia	68N58	033E05
21.6	3SB	28.0	DHO 3SB	VTI	Bombay	India	19N00	073E00
21.8	TBA	28.5	NAU NPL	VTX	Vijayanagaram	India	08N26	077E44
21.9	JJI	28.6	RAM					
22.2	JJI	29.0	3SA					
22.3	NWC NAA NLK NPC NSS NPM	30.0	UNW					
22.6	GBR							

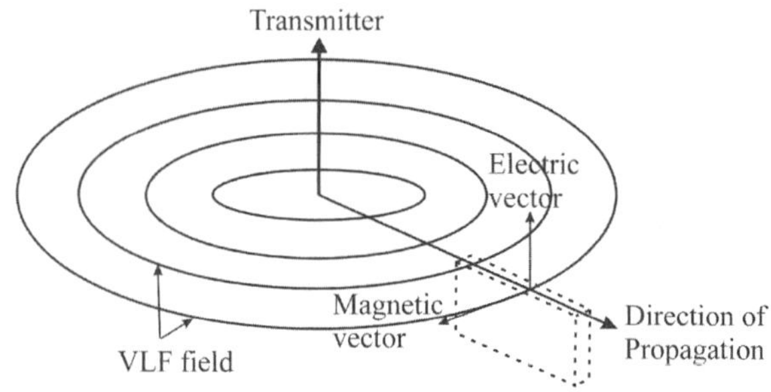

Fig. 3 Principle of VLF EM method. *Dashed lines* show a conductor striking towards the transmitter which is cut by the magnetic vector of the EM field

3 VLF and VLF-R Anomalies

3.1 E-Polarization

If y-axis is the strike direction and VLF transmitter is located in y-direction, then tilt angle α, which is the inclination of the major axis of the polarization ellipse, and the ellipticity e, which is the ratio of the minor to the major axis of the ellipse, are calculated by the formulae (Smith and Ward 1974)

$$\tan 2\alpha = \pm \frac{2(H_z/H_x)\cos\Delta\phi}{1-(H_z/H_x)^2} \tag{1}$$

and

$$e = \frac{H_z H_x \sin\Delta\phi}{H_1^2}. \tag{2}$$

where H_z and H_x are the amplitudes, and the phase difference $\Delta\phi = \phi_z - \phi_x$, in which ϕ_z is the phase of H_z, and ϕ_x is the phase of H_x, and $H_1 = |H_z e^{i\Delta\phi}\sin\alpha + H_x\cos\alpha|$. The tangent of the tilt angle and the ellipticity are good approximations to the ratio of the real component of the vertical secondary magnetic field to the horizontal primary magnetic field, and to the ratio of the quadrature component of the vertical secondary magnetic field to the horizontal primary field, respectively (Paterson and Ronka 1971). These quantities are called the real (=tan $\alpha \times 100\%$) and imaginary ($= e \times 100\%$) anomalies, respectively, and are normally expressed in percentage.

A VLF-R measurement records the horizontal electric field component and an orthogonal horizontal magnetic field. It is possible to use both the E- and H-polarization. In E-polarization mode, the electric field is measured in a direction parallel to the geological strike. The apparent resistivity ρ_a and phase angle φ computed from horizontal electric field E_y and magnetic field H_x are given by the formulae (e.g., Kaikkonen 1979):

$$\rho_a = \frac{1}{\omega\mu}\left|\frac{E_y}{H_x}\right|^2 \tag{3}$$

and

$$\phi = \arctan\left[\frac{\mathrm{Im}\left(\frac{E_y}{H_x}\right)}{\mathrm{Re}\left(\frac{E_y}{H_x}\right)}\right] \tag{4}$$

3.2 H-Polarization

In H-Polarization mode, real and imaginary VLF anomalies do not exist and they are considered as zero.

For H-polarization measurement, the VLF transmitter should be located in a direction perpendicular to the geological strike, the electric field is measured in a direction perpendicular to the geological strike, and the magnetic field in the direction of the strike. Expressions for apparent resistivity and phase is given by

$$\rho_a = \frac{1}{\omega\mu}\left|\frac{E_x}{H_y}\right|^2 \tag{5}$$

$$\phi = \arctan\left[\frac{\mathrm{Im}\left(\frac{Ex}{Hy}\right)}{\mathrm{Re}\left(\frac{Ex}{Hy}\right)}\right] \tag{6}$$

For theoretical computations, the field components *Ex* and E_y are obtained by applying the finite element technique with the Galerkin process directly to the Maxwell's equations, and other field components can be computed from Maxwell's equations by numerical differentiation. To solve the forward problem the magnetotelluric finite element program of Wannamaker et al. (1987) has been used after a slight modification for VLF purposes. For the details of the finite element modeling and VLF computation, papers by Kaikkonen (1979) and Wannamaker et al. (1987) are referred. Cubic spline interpolation (Press et al. 1992) is used to obtain response at the desired observation locations.

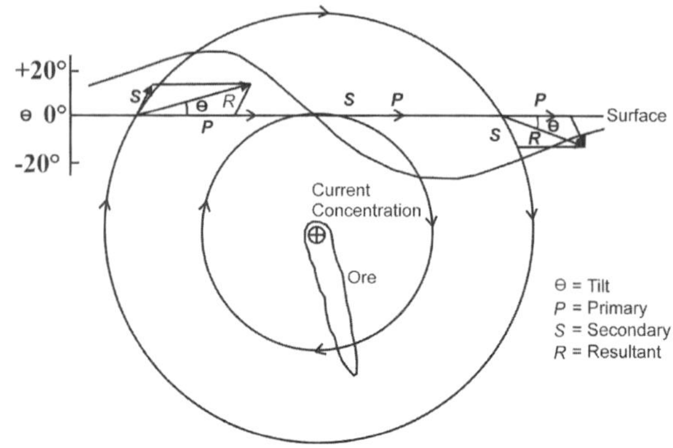

Fig. 4 Tilt-angle profiles resulting due to presence of a vertical conductor. (After Parasnis 1986)

Depth of Penetration or Skin Depth

In general, depth of penetration for EM fields is assumed to be equal to a skin depth value δ. The skin depth is the depth at which EM fields are attenuated to 1/e of its surface amplitude during the propagation through an isotropic earth. The value of skin depth depends on frequency f (in Hz) and effective resistivity ρ (in Ωm) which is a sort of average resistivity of an equivalent homogeneous subsurface to the actual subsurface. The expression for skin depth in meter is:

$$\delta = 503\sqrt{\frac{\rho}{f}}. \tag{7}$$

It is evident from Eq. (7) that the depth of penetration will be greater in a more resistive subsurface at the same frequency of EM fields and lower frequencies will achieve greater depth of penetration in same resistive subsurface. In a typical resistive medium (1,000 Ωm) the depth of investigation is less than 150 m at VLF frequencies.

4 Interpretation Procedures

In the presence of a single vertical conductive body, VLF real and imaginary anomalies change from negative to positive (or positive to negative depending on the convention of the sign used in the instrument) with a zero crossover exactly above the location of the vertical conductor (Fig. 4). Paterson and Ronka (1971) have discussed general nature of the tilt angle and ellipticity data in various situations. They demonstrated by numerical calculations that tilt angle and ellipticity data will be of opposite sign with approximately the same numerical value if a good conductor lies in a weakly conductive ground. However, tilt angle and ellipticity data will be of same sign with lower numerical value of ellipticity in comparison to tilt angle if a poor conductor lies in a non-conductive ground or on the surface.

It is easy to make qualitative interpretation of the location of conductive body in such simple cases, when a single conductor is present. The problem arises when more than one conductor are present along the same profile line. When many conductors are present in the subsurface then responses corresponding to these conductors couple together and it is difficult to decide about the location of each conductor separately.

Some researchers (Chouteau et al. 1996; Gharibi and Pedersen 1999; Becken and Pedersen 2003) have presented the mathematical techniques to transform the VLF data (real and imaginary anomalies) into VLF-R data (apparent resistivity and phase) for the interpretation. However, most common techniques to interpret the VLF data qualitatively are filtering techniques presented by Fraser (1969) and Karous and Hjelt (1983). Moreover, these filtering techniques do not provide exact depth information of the conductor. A more advanced filtering technique has been discussed by Pedersen and Becken (2005) to construct corresponding equivalent current density distribution images which also provides information about the depth of the conductor quantitatively. However, a fully quantitative interpretation in terms of resistivity and depth together can be obtained by various linearized and nonlinearized inversion schemes of VLF and VLF-R data.

4.1 Fraser Filtering Technique

Fraser (1969) presented a very simple filtering technique which performs phase shift of the tilt angle data by 90°, such that crossovers are transformed into peaks. The filtering process is simply performed by summing the observations at two consecutive data stations and then subtracting from the sum at the next two consecutive data stations. Mathematical formulation for the Fraser filtering is written as:

$$F(0) = (H_{-2} + H_{-1}) - (H_1 + H_2), \tag{8}$$

where H_{-2}, H_{-1}, etc., are the measured data (real or imaginary part of the magnetic transfer function) at consecutive stations with station interval Δx. $F(0)$ is filtered data obtained, corresponding to center of H_{-2} and H_2. If the VLF data is collected along parallel profile lines of same length and for the same transmitter, then filtered data can be contoured to locate the extension of the conductor in the survey area.

4.2 Karous–Hjelt Filtering Technique

Karous and Hjelt (1983) developed a filtering technique to calculate the equivalent current density which produces a magnetic field identical to the measured field. They applied the concept of linear filter theory to calculate apparent current density. According to the Biot–Savart's law, a magnetic field is produced by a current source. In case of VLF, we measure the secondary magnetic field which is produced by the eddy current set in the conductor by induction phenomenon. Therefore, Karous and Hjelt (1983) used this analogy and developed an efficient filter to calculate equivalent current density from the observed magnetic field. A number of filters of various lengths have been examined by Karous and Hjelt (1977), and a six point filter coefficient was found to be the shortest and most suitable within 8% accuracy for the interpretation of the field of a single current source.

The application of this filtering technique is rather simple. The filter coefficients are multiplied with equi-spaced and consecutive observation points and then added to give the equivalent current density value at the centre of those observation points. This procedure is repeated by shifting filter coefficients to the next observation points until the last observation point along the profile. Therefore, first equivalent current density value is obtained for a location at the centre of second and third observation points, and the last equivalent current density value is obtained for a location at the centre of second and third observation points from the last station on the profile. The expression for the filtering technique is given as:

$$\frac{\Delta z}{2\pi} I_a(0) = -0.102H_{-3} + 0.059H_{-2} - 0.561H_{-1} + 0.561H_1 - 0.059H_2 + 0.102H_3 \quad (9)$$

where $I_a(0) = 0.5\left[I(\Delta x/2) + I(-\Delta x/2)\right]$ is the equivalent current density value. This filter is designed to be applied on equi-spaced data with the assumption that transformed current density values corresponds to the depth Δz, equal to the data spacing at which the current densities are to be calculated. Thus, it is possible to calculate the current distribution at different depths and to construct the equivalent current density cross-section from VLF data. However, it is important to note that the equivalent current density cross-section is a pseudo section which does not correspond to the actual current distribution. The relationship between the pseudo depth and the data interval Δx is a convention only, not a mathematical relationship.

Due to symmetry of the filter coefficients, it produces almost same features when applied on the real and imaginary components of the magnetic transfer function (Ogilvy and Lee 1991), therefore, generally current density cross-section generated by real component is used in the interpretation. A higher value of the current density shows presence of the conductor exactly below that location in the cross-section. The Karous–Hjelt filter is used for interpretation of VLF data collected from different regions.

4.3 Inversion of VLF and VLF-R Data

A quantitative interpretation of VLF or VLF-R data for depth and resistivity distribution can be carried out by applying various inversion schemes. In principle, approaches developed for the inversion of magnetotelluric (MT) data can be used to invert VLF and VLF-R data with slight modification in the forward modeling for computation of VLF responses.

Beamish (1994) used regularized inversion approach well known as OCCAM inversion to invert VLF-R data. The OCCAM inversion is developed by deGroot-Hedlin and ConsTable (1990) for inversion of 2-D MT data. Further, Beamish (2000) used 2-D MT inversion code developed by Rodi and Mackie (2001). Sharma and Kaikkonen (1998a, 1998b) have used very fast simulated annealing for inversion of VLF-R data. Kaikkonen and Sharma (1998) have also jointly inverted VLF and VLF-R data using very fast simulated annealing. A comparison of local and global inversion approaches for VLF and VLF-R data is discussed by Kaikkonen and Sharma (2001).

There are few literatures available on the inversion of VLF data. A 2-D OCCAM approach (deGroot-Hedlin and Constable 1990) works very well in a stable manner for inversion of VLF-R data. A further efficient variant of OCCAM inversion called REBOCC inversion developed by Siripunvaraporn and Egbert (2000) for MT data is used for inversion of VLF data (Oskooi and Pedersen 2005) and for the inversion of airborne VLF data (Oskooi and Pedersen 2006; Pedersen and Oskooi 2004). Monteiro Santos et al. (2006) have used a 2-D regularized inversion based on the approach of Sasaki (1989) to interpret single frequency VLF data.

5 Applications

5.1 Groundwater Exploration

Groundwater exploration in hard rock area is often challenging. Worldwide, hard rock areas are more problematic for the occurrence of groundwater than the soft rock areas. Groundwater movement in hard rock area takes place through fine fractures, and it is rather difficult to detect these fractures using a conventional direct current (DC) resistivity technique. Groundwater movement in hard rock areas forms vertical as well as dipping conductors and it could be delineated by VLF EM surveys. Therefore, VLF survey can be used to find the appropriate location and subsequently, resistivity survey can assist in verifying the suitability of the location.

An example for groundwater investigation from Purulia (WB), India is presented. The area belongs to Chhotanagpur Granite Gneiss Complex (CGGC). The area is characterized by gently dipping metamorphic rocks striking approximately in east–west direction. A ridge type structure with its axis approximately perpendicular to the strike of the formations, characterizes the area. The rock types in the area are granite gneiss, amphibolite, mica schist, quartzite, quartz vein, and calc-silicate rocks with interbanded crystalline limestone. A thin soil cover forms the upper surface of the study area which is followed by crystalline massive metamorphic rocks of very high resistivity. Exposure of hard rock can be seen at several isolated locations in the entire area. The surface exposure shows the strike of the formation approximately in the east–west direction and it is gently dipping. The most common rocks in the Purulia district are granites and granite gneiss in which metabasics occur as intrusives.

Integrated electrical and EM surveys were carried out in the hard rock areas of Purulia district (West Bengal), India, for delineation of groundwater-bearing zones that would be suitable for construction of deep tube-wells for large amounts of water. A detailed survey of the area was done using a VLF-WADI instrument and appropriate locations were selected for further study using Schlumberger resistivity sounding. Hence, the entire area was surveyed in a relatively short time by the combined use of resistivity and EM surveys.

Figure 5 shows the area map with detailed VLF profiles. From the various VLF profiles suitable locations were marked and subsequently resistivity surveys were performed. The VLF profile 0300E reveals the maximum amplitude in VLF anomaly. The real anomaly and apparent current density along this profile is shown in Fig. 6. It is important to mention that the entire area is almost flat and dry with hard rock exposure. However, the fracture has been depicted which is also confirmed with other electrical resistivity measurements such as profiling, self-potential measurement, as well as resistivity sounding (Sharma and Baranwal 2005). This location on VLF profile, 0300E, was drilled and a deep tube well is working successfully.

5.2 Mineral Exploration

Chromite Investigation

Chromite is a high density metallic mineral and it is often associated with mafic/ultramafic rock. Due to its high density, gravity method of prospecting is the best approach for its investigation. However, due to non-uniqueness in the interpretation, gravity method alone cannot be fully reliable. Since chromite is a metallic mineral, it possesses good electrical conductivity. Electrical and EM methods that deals with electrical conductivity of the subsurface are used in integrated study for the chromite investigation. Such an integrated study is performed around Tengarapada (Orissa), India (Fig. 7).

Major rock type present in the area sheared granite, quartzofeldspathic gneiss (QFG), and mafic/ultramafic rocks. Mafic/ultramafic rocks occur as pods that are aligned approximately parallel to a mylonitic shear zone that cuts through the QFG unit. Chromite is associated with these mafic/ultramafic rocks. The densities of chromites as well as mafic/ultramafic rocks are high. Therefore, mafic/ultramafic rock also can be interpreted as chromite from the gravity anomaly. However, chromite is electrically conducting but mafic/ultramafic rocks are highly resistive. Therefore, a VLF EM method is used to distinguish between these two high density materials.

The upper part of Fig. 8 shows measured VLF real and imaginary anomaly, and the lower part shows the current density section. We can see that real and imaginary anomaly reverses polarity around 300 m location. About 200 m wide (200–400 m location) bowl shaped high conducting zone can be seen from the current density section. This anomaly is also characterized by high gravity anomaly (Mohanty et. al

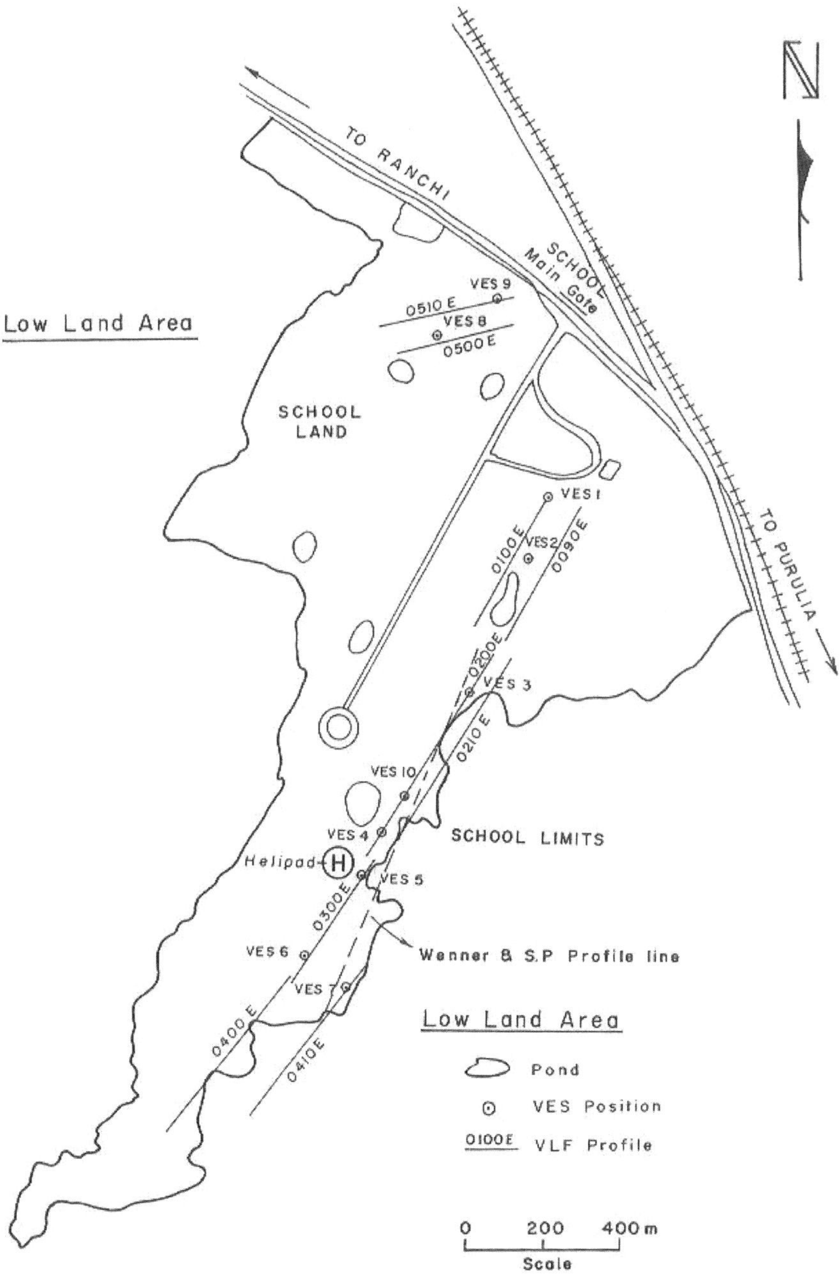

Fig. 5 Study area showing VLF profiles and other geophysical measurements. (After Sharma and Baranwal 2005)

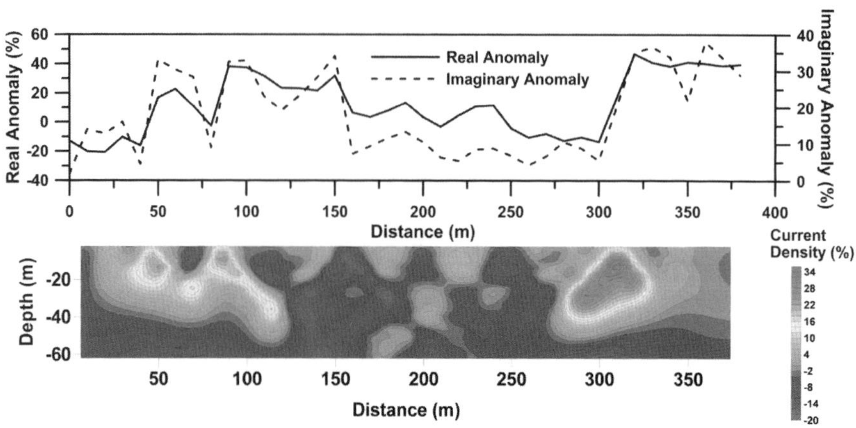

Fig. 6 VLF real anomaly and apparent current density cross-section. (Sharma and Baranwal 2005)

2011). Hence, high density and high conducting structure in the present study area reveals chromite formation. Therefore, VLF and gravity methods can be used for rapid investigation of high density metallic mineral deposit whether it is chromite or any other mineral.

Uranium Investigation

Uranium mineralization could occur in different shapes (vertical or horizontal sheet) and sizes depending on the depositional environment. The VLF EM method that suits well for delineation of vertical and gently dipping structures could also be helpful in identifying the conducting structures associated with shear zones. South Purulia Shear Zone (SPSZ), an east-west trending and almost 150 km in length, is a probable source of uranium mineralization. Beldih is an important location for apatite mining, and is exactly located along South Purulia Shear Zone (SPSZ). Particularly, the mine area consists of quartz-magnetite-apatite rocks, kaolin rocks, granite, quartzite, carbonatite, syenites, ultramafics, and some other rocks of early mesoproterozoic age. VLF survey is performed to find the location of the conductor associated with uranium mineralization, and subsequently, the extension of the body in lateral direction.

Figure 9 shows the real and imaginary VLF anomaly and current density section along a known uranium mineralization near Beldih mine. It reveals two prominent conducting features, first near −110 m location and other near +50 m location. The first conductor is shallow and anomaly is very sharp. This conductor is shallow due to power line disturbance. Second conductor coincides well with the uranium mineralization. Several VLF profiles were carried out on either side of this profile and continuation of the possible mineralization zone is depicted. It is important to highlight that only VLF survey is not enough to confirm the uranium mineralization.

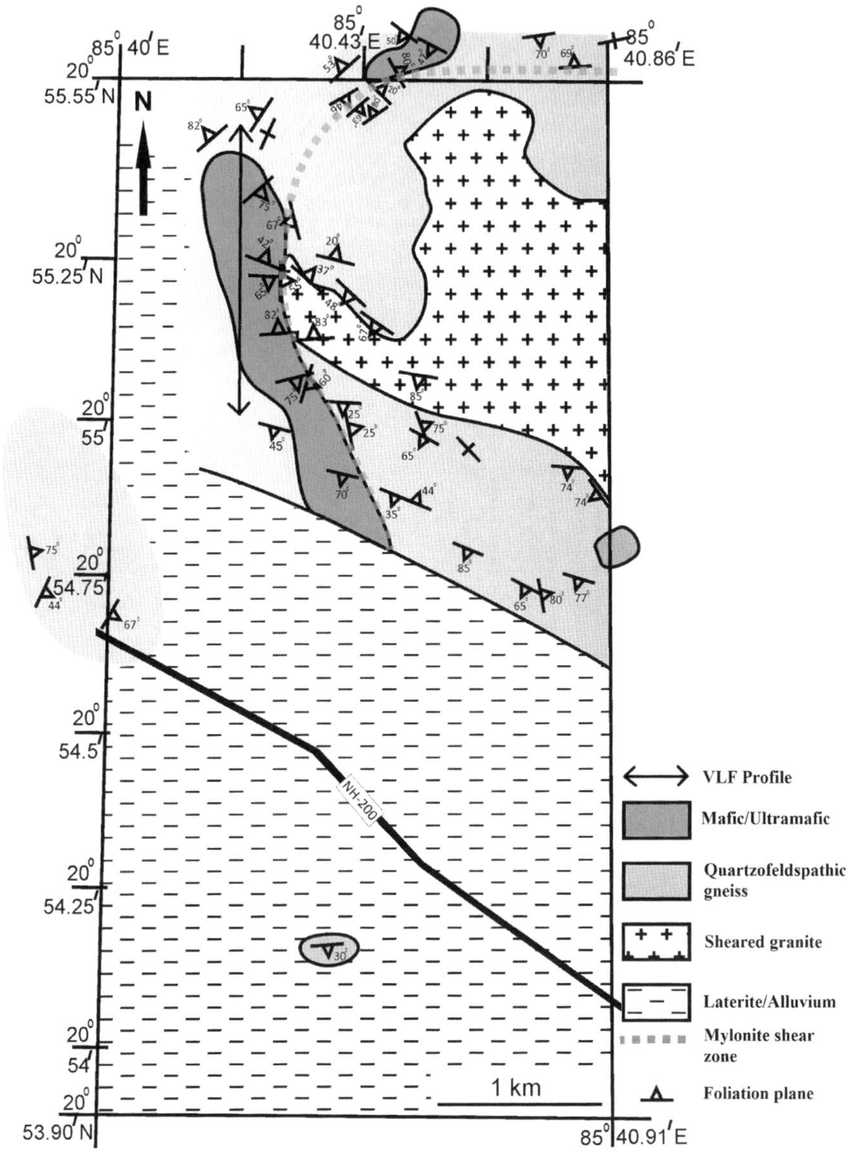

Fig. 7 Geological map of the study area. (After Mohanty et al. 2011)

Geophysical surveys such as gravity, magnetic, electrical, are necessary to confirm the extension of mineralization. This is because conductor in the area may be due to groundwater filled fractures or due to mineralization. VLF method will not be able to discriminate between these two types of conductor.

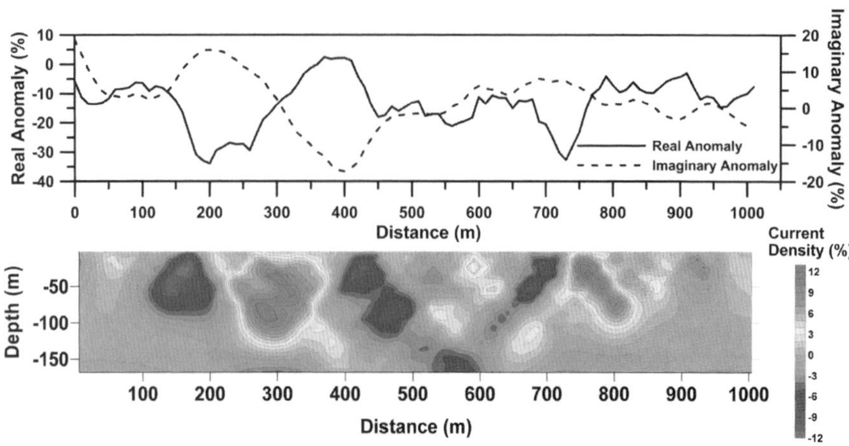

Fig. 8 Real and imaginary anomaly and current density along a profile over chromite deposit

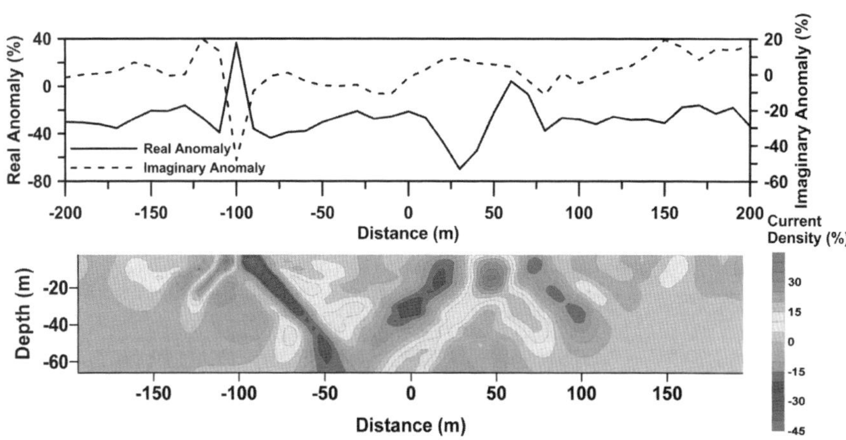

Fig. 9 Real and imaginary anomaly and current density along a profile over uranium deposit

Graphite Investigation

Graphite is an important non-metallic mineral which possesses an excellent electrical conductivity. Graphite also occurs in thin vertical as well as dipping sheet type structures. CGGC also hosts a number of graphite deposits. VLF survey was carried out over an existing graphite mine near Daltanganj (Jharkhand) India.

The study area exhibits a number of graphite mines. Mining in the area is being done mostly with open cast approach. Graphite veins are exposed in the area and mining companies are taking clue for it and digging randomly. This often results in failure and also getting poor quality graphite. It is necessary to do surveys and plan mining accordingly.

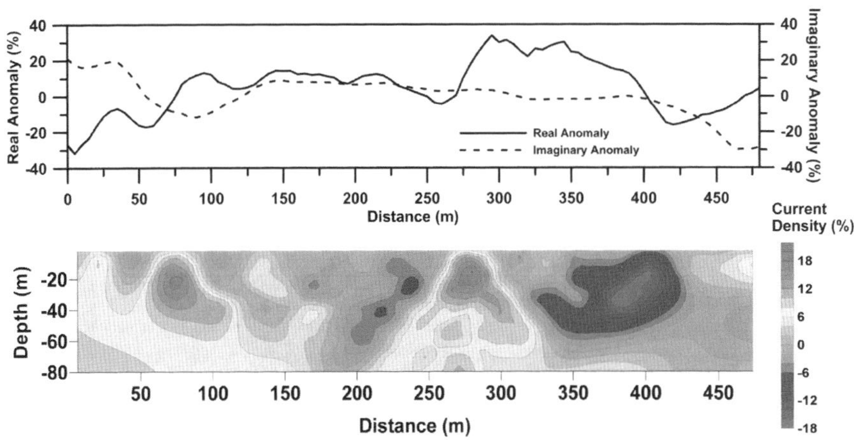

Fig. 10 Real and imaginary anomaly and current density along a profile over graphite deposit

Unplanned mining has resulted in bad quality graphite as well as loss of money due to very limited amount of graphite at several places. Figure 10 shows a VLF profile that passes nearly three exposed veins at 20, 80, and 150 m locations. Clearly, only the anomaly at the 80 m location is worth mining using open cast approach. We can see one prominent VLF anomaly near the 280 m location and one small anomaly near the 330 m location. These two bodies are not exposed on the surface but are located very close to the surface. Body at the 280 m location is the best on this profile and can be mined using open cast approach. A number of VLF profiles can help in the proper identification and to use the graphite deposit judiciously with minimum destruction to the nature.

Once again it is important to highlight that graphite deposits also exhibit strong self-potential anomaly. Therefore, self-potential survey is required to confirm the VLF anomaly due to graphite or simply due to water at the contact of hard rocks present in the area. Presence of good quality graphite, 38 %, is established for the 280 m target.

5.3 *Landslide Studies*

The VLF method can also be applied for landslide studies and monitoring. A VLF survey is carried out at Lantakhola landslide, Sikkim, India. Lantakhola is located approximately 74 km from the state capital Gangtok on the North Sikkim Highway, and is one of the most dangerous landslides on this highway. A number of profiles are selected on the hill slope, and the VLF measurements were carried out. It is important to mention that water flowing channels disappear on the way from hill scarf to the road level. It was important to depict the location of these channels in the subsurface for any mitigation work. Figure 11 clearly depicts the location of the

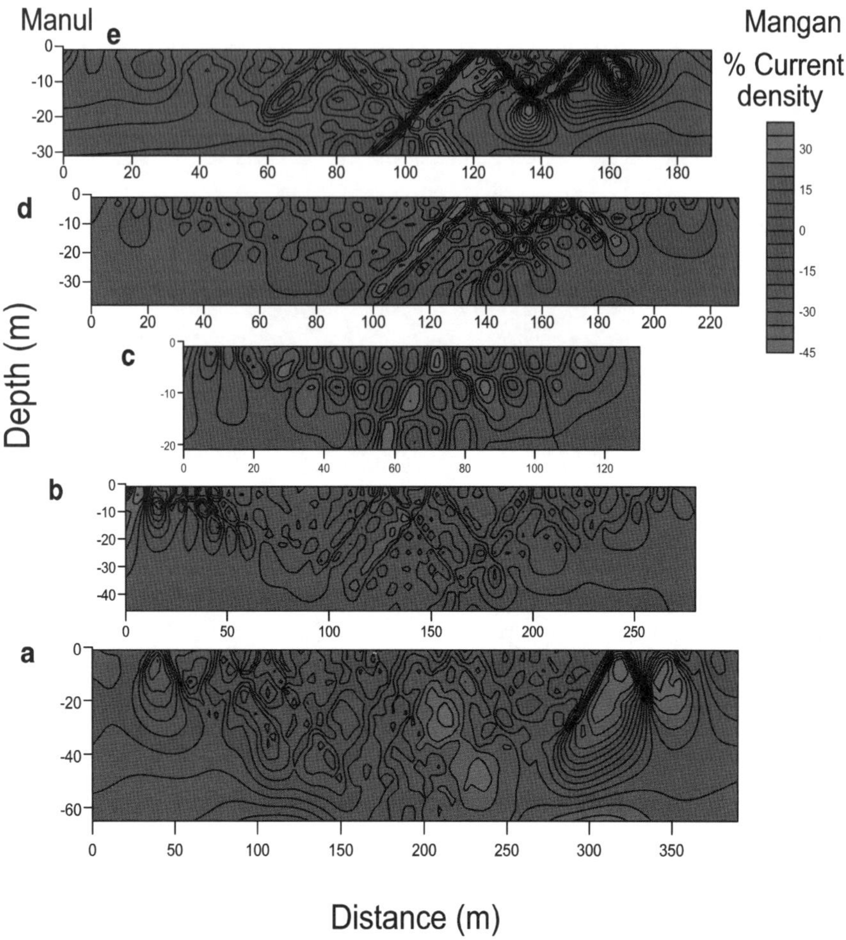

Fig. 11 VLF current density section along five profiles on Lantakhola landslide. (Sharma et al. 2010)

water channel passing below each profile. Figure 11a, which shows the results of VLF profile at the road level, also depicts the two stable regions on either side of the main landslide which can be used to build some stable structures such as a bridge for uninterrupted traffic movement on this strategic highway.

5.4 Subsurface Pollution Studies

Natural or man-made hazardous substances leaching through the subsurface fractures contaminate the underground fresh aquifer systems. Detection and regular monitoring of the area around such hazardous zones using VLF EM method is pos-

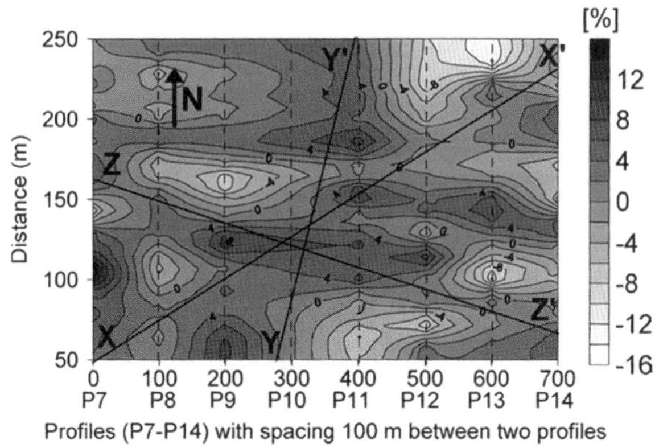

Fig. 12 Plan view of current density at 20 m depth near Tarbalu hot spring. (After Baranwal and Sharma 2006)

sible to prevent the adverse effect of subsurface pollution. Two case studies have been presented to demonstrate the efficacy of VLF EM method in such studies.

The first example consists of fluoride contamination resulting from Tarbalu hot spring, Orissa, India. Hot water gets continuously discharged though several natural openings in this area. The hot water is enriched with fluorite and also moved through subsurface fractures towards the surrounding villages. A detailed VLF study was performed around this hot spring. Plan view of current density at 20 m depth along various profiles (P7-P14) was contoured. The hot spring is located 50 m away from the bottom left corner of the Fig. 12.

It is interesting to note that high current density indicates that the fractures are filled with fluoride-contaminated groundwater. The villages marked in the direction of solid lines are affected by a high concentration of fluoride. Nearly 12 ppm fluoride has been detected in this area. Once the direction is delineated, appropriate measures can be taken to stop the contamination or to drill tube wells in another safe region.

The second example related to subsurface pollution deals with uranium mine tailings pond near Jaduguda, Jharkhand, India. Jaduguda uranium mine processes uranium ore from various uranium mines located in this region, and dumps the tailing in the nearby hills by designing suitable storage ponds. VLF survey was performed over a tailing pond to know the subsurface characteristics and any leaching of contaminated groundwater in the neighboring villages.

VLF measurements were performed along five profiles over a uranium tailing pond which is completely filled and capped with soil cover, since rainwater comes in this area from adjoining hills and runs through the tailing. Rainwater movement as well as seepage from nearby hills through tailings may contaminate groundwater in this region. Several fractures have been delineated in this area (Fig. 13). A plan view at 10 and 20 m depth was also prepared over the pond as shown in Fig. 14. It reveals that rainwater moves from the upper surface only. Current densities at 20 m depth suggest that the pond is safe and contamination is not taking place at depth. A suitable check dam can be made to avoid the flow at shallower level only.

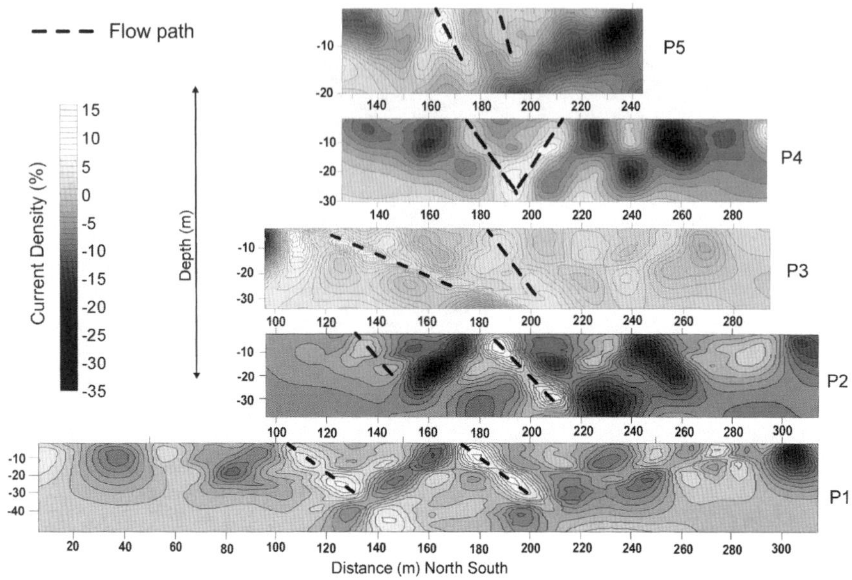

Fig. 13 Current density along various VLF profile over a tailing pond

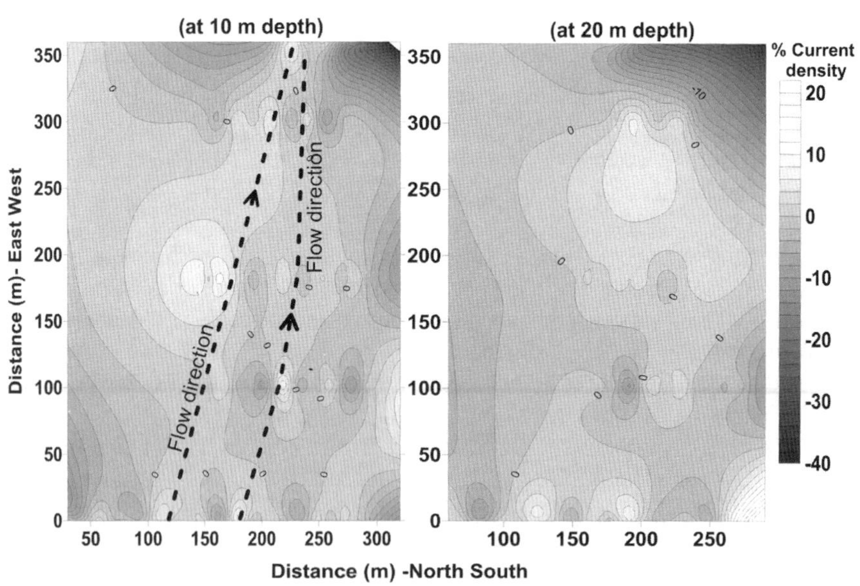

Fig. 14 Plan view of current density at 10 and 20 m depth, respectively

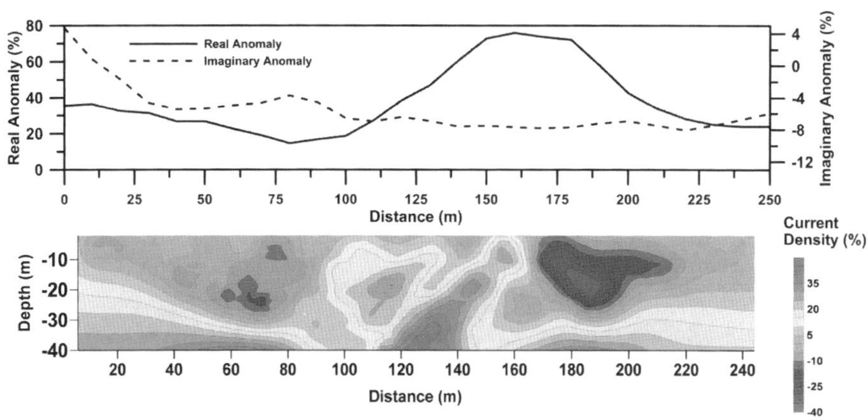

Fig. 15 VLF anomaly and current density near Bhangur chromite mine

5.5 Underground Mining Seepage

Massive seepage of groundwater in underground mines is one of the serious problems faced by mining industries. Several accidents are reported worldwide regarding flooding in the underground mines. Dewatering and discharging water to nearby big ponds are also dangerous, if these ponds are connected with fractures leading toward the underground mine. A VLF study is presented over a mine dealing with chromite mining activity near Bhangur, Orissa, India. Major fracture zones are often related with chromite formation. These fractures can be easily mapped using VLF EM method and mining planning should be done accordingly.

A VLF survey was performed near Bhangur mine in the anticipation that groundwater entering the mine is coming from a nearby big storage pond managed by another mining company. The survey was performed along a profile passing between Bhangur mine and water storage pond. Figure 15 shows the measured anomaly and inferred current density section. There is a massive fracture that connects the pond and Bhangur mine. The dip of the fracture is roughly 45° and it matches well with the dip seen in the underground mine. The study concludes that this pond will continue to create problems for Bhangur Mine.

The study demonstrates that such studies are required before the execution of any mining activity.

6 Conclusions

VLF EM method is a rapid investigation tool for shallow subsurface investigation. It has versatile applications in various explorations as well as environmental applications related with subsurface studies. Its applications have been demonstrated for groundwater exploration, mineral investigation, landslide studies, subsurface contamination studies and monitoring, and underground mine seepage studies. Even though VLF is a rapid technique for subsurface investigation, use of complementary geophysical methods such as gravity, DC resistivity, self potential, radiometric, etc., reduces the ambiguity in the interpretation and yields reliable subsurface information.

References

Baranwal VC, Sharma SP (2006) Integrated geophysical studies in the East-Indian geothermal province. Pure Appl Geophys 163:209–227

Beamish D (1994) Two-dimensional, regularised inversion of VLF data. J Appl Geophys 32:357–374

Beamish D (2000) Quantitative 2D VLF data interpretation. J Appl Geophys 45:33–47

Becken M, Pedersen LB (2003) Transformation of VLF anomaly maps into apparent resistivity and phase. Geophysics 68:497–505

Chouteau M, Zhang P, Chapellier D (1996) Computation of apparent resistivity profiles from VLF-EM data using linear filtering. Geophys Prospect 44:215–232

deGroot-Hedlin C, Constable S (1990) Occam's inversion to generate smooth, two-dimensional models from magnetotelluric data. Geophysics 55:1613–1624

Fraser DC (1969) Contouring of VLF-EM data. Geophysics 34:958–967

Gharibi M, Pedersen LB (1999) Transformation of VLF data into apparent resistivities and phases. Geophysics 64:1393–1402

Kaikkonen P (1979) Numerical VLF modeling. Geophys Prospect 27:815–834

Kaikkonen P, Sharma SP (1998) 2-D nonlinear joint inversion of VLF and VLF-R data using simulated annealing. J Appl Geophys 39:155–176

Kaikkonen P, Sharma SP (2001) A comparison of performances of linearized and global nonlinear 2-D inversions of VLF and VLF-R electromagnetic data. Geophysics 66:462–475

Karous M, Hjelt SE (1977) Determination of apparent current density from VLF measurements. Department of Geophysics, University of Oulu, Contribution No 89, pp 1–81

Karous M, Hjelt SE (1983) Linear filtering of VLF dip-angle measurements. Geophys Prospect 31:782–794

McNeill JD, Labson VF (1991) Geological mapping using VLF radio fields. In: Nabighian MN (ed) Electromagnetic methods in applied geophysics II. Society of Exploration Geophys, pp 521–640.

Mohanty WK, Mandal A, Sharma SP, Gupta S, Misra S (2011) Integrated geological and geophysical studies for delineation of chromite deposits: a case study from Tangarparha, Orissa, India. Geophysics 76:B173–B185

Monteiro Santos FA, Mateus A, Figueiras J, Gonçalves MA (2006) Mapping groundwater contamination around a landfill facility using the VLF-EM method—a case study. J Appl Geophys 60:115–125

Ogilvy RD, Lee AC (1991) Interpretation of VLF-EM in-phase data using current density pseudosections. Geophys Prospect 39:567–580

Oskooi B, Pedersen LB (2005) Comparison between VLF and RMT methods: a combined tool for mapping conductivity changes in the sedimentary cover. J Appl Geophys 57:227–241

Oskooi B, Pedersen LB (2006) Resolution of airborne VLF data. J Appl Geophys 58:158–175

Parasnis DS (1986) Principles of applied geophysics. Chapman and Hall, London

Paterson NR, Ronka V (1971) Five years of surveying with the very low frequency electromagnetic method. Geoexploration 9:7–26

Pedersen LB, Becken M (2005) Equivalent images derived from very-low-frequency (VLF) profile data. Geophysics 70:G43–G50

Pedersen LB, Oskooi B (2004) Airborne VLF measurements and variations of ground conductivity: a tutorial. Surv Geophys 25:151–181

Press WH, Teukolsky SA, Vetterling WT, Flannery BP (1992) Numerical recipes in C: the art of scientific computing. Cambridge University, New York

Rodi WL, Mackie RL (2001) Nonlinear conjugate gradient algorithm for 2-D magnetotelluric inversion. Geophysics 66:174–187

Sasaki Y (1989) Two-dimensional joint inversion of magnetotelluric and dipole-dipole resistivity data. Geophysics 54:254–262

Sharma SP, Baranwal VC (2005) Delineation of groundwater bearing fracture zones in a hard rock area integrating very low frequency electromagnetic and resistivity data. J Appl Geophys 57:155–166

Sharma SP, Anbarasu K, Gupta S, Sengupta A (2010) Integrated very low frequency EM, electrical resistivity and geological studies on the Lanta Khola landslide, North Sikkim, India. Landslides 7(1):43–53

Sharma SP, Kaikkonen P (1998a) Two-dimensional nonlinear inversion of VLF-R data using simulated annealing. Geophys J Int 133:649–668

Sharma SP, Kaikkonen P (1998b) An automatic finite element mesh generation and element coding in 2-D electromagnetic inversion. Geophysica 34:93–114

Siripunvaraporn W, Egbert G (2000) An efficient data-subspace inversion method for 2-D magnetotelluric data. Geophysics 65:791–803

Smith BD, Ward SH (1974) On the computation of polarization ellipse parameters. Geophysics 39:867–869

Wannamaker PE, Stodt JA, Rijo L (1987) A stable finite element solution for two-dimensional magnetotelluric modeling. Geophys J R Astron Soc 88:277–296

Treatment of Industrial Wastewater

A. K. Gupta and C. Sahoo

Abstract Textile wastewater is one among the most complex industrial wastewaters. Photocatalytic degradation process has been effectively used to degrade real and synthetic wastewater. Response surface modelling has been used to optimize many physical and chemical processes. This chapter presents a case study on the application of response surface modelling to optimize the photocatalytic degradation of real textile wastewater from a fabric dyeing and finishing industry using 1% Ag^+ doped TiO_2 in a batch reactor which was irradiated by UV rays of wavelength 254 nm. Response surface modelling (RSM) was used to evaluate the individual and interaction effects of catalyst dose, initial concentration of textile wastewater and reaction pH, on two responses: colour removal and chemical oxygen demand (COD) removal. These variables were found to impart maximum effect on the degradation process. Box–Behnken design of experiment was used to design 15 sets of experiments, to be used in RSM optimization, with the help of Design Expert software version 8.0.6.1. After the optimization under three imposed constraints, the values for the process variables giving optimum results were: catalyst dose, 1.5 g/L, real textile wastewater concentration, 60% and reaction pH, 7. The results of the validation experiment under the optimized conditions were colour removal and COD removal of 90.46% and 71.72%, respectively after UV irradiation for 180 min. The resulting polynomial equations for real textile wastewater were also compared with the equations developed for synthetic wastewater. The regression analysis showed that the developed model is adequate to predict the individual and interaction effects of the process variables.

Keywords Response · Degradation · Mineralization · ANOVA · Textile wastewater

---

C. Sahoo is currently a research scholar at the Department of Civil Engineering, IIT Kharagpur.

A. K. Gupta (✉)
Environmental Engineering Division, Department of Civil Engineering,
Indian Institute of Technology, Kharagpur 721302, India
e-mail: agupta@civil.iitkgp.ernet.in

C. Sahoo
Department of Civil Engineering, Indira Gandhi Institute of Technology,
Sarang, Odisha 759146, India
e-mail: crsahoo65@yahoo.co.in

D. Sengupta (ed.), *Recent Trends in Modelling of Environmental Contaminants*,
DOI 10.1007/978-81-322-1783-1_6,

1 Introduction

Water is the most precious vehicle of nature which helps us to sustain our lives on earth. Human civilization is starved of pure water due to overexploitation by ever-increasing world population. To couple with it, water is getting polluted by indiscriminate wastewater discharge from domestic, agricultural and industrial sources. As the industries are major water polluting sources, the emphasis is now on zero discharge by means of recycle and reuse.

1.1 Industrial Wastewater

Industrial wastewater comprises process water, water used for transporting raw materials, carrying waste products, heating and cooling, gardening and road sprinkling, and water used in toilets and canteens. Water from all these sources gets mixed up and ends up as industrial wastewater. The composition of industrial wastewater varies with the nature of the industry. It mostly contains various inorganic pollutants like acids, bases, salts of metals (like zinc, chromium, copper, nickel, cadmium, silver etc.), anions (like sulphates, phosphates, cyanides, chlorides, nitrates and nitrites), cations (like sodium, potassium, calcium, magnesium, iron, manganese, mercury, arsenic etc.) and high molecular weight organic pollutants (oils and fats, proteins, carbohydrates, hydrocarbons, detergents, phenolic substances, organic acids and their salts, dyes and dye intermediates etc.; Mahajan 1998). Some of the pollutants are toxic and recalcitrant in nature. Hence, they should either be completely removed or degraded to biodegradable form in order to make the wastewater suitable for treatment using conventional biological methods.

Industrial wastewater can be treated by various physical, chemical and biological processes. Physical processes include adsorption, coagulation, membrane filtration, reverse osmosis, ion exchange, electrocoagulation etc. The chemical processes include oxidative processes like ferrous iron/H_2O_2 (Fenton process), ferrous iron/H_2O_2/UV (Photo Fenton process), ozonation, H_2O_2/ozone, H_2O_2/UV, H_2O_2-peroxidase, action of NaOCl etc. Recently, heterogeneous photocatalytic methods involving semiconductor photocatalysts like TiO_2, ZnO, CdO etc. and UV light are becoming popular as they not only degrade the contaminants but also cause their mineralization to CO_2 and water. Biological processes include mainly biosorption, degradation by white rot fungi, aerobic and anaerobic biodegradation processes and enzymatic processes. A combination of the aforementioned processes can also be employed by arranging them in series. Such combined systems may work with higher efficiency.

2 Textile Wastewater

Textile industry is one of the most polluting industries. It is a water-intensive industry and mostly located in South Asian counties like India, China, Bangladesh, Pakistan etc.

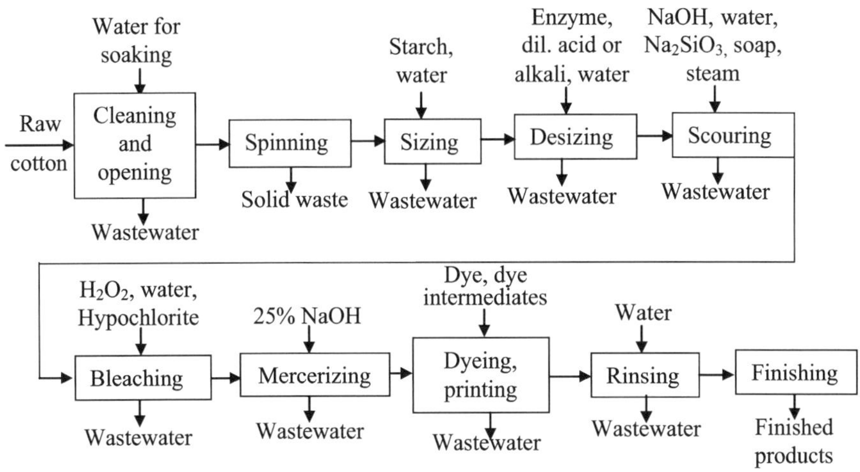

Fig. 1 Flowchart showing wet processing units in a composite cotton textile mill

A flowchart showing different textile wet processing units in a composite cotton textile mill is shown in Fig. 1. The process carried out in each unit involves the addition of various chemicals and produces varying quantities of wastewater. The wastewater produced by all the units mentioned in the flowchart combines to produce textile wastewater. Sizing involves the addition of sizing agents like starch and cellulose to stiffen the fabric. Desizing involves the removal of excess sizing agents by adding enzymes, dilute acids or alkalis. Scouring involves the addition of mild soap or detergent to remove impurities like oil and grease. In the bleaching process, the natural colour of the fabric is removed by adding bleaching agents like sodium hypochlorite and H_2O_2. Mercerizing involves the addition of sodium hydroxide to improve strength, lustre and dye affinity of the fabric. The excess alkali is then removed by acid washing. Mercerizing is done in case of 100 % cotton fabric. Dyeing process involves the addition of various dyes and dye intermediates to the fabric in order to impart colour and make it colourfast and weather resistant. The wastewater emerging from this section is very complex and most difficult to treat by conventional biological processes, as it contains toxic and refractory organic compounds. In the rinsing step, the fabric is washed with excess water to remove the unfixed dyes from the fabric. The general characteristics of wastewater generated by different units are shown in Table 1.

Textile wastewater poses a threat to the environment, as it contains a large amount of chemically different dyes, many of which are poisonous and prone to produce cancer (Chung and Cerniglia 1992). They are not affected by the attack of sun, rain or adverse chemicals. Hence, there is a need to treat textile wastewater satisfactorily before disposing it into the water bodies. Many researchers have

Table 1 General characteristics of wastewater from different wet processing units. (Source: Sharma 1989)

Unit	Contaminants	Characteristics of wastewater
Sizing, desizing	Starch, cellulose, fats and wax	High BOD
Scouring	Caustic soda, waxes, oil and grease, soda ash, sodium silicate, fibrous matter	Strongly alkaline, high BOD, dark brown in colour
Bleaching	Hydrogen peroxide, sodium hypochlorite, sodium silicate, caustic soda, acids, chlorine	Alkaline, low BOD
Mercerizing	Caustic soda	Strongly alkaline low BOD
Dyeing	Dyes, dye intermediates, soap	Strongly coloured with varying colours, low BOD, high COD
Printing	Colours, thickeners, auxiliaries	Highly coloured, low BOD, high COD

BOD biochemical oxygen demand, *COD* chemical oxygen demand

proposed physical, chemical and biological treatment processes to remove colour from synthetic and real textile wastewater. Physical methods such as adsorption, coagulation, membrane filtration etc. (Li et al. 2010; El-Gohary and Tawfik 2009; Akbari et al. 2007) are very efficient in colour removal and are being used in many industries. Dyes are not degraded by these methods, but they are either concentrated or transferred to the solid medium. Chemical methods (Kazmi and Thul 2007; Khataee et al. 2009; Khadhraoui et al. 2009) are very efficient, but may not be cost effective. Many produce large amount of sludge. Biological methods (Buitron et al. 2004; Zee and Villaverde 2005; Erkurt et al. 2007) are slow and show low efficiency, as many dyes are difficult to biodegrade due to their complex aromatic ring structures. However, they are environment friendly and cheap.

Photocatalytic degradation, mediated by large band gap semiconductors like TiO_2, has gained importance in the last few decades (Arabatzis et al. 2003; Gupta et al. 2006; Zhang et al. 2008; Sahoo et al. 2012b). TiO_2 is used because it is cheap and easily available, photo stable and nontoxic. It can degrade the contaminant ultimately to CO_2, water and salts of mineral acids, by successive attack of OH· radical (Sleiman et al. 2007). Hence, the complete mineralization of the contaminants is possible if the reaction is performed under optimum conditions in a controlled manner for a longer duration. The limiting steps in the reaction are recombination of electron hole pair and difficulty in settling of the fine photocatalyst particles. These limitations are removed by immobilizing TiO_2 on various supports (Alinsafi et al. 2007; SubbaRao et al. 2003) or doping it with transition metals or metal ions in very small proportions (Vinu and Madras 2008; Sathishkumar et al. 2011). After immobilization, the photocatalyst becomes less efficient because of the decrease in exposed surface area, while it becomes more efficient after doping with metal or metal ion and becomes more settleable (Sahoo et al. 2005b).

3 Case Study—Optimization of Photocatalytic Degradation of Real Textile Wastewater Applying Box–Behnken Experimental Design and Research Surface Modelling (RSM) Approach

3.1 Background of the Study

The photocatalytic degradation process mainly depends upon the initial concentration of the contaminant, the catalyst dose and the pH. The efficiency of the photocatalyst in removing the contaminants can be maximized by optimizing the influential parameters on which the process depends. In the classic optimization process, a large number of experiments are performed, making the process very time consuming and less cost effective. It does not show how the process parameters interact with each other and the effect of their interaction on the responses (Sahu et al. 2009). The statistical design of experiment method helps in overcoming the aforementioned limitations by reducing the number of experiments required to be performed, thus consuming less time. It develops two-dimensional polynomial models to optimize the process which can evaluate the effect of interaction among the variables on the responses. Recently, many studies have reported the use of experimental design methods to optimize multivariable chemical processes (Alam et al. 2007; Aleboyeh et al. 2008). The independent and interaction effects of many variables on the outcome of the process are effectively studied using RSM. This is achieved by simultaneously varying the independent variables and conducting very few experiments. There are very few studies in the literature which have applied statistical experimental design and RSM modelling approach to optimize the photocatalytic degradation of real and synthetic wastewater (Pang et al. 2011; Sahoo and Gupta 2012).

The present study is the first attempt to optimize the photocatalytic degradation of real textile wastewater using silver ion doped TiO_2 as a catalyst. The real textile wastewater used in the study was collected from a fabric dyeing and finishing industry located in Rourkela, India. Initial information about the ranges of influential process variables were gathered from the preliminary experiments. The number of experiments required to be performed for the optimization was determined using the Box–Behnken design of experiment and the independent and interaction effects of the three influential process variables: catalyst dose, textile wastewater concentration and reaction pH, on the dependent variables. Colour removal and COD removal (mineralization), was ascertained using RSM:

3.2 Materials and Methods

Materials and Chemicals

Micro sized titanium dioxide used as photocatalyst in the study was obtained from Merck India. Silver ion doping was done using silver nitrate (99.9% pure) purchased from S.D. Fine Chem. The procedure was reported elsewhere (Sahoo et al. 2005a). pH of the reaction mixture was adjusted using NaOH and HNO_3 purchased from Merck India. $K_2Cr_2O_7$, H_2SO_4, $HgSO_4$ and Ag_2SO_4 obtained from Merck India were used to analyse the COD. All the discussed reagents used in the study were of analytical reagent grade and triple distilled water prepared in the laboratory was used to prepare the aqueous solutions. The instruments used in the study were reported elsewhere (Sahoo and Gupta 2012).

Characterization of Real Textile Wastewater

Real textile water was collected from the cotton dyeing bath of New India Dyers and Finishers in Rourkela, India, which is a fabric dyeing and finishing industry. They dye cotton, synthetic and woollen fabric. Sodium chloride is used to impart colourfastness and for better bonding during cotton dyeing. The industry uses reactive, disperse and direct dyes. Before dyeing, the fabric is subjected to processes like sizing, desizing, scouring, bleaching and mercerizing. After dyeing, the fabric is washed with a large amount of water to remove excess colour and is subjected to other finishing operations. The flowchart for the processes involved in wet processing of fabric, in the industry, was shown elsewhere (Sahoo et al. 2012a). The composition of real textile wastewater is shown in Table 2. The wastewater was found to contain high total solids, dissolved and colloidal solids, fixed solids, COD and chloride. The low BOD/COD ratio shows the recalcitrant nature of the wastewater.

Experimental Setup

The experiments designed by Box–Behnken experimental design were conducted in a batch reactor which was set up by placing a 500 mL Borosil beaker over a magnetic stirrer and mounting the whole assembly on a wooden frame. The light source used was a 15 W UV fluorescent lamp emitting UV light of wavelength 254 nm, which was placed over the beaker, on the wooden frame. The real textile wastewater concentrations in the reaction mixture (60 mL) were 20, 60 and 100% respectively. The degradation was carried out with silver ion doped TiO_2 at doses of 1, 1.5 and 2 g/L, and the reaction pH was maintained at 4, 7 and 10, respectively. The reaction mixture was UV irradiated for a fixed time of 180 min after stirring it in the dark for 30 min to allow for adsorption. Samples were taken at adequate intervals and analysed for the decolourization and mineralization of real textile wastewater after centrifugation.

Table 2 Characteristics of real textile wastewater

Parameters	Values
pH	7.47
Temperature	33 °C
Total solids	2,153 mg/L
Suspended solids	177 mg/L
Dissolved and colloidal solids	1,976 mg/L
Volatile solids	439 mg/L
Fixed solids	1,714 mg/L
Turbidity	85.2 NTU
Conductivity	32,400 μS
COD	423 mg/L
TOC	157 mg/L
BOD	73 mg/L
Chloride	1,645 mg/L
Nitrate N	23.2 mg/L
NH_4 N	31.4 mg/L
Sulphate	278 mg/L
Surfactants	87 mg/L

BOD biochemical oxygen demand, *COD* chemical oxygen demand, *TOC* total organic carbon

Analytical Procedure

Spectral analysis of real textile wastewater was done using Genesys 20 spectrophotometer procured from Thermo Spectronic, USA and its λ_{max} value was found to be 515 nm. The colour removal of real textile wastewater was monitored at this λ_{max} value. The colour removal was determined as:

$$\%\,Colourremoval = \left(\frac{A_0 - A_t}{A_0}\right) \times 100 \quad (1)$$

where

A_0 is the absorbance of real textile wastewater at zero time
A_t is the corresponding absorbance at any time t.

The absorbance was taken instead of concentration in the equation because the concentration of the real textile wastewater was unknown, and according to Beer–Lambert's law, concentration varies directly with absorbance. The degree of mineralization of real textile wastewater was determined by ascertaining the COD removal using closed reflux colorimetric method (5220D) (Clesceri et al. 2005). The COD removal (mineralization) was calculated as:

$$\%\,CODremoval = \left(\frac{COD_0 - COD_t}{COD_0}\right) \times 100 \quad (2)$$

Table 3 Ranges of variables considered in Box–Behnken statistical experimental design

Independent variables	Symbol	Coded level of variable		
		Low	Centre	High
		−1	0	+1
Catalyst dose (g/L)	A	1	1.5	2
Textile wastewater (%)	B	20	60	100
Reaction pH	C	4	7	10

where

COD_0 initial COD in ppm
COD_t COD at any time t in ppm.

Response Surface Modelling and Experimental Design

Initially, a few experiments were conducted by changing one factor at a time while keeping the other factors constant in order to find the process variables which influence the photocatalytic degradation of real textile wastewater most, in a bid to arrive at the most optimized condition for the reaction. Catalyst dose, initial concentration of real textile wastewater and reaction pH, found as the most influential process variables from the single factor study, were optimized using RSM (Zhang et al. 2010).

The process was optimized for the three selected process variables obtained from the initial experiments using RSM. The process variables were taken as independent variables and % colour removal and % COD removal (mineralization) were taken as two response variables. Box–Behnken experimental design procedure was used to design 15 experiments to evaluate the independent and interaction effects of the three independent process variables on the two responses. Box–Behnken design was used because of its efficiency and as it does not contain any point at the vertices of the cubic region formed by the upper and lower extremes of the variables (Montgomery 2004; Ferreira et al. 2007). Table 3 compiles the ranges and levels of independent process variables used in the design. The following equation is used to calculate the number of experiments required for the design:

$$N = n^2 + n + c \tag{3}$$

where

N is the number of experiments required
n is the number of process variables
c is the number of central runs (Edrissi et al. 2008).

The response surface model is usually expressed as:

$$Y = f\left(x_1, x_2, x_3, x_4, x_4, \ldots\ldots, x_n\right) \pm e \tag{4}$$

where

Y represents the response variable
f is the response function
x_i are the independent variables
e represents the experimental error (Singh et al. 2011).

The response function is constructed based on the relationship between the process variables and the response variable. RSM is used to approximate the response function f by a suitable quadratic polynomial in some region of the space comprising independent process variables (Singh et al. 2011). Thus in the present study, the results of the experimental design were fitted to the following empirical quadratic polynomial model for three process variables:

$$Y = \beta_0 + \beta_1 A + \beta_2 B + \beta_3 C + \beta_{11} A^2 + \beta_{22} B^2 + \beta_{33} C^2 + \beta_{12} AB + \beta_{23} BC + \beta_{31} CA \quad (5)$$

where

Y represents the response variable
β_0 is the intercept
$\beta_1, \beta_2, \beta_3$ are the coefficients of the independent process variables
$\beta_{11}, \beta_{22}, \beta_{33}$ are quadratic coefficients
$\beta_{12}, \beta_{23}, \beta_{31}$ are interaction coefficients
A, B, C are the independent process variables studied

The statistical adequacy of the developed model was tested by performing multivariate regression analysis and analysis of variance (ANOVA) using Fisher's F-statistics. The RSM optimization was performed with the help of Design Expert software version 8.0.6.1, Stat Ease Inc., USA. The predicted and experimental values of the response variable y were used to calculate the statistical tools such as the coefficient of determination (R^2), the root means square error of prediction (RMSEP) and the relative standard error of prediction (RSEP). The correlation between the experimental and predicted values indicates that the model fits well to the experimental results, while the RMSEP and RSEP values evaluate the predictive ability of the selected model (Singh et al. 2011). The RMSEP and RSEP were computed as follows:

$$RMSEP = \sqrt{\frac{\sum_{i=1}^{N} \left(y_{\text{pred},i} - y_{\text{exp},i} \right)^2}{N}} \quad (6)$$

$$RSEP = \sqrt{\frac{\sum_{i=1}^{N} \left(y_{\text{pred},i} - y_{\text{exp},i} \right)^2}{\sum_{i=1}^{N} \left(y_{\text{exp},i} \right)^2}} \times 100 \quad (7)$$

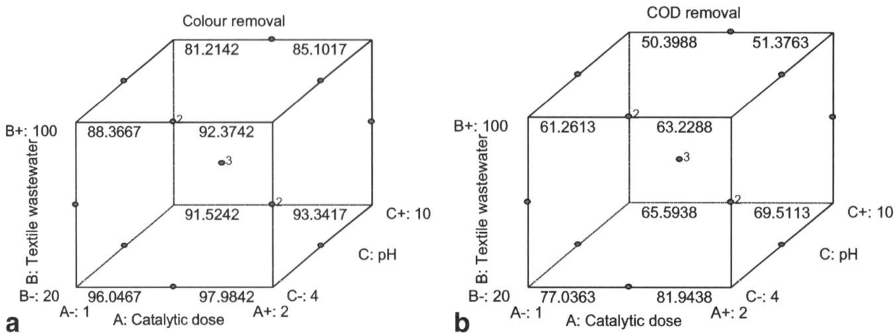

Fig. 2 Cube plots of Box–Behnken designs for colour removal (**a**) and COD removal (**b**) of real textile wastewater

where

$y_{pred,i}$ and $y_{exp,i}$ represent the predicted and experimental values of the response variable y, respectively

N represents the number of experiments conducted (Singh et al. 2011).

Statistical Analysis

The data obtained from the experiments, designed using Box–Behnken experimental design methodology, were analysed statistically using Design Expert software version 8.0.6.1 from Stat Ease Inc., USA. Analysis of variance (ANOVA) of the model as a whole and its coefficients was performed and the values obtained for which p values <0.05 were considered significant. Various descriptive statistical tools such as p value, F value, degree of freedom (df), determination coefficient (R^2), adjusted determination coefficient (R^2_{adj}), sum of squares (SS), mean sum of squares (MSS) are calculated to demonstrate the statistical significance of the developed quadratic model (Singh et al. 2011). The process was optimized by conducting three-dimensional response surface analyses of the independent process variables and dependent response variables (Sahoo and Gupta 2012). The designed experiments along with experimental and predicted values of the two response variables are shown in Table 4 and the cube diagrams for the designs of the two response variables are shown in Fig. 2.

3.3 Results and Discussion

Initial Experiments

In order to fix the effective ranges of the three process variables, catalyst dose, concentration of real textile wastewater and reaction pH, to be used in the Box–Behnken

Table 4 Box–Behnken experimental design along with actual and predicted values of the response variable

Run	A, Catalytic dose (g/L)	B, Textile wastewater (%)	C, pH	Y_1, Colour removal (%)		Y_2, COD Removal (%)	
				Actual	Predicted	Actual	Predicted
1	1	20	7	94.57	94.53	75.61	75.78
2	2	20	7	96.38	96.41	80.49	80.19
3	1	100	7	85.57	85.54	60.00	60.30
4	2	100	7	89.45	89.49	61.94	61.77
5	1	60	4	91.50	91.60	68.32	68.22
6	2	60	4	94.54	94.57	71.29	71.66
7	1	60	10	85.79	85.76	57.43	57.07
8	2	60	10	88.71	88.61	59.41	59.51
9	1.5	20	4	97.10	97.04	82.93	82.86
10	1.5	100	4	90.46	90.39	65.81	65.62
11	1.5	20	10	92.39	92.46	70.73	70.92
12	1.5	100	10	83.12	83.18	54.19	54.26
13	1.5	60	7	90.74	90.91	71.29	71.95
14	1.5	60	7	90.99	90.91	72.28	71.95
15	1.5	60	7	90.99	90.91	72.28	71.95

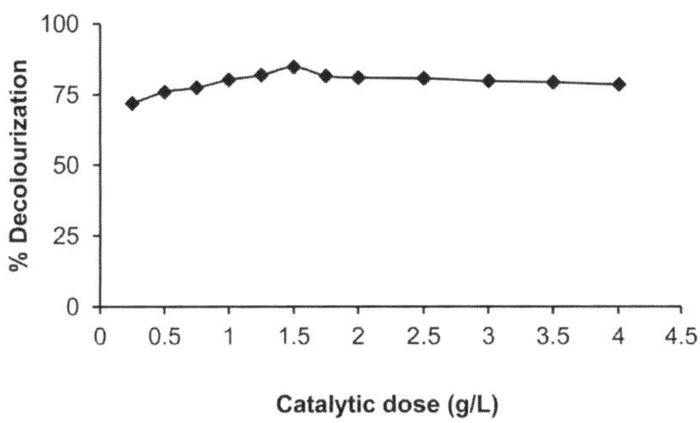

Fig. 3 Effect of catalyst dosage on the decolourization rate of real textile wastewater; concentration of textile wastewater: 50 %; pH: 7.47; temperature: 25±2 °C; time: 1 h

experimental design, initial experiments were performed using the batch reactor. The percent decolourization of real textile wastewater (50 %) at different catalyst doses after UV irradiation for 60 min is represented in Fig. 3. Figure 3 demonstrates a gradual increase in the decolourization rate of real textile wastewater up to a catalyst dose of 1.5 g/L and then it slowly decreased. The increase in decolourization rate may be due to increase in production rate of oxidising species like hydroxyl ions and super oxide ions in the reaction medium, with increase in catalyst dosage. The reaction mixture becomes turbid at higher catalytic dosage, impeding light penetration on to the catalyst surface, and hence decreasing the decolourization rate.

The effect of the concentration of real textile wastewater on its decolourization rate was then investigated by UV irradiation of different concentrations of real textile wastewater for 60 min with 1.5 g/L silver ion doped TiO_2, and the results

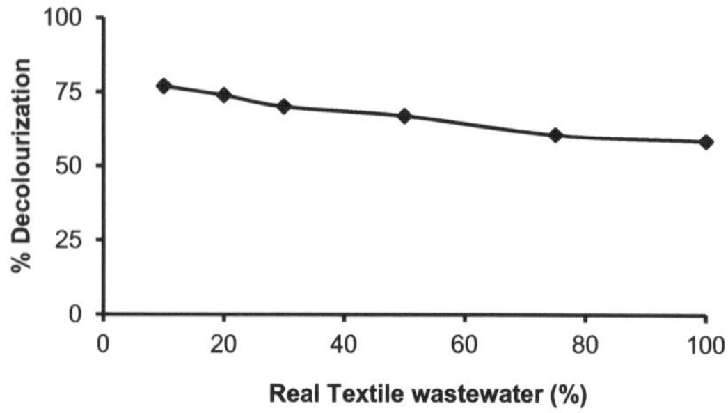

Fig. 4 Effect of initial wastewater concentration on the decolourization rate of real textile wastewater; silver ion doped TiO_2: 1.5 g/L; pH: 7.47; temperature: 25 ± 2 °C; time: 1 h

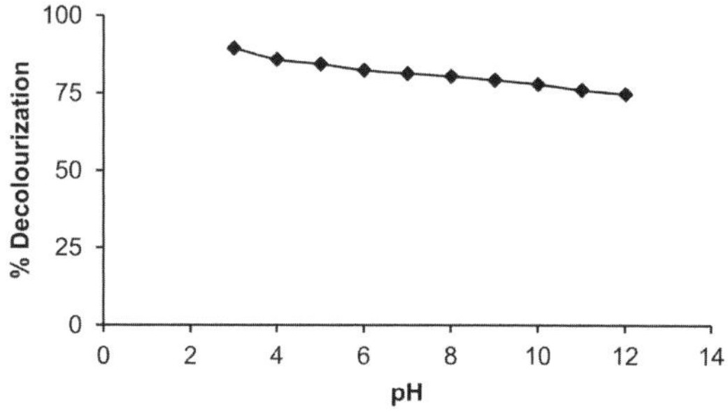

Fig. 5 Effect of reaction pH on the decolourization rate of real textile wastewater; silver ion doped TiO_2: 1.5 g/L; concentration of real textile wastewater: 50 %; temperature: 25 ± 2 °C; time: 1 h

are represented in Fig. 4. A gradual decrease in the rate of decolourization of real textile wastewater was observed with increase in its concentration as photocatalytic decolourization is a light dependent process. The depth of penetration of UV light in to the reaction mixture decreased with increase in the concentration of real textile wastewater as it did not reach the catalyst surface. The requirement of catalyst surface needed for the decolourization increased with increase in initial concentration of real textile wastewater. As illumination time and catalyst dose were constant, the amount of hydroxyl radical (primary oxidant) formed on the surface of the photocatalyst was also constant. The rate of decolourization of real textile wastewater decreased as the relative number of free radicals attacking the real textile wastewater molecules decreased (Tayade et al. 2009).

The effect of reaction pH on the decolourization rate of real textile wastewater was studied in the pH range 3–13 after UV irradiation of real textile wastewater (50 %) for 60 min with 1.5 g/L silver ion doped TiO_2 and represented in Fig. 5. From the figure, it is observed that there was a higher decolourization rate in the acidic pH range and lower decolourization rate in the basic pH range. This may be because the pH_{zpc} of silver ion doped TiO_2 is approximately 6.4 and at a pH less than pH_{zpc}, the surface of the catalyst is positively charged, whereas at a pH more than pH_{zpc}, it is negatively charged. The real textile wastewater may contain some anionic dye which is attracted by positively charged catalytic surface at low pH and vice versa (Sahoo et al. 2005a).

Statistical Analysis and Model Fitting

The photocatalytic degradation process of real textile wastewater using silver ion doped TiO_2 under UV irradiation was optimized using one of the very efficient designs of experiment called Box–Behnken design. It contains all the points lying on a sphere whose radius is $\sqrt{2}$, and hence it is called a spherical design. It does not consider any of the extreme points, i.e. the points formed by the combinations of all maximum or all minimum values of the variables on the vertices of the cubic region (Montgomery 2004; Ay et al. 2009). The design combines a two-level (full or fractional) factorial design to an incomplete block design. The design consists of all combinations of a few factors used in a factorial design, while the central values of some other factors are also included (Ay et al. 2009). In the statistical experimental designs, the number of experiments and the number of coefficients in the quadratic model are in the ratio range 1.5–2.6 and in case of Box–Behnken design of experiment with three process variables the ratio is 1.67 (Ay et al. 2009).

Design Expert software version 8.6.0.1 from Stat Ease Corporation, USA was used to develop two empirical polynomial equations of second order for the two response variables % colour removal and % COD removal using three independent process variables and the experimental design represented by Table 4.

$$Y_1 = 90.91 + 1.46A - 3.89B - 2.95C + 0.52AB - 0.03AC - 0.66BC - 0.023A^2 + 0.61B^2 - 0.75C^2 \quad (8)$$

$$Y_2 = 71.95 + 1.47A - 8.48B - 5.82C - 0.73AB - 0.25AC + 0.14BC - 3.37A^2 + 0.93B^2 - 4.47C^2 \quad (9)$$

where

Y_1 and Y_2 represent the percentage colour and COD removal of real textile wastewater, respectively after UV irradiation for 180 min

A, B and *C* are the coded values of catalyst dose, concentration of real textile wastewater and reaction pH, respectively.

ANOVA was performed for the experimental data for both the responses colour and COD removal, and the results are presented in Tables 5 and 6. The model was highly significant as found from the ANOVA of the second-order quadratic polynomial model for both the responses demonstrated by high F values of 1,465.05 and 433.16 and low p values <0.0001. It means that there is only 0.01 % chance of occurrence of such high model F values due to noise. The lack of fit F values of 0.67 and 0.62 are found to be insignificant with p values >0.05. The ANOVA results for the coefficients of the polynomial regression models for the responses colour removal and COD removal are presented in Tables 7 and 8.

For the response colour removal, all the independent variables represented by *A, B* and *C* are significant. Along with them, the effect of interaction of catalyst dose (*A*) and real textile wastewater concentration (*B*), *AB,* the effect of interaction of

Table 5 ANOVA table for the response surface quadratic polynomial model generated for the response variable colour removal

Source	Sum of squares	Degree of freedom	Mean square	F value	p value
Model	219.77	9	24.42	1465.05	<0.0001
Residual	0.083	5	0.017		
Lack of fit	0.042	3	0.014	0.67	0.6464
Pure error	0.042	2	0.021		
Total	219.86	14			
	$R^2=0.9996$	Adj $R^2=0.9989$	Ped $R^2=0.9965$		

Table 6 ANOVA table for the response surface quadratic model generated for the response variable COD removal

Source	Sum of squares	Degree of freedom	Mean square	F value	p value
Model	982.08	9	109.12	433.16	<0.0001
Residual	1.26	5	0.25		
Lack of fit	0.61	3	0.20	0.62	0.6661
Pure error	0.65	2	0.33		
Total	983.34	14			
	$R^2=0.9987$	Adj $R^2=0.9964$	Ped $R^2=0.9886$		

Table 7 ANOVA table for the coefficients of quadratic model developed for the response variable colour removal

Factor	Coefficients	Degree of freedom	Standard error	F value	95% confidence interval low	95% confidence interval high	p value
Intercept	90.91	1	0.075	–	90.72	91.10	–
A	1.46	1	0.046	1017.82	1.34	1.57	<0.0001
B	−3.98	1	0.046	7602.63	−4.10	−3.86	<0.0001
C	−2.95	1	0.046	4173.24	−3.07	−2.83	<0.0001
AB	0.52	1	0.065	64.27	0.35	0.68	0.0005
AC	−0.030	1	0.065	0.22	−0.20	0.14	0.6617
BC	−0.66	1	0.065	103.74	−0.82	−0.49	0.0002
A^2	−0.023	1	0.067	0.12	−0.20	0.15	0.7425
B^2	0.61	1	0.067	82.20	0.44	0.78	0.0003
C^2	−0.75	1	0.067	124.05	−0.92	−0.58	0.0001

real textile wastewater concentration (B) and reaction pH (C), BC, and the quadratic terms B^2 and C^2 are significant with p value <0.05. The independent variables affect the rate of decolourization of real textile wastewater in the order initial concentration of real textile wastewater (B) > reaction pH (C) > catalyst dose (A). B has a higher negative effect while A has lower positive effect and C has lower negative effect.

Table 8 ANOVA table for the coefficients of quadratic model developed for the response variable COD removal

Factor	Coefficients	Degree of freedom	Standard error	*F* value	95% confidence interval low	95% confidence interval high	*p* value
Intercept	71.95	1	0.29	–	71.21	72.69	–
A	1.47	1	0.18	68.74	1.02	1.93	0.0004
B	−8.48	1	0.18	2282.29	−8.93	−8.02	<0.0001
C	−5.82	1	0.18	1077.06	−6.28	−5.37	<0.0001
AB	−0.73	1	0.25	8.58	−1.38	−0.090	0.0327
AC	−0.25	1	0.25	0.97	−0.89	0.40	0.3693
BC	0.14	1	0.25	0.33	−0.50	0.79	0.5884
A^2	−3.37	1	0.26	166.58	−4.04	−2.70	<0.0001
B^2	0.93	1	0.26	12.71	0.26	1.60	0.0161
C^2	−4.47	1	0.26	292.37	−5.14	−3.79	<0.0001

For the response COD removal (mineralization), all the independent variables of the model, *A, B* and *C,* the effect of interaction between catalyst dose (*A*) and concentration of real textile wastewater (*B*), AB, and the quadratic terms A^2, B^2 and C^2 are significant with p value <0.05. The independent variables affect the rate of mineralization of real textile wastewater in the order initial concentration of real textile wastewater (*B*) > reaction pH (*C*) > *catalyst dose* (*A*).

B has a higher negative effect while *A* has lower positive effect and *C* has lower negative effect. The initial concentration of real textile wastewater (*B*) shows a pronounced negative effect, whereas catalyst dose shows low positive effect and the reaction pH shows low negative effect on both the response variables decolourization and mineralization of real textile wastewater. The aforementioned observations are in good agreement with the experiments performed initially to fix the ranges of the independent variables. Thus, the initial concentration of real textile wastewater affect its photocatalytic degradation using silver ion doped TiO_2 in a major way. At higher real textile wastewater concentration, the depth of penetration of UV light becomes low, preventing it from reaching the surface of major portion of photocatalyst. This reduces the production of hydroxyl, superoxide and other radicals responsible for the photocatalytic degradation of real textile wastewater.

Adequacy Check of the Model

In order to measure whether the statistical model is adequate to predict the responses or not, the coefficient of determination (R^2), adequate precision, prediction error sum of square (PRESS), RMSEP and RSEP values were determined for the model. The R^2_{pred} values of 0.9965 and 0.9886 for decolourization and mineralization respectively are very close to the corresponding R^2_{adj} values of 0.9989 and 0.9964. Adequate precision measures the adequacy of the signal and is represented

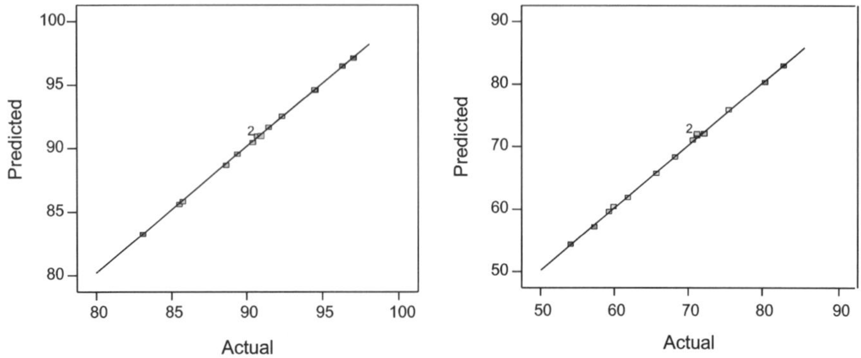

Fig. 6 (**a**) Plot of the actual and predicted values for colour removal (**b**) and COD removal

by the ratio of signal and noise. The signal is adequate for adequate precision values greater than 4. The adequate precision values of 131.457 and 69.795 for the two responses decolourization and mineralization, respectively, indicate adequate signal. PRESS shows how effectively the model developed for the experiment can be used to predict the responses for a similar experiment. Small PRESS values of 0.76 and 11.17 for decolourization and mineralization show adequacy of the predicted values. The values of R^2 obtained show the effectiveness of the prediction of decolourization and mineralization of real textile wastewater using the developed quadratic polynomial equations in the range of the experimental (Ravikumar et al. 2007). The values of RMSEP for decolourization and mineralization of real textile wastewater were 0.076 and 0.29, and the corresponding values of RSEP were 0.083 and 0.422, respectively. The values are considerably low showing the adequacy of the fitted model.

The plots of the predicted values of colour and COD removal against their observed values are represented in Fig. 6a and b. The plots reveal the closeness of the predicted values to their experimental values. The normal probability plots for the internally studentized residuals for colour and COD removal represented in Fig. 7a and b are used to detect and explain the correctness of the assumptions of errors following normal distribution, their independence over one another and the homogeneity of error variance. Residuals show the degree of lack of fit of the developed model (Singh et al. 2011). From Fig. 7, it evident that the assumptions made are not seriously violated and the normality of errors and independence of residuals are true (Sen and Swaminathan 2004).

RSM Analysis

The response surface plots in three dimensions and the corresponding contour plots in two dimensions, generated using response surface methodology, represented the independent and interaction effects of the process variables (catalyst dose, initial con-

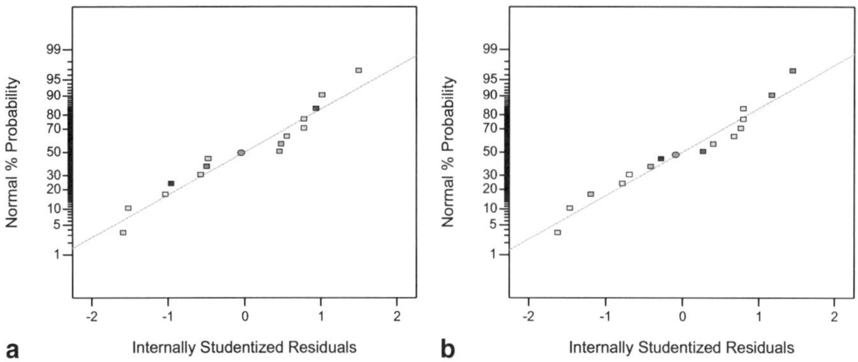

Fig. 7 Plot of the normal probability and internally studentized residuals for colour removal (**a**) and COD removal (**b**)

centration of real textile wastewater and reaction pH) on the decolourization and mineralization of real textile wastewater graphically. The plots were used to predict the responses and to optimize the values of the process variables for decolourization and mineralization of real textile wastewater (Clesceri et al. 2005; Tavares et al. 2009).

The effect of interaction of initial concentration of real textile wastewater and the dosage of silver ion doped TiO_2 on the decolourization rate of real textile wastewater is shown in Fig. 8a and that on its mineralization rate is shown in Fig. 9a. Figure 8a shows a gradual decrease in the decolourization rate of real textile wastewater with increase in its concentration independent of the catalyst dose. The contour plot showed that the maximum degradation occurred at 20% concentration of real textile wastewater. The rate of decolourization of real textile wastewater increased continuously till the catalyst dose increased to 1.5 g/L and then remained nearly constant independent of its initial concentration. The contour plots showed the maximum decolourization to be in the catalyst dose range of 1.4–2 g/L. From Fig. 9a, a similar trend as in case of decolourization rate, was observed for the effects of interaction of initial concentration of real textile wastewater and catalyst dose on the mineralization rate. The maximum ranges were real textile wastewater concentration 20% and catalyst dose 1.4–2 g/L. The observations, as discussed, obeyed the results of initial tests.

The effect of interaction of reaction pH and catalyst dose on the degree of decolourization of real textile wastewater is represented in Fig. 8b and that on the degree of mineralization is represented in Fig. 9b. Figure 8b demonstrates the slow decrease in decolourization rate as the reaction pH increases independent of the catalyst dose. The degree of decolourization increased as the catalyst dose increased up to a dosage of 1.5 g/L and then became nearly constant independent of the reaction pH. The contour plot demonstrated the maximum decolourization to be for catalyst dose in the range 1.4–2 g/L. The interaction of reaction pH and catalyst dose affected the degree of mineralization of real textile wastewater in a similar way as in case of decolourization as demonstrated by Fig. 9b. Maximum mineralization occurred in the reaction pH range 4–6 and catalyst dose range 1.4–2 g/L.

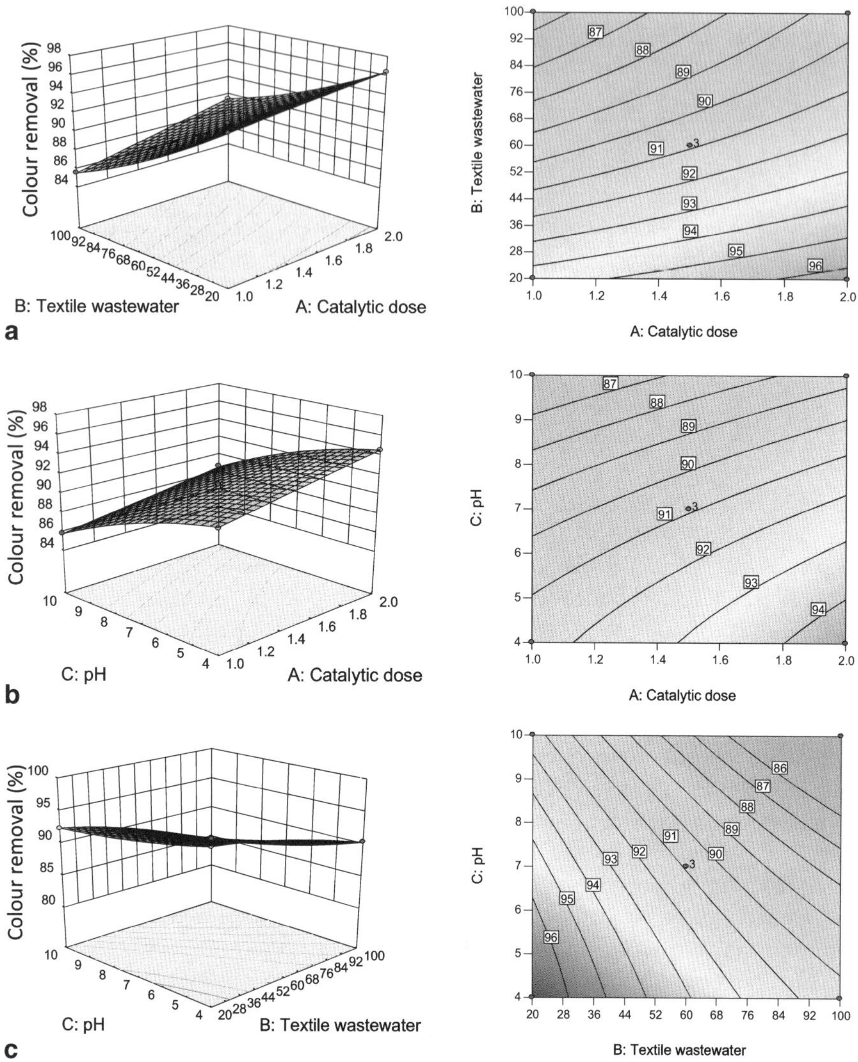

Fig. 8 Effects of catalyst dose, initial concentration and reaction pH on decolourization of real textile wastewater. **a** Reaction pH constant at 7. **b** Concentration of textile wastewater constant at 60 %. **c** Catalyst dose constant at 1.5 g/L

The effect of interaction of reaction pH and the initial concentration of real textile wastewater on its degree of decolourization is demonstrated in Fig. 8c and that on its degree of mineralization is demonstrated in Fig. 9c. Figure 8c demonstrates that the degree of decolourization decreased slowly with the increase in reaction pH independent of the real wastewater concentration. Maximum decolourization was in the reaction pH range 4–6. It also demonstrates that the degree of decolourization decreased as the initial concentration of real textile wastewater increased with the

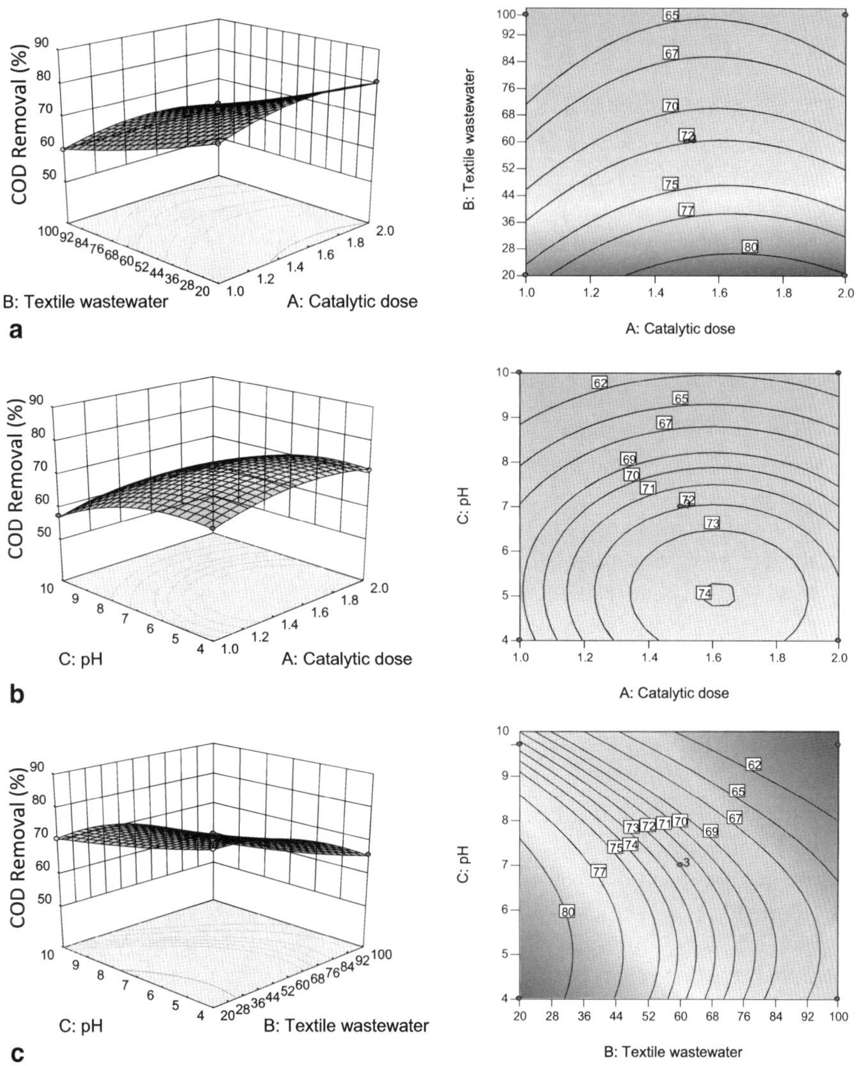

Fig. 9 Effects of catalyst dose, initial concentration and reaction pH on mineralization of real textile wastewater. **a** Reaction pH constant at 7. **b** Concentration of textile wastewater constant at 60 %. **c** Catalyst dose constant at 1.5 g/L

maximum in the range of 20–30 % real textile wastewater. Figure 9c shows that the degree of mineralization decreased slowly as reaction pH increased independent of the initial concentration of real textile wastewater while it decreased as the initial concentration of real textile wastewater increased, with the maximum at reaction pH 4–6 and real textile wastewater concentration 20–30 %. All the observations were in accordance with results of the tests performed in the beginning to fix the process parameters.

Optimization of the Independent Process Variable

The process of finding out the optimum operating conditions for any physical or chemical process which will produce best possible results is called optimization. Most of the researchers aim at maximizing the results while optimizing a process without considering economy and ecological balance. Furthermore, they do not consider that the degraded products can be subsequently treated. So in the present study, instead of stressing on maximization of results, optimization was done under three predefined constraints. The pH was kept close to the pH of the reaction mixture, the use of photocatalyst was minimized and the concentration of real textile wastewater was maximized while optimizing. The process was optimized numerically using Design Expert version 8.0.6.1 for the three independent process variables and two responses under the aforementioned constraints. As a result of optimization, the maximum possible decolourization and mineralization of real textile wastewater under the applied constraints were at 1.5 g/L of silver ion-doped TiO_2, concentration of real textile wastewater 60 % and pH of medium 7. Under the optimum conditions, the colour removal and COD removal rates of real textile wastewater were predicted by the model to be 90.91 % and 71.95 % after 180 min of UV irradiation. These results demonstrated the successful use of RSM to optimize the photocatalytic decolourization and mineralization of real textile wastewater using silver ion-doped TiO_2 (Clesceri et al. 2005).

Validation of the Optimization Result

Validation experiments were conducted in the batch reactor by degrading the real textile wastewater photocatalytically under the optimum conditions obtained from RSM. This provided a check on the accuracy of the optimization by RSM. The decolourization and mineralization rates were found out to be 90.46 and 71.72 %, respectively, after 180 min which are very close to the predicted values. This validated the set of optimum conditions arrived at by RSM optimization under predetermined constraints and proved that the photocatalytic degradation process of real textile wastewater using silver ion doped TiO_2 can be optimized using RSM.

Comparison with Synthetic Wastewater

The polynomial equations developed for colour and COD removal of real textile wastewater were compared with the corresponding equations developed for methyl blue reported in our earlier paper (Sahoo and Gupta 2012). Equations 8 and 9, developed for colour and COD removal of real textile wastewater, show that concentration of real textile wastewater and its pH have a negative effect on both the responses while catalytic dose has a lower positive effect. Corresponding equations developed for methyl blue showed a high negative effect of initial dye concentration while the other two variables have low positive effect on both the responses.

Presence of other interfering ions may be the cause of negative effect of pH on the decolourization and mineralization of real textile wastewater. The negative effect of concentration of real textile wastewater on the responses was lower than that in case of methyl blue, maybe due to the low colour intensity of textile wastewater. The nature of the equations developed was similar and they were used successfully for the optimization of the process.

4 Conclusions

Treatment and safe disposal/reuse of industrial wastewater is very important in the present scenario of water scarcity. As witnessed from the case study, advanced oxidation processes like photochemical oxidation hold future in the breakdown and removal of nonbiodegradable organic contaminants.

In the case study, an RSM approach was successfully employed to optimize the photocatalytic decolourization and mineralization of real textile wastewater using Ag^+ doped TiO_2. Box–Behnken spherical design was used to design 15 experiments, and using them, the effects of three process parameters catalytic dose, initial concentration of real textile wastewater and reaction pH on the photocatalytic decolourization and mineralization of real textile wastewater were studied. The results obtained were used to develop two empirical quadratic polynomial models using RSM. Initial concentration of real textile wastewater was the most influential parameter while the other parameters also influenced the process but to lesser extent. The high R^2 values and low RMSEP and RSEP values showed goodness of fit of the model with the experimental data. The optimization process showed that both decolourization and mineralization of real textile wastewater were simultaneously optimum at catalytic dose, 1.5 g/L, textile wastewater concentration, 60 %, and reaction pH, 7. Under the optimum condition, the predicted values of decolourization and mineralization of real textile wastewater were 90.91 % and 71.95 %, respectively, after 180 min. The experimental values obtained from the validation experiments were very close to the predicted values.

References

Akbari A, Desclaux S, Rouch JC, Remigy JC (2007) Application of nanofiltration hollow fibre membranes, developed by photografting, to treatment of anionic dye solutions. J Membrane Sci 297:243–252

Alam Z, Muyibi SA, Toramae J (2007) Statistical optimization of adsorption processes for removal of 2 4-dichlorophenol by activated carbon derived from oil palm empty fruit bunches. J Environ Sci 19:674–677

Aleboyeh A, Daneshvar N, Kasiri MB (2008) Optimization of C.I. Acid Red 14 azo dye removal by electrocoagulation batch process with response surface methodology. Chem Eng Process 47:827–832

Alinsafi A, Evenou F, Abdulkarim EM, Pons MN, Zahraa O, Benhammou A, Yaacoubi A, Nejmeddine A (2007) Treatment of textile industry wastewater by supported photocatalysis. Dye Pigment 74:439–445

Arabatzis IM, Stergiopoulos T, Andreeva D, Kitova S, Neophytides SG, Falaras P (2003) Characterization and photocatalytic activity of Au/TiO_2 thin films for azo-dye degradation. J Catal 220:127–135

Ay F, Catalkaya EC, Kargi F (2009) A statistical experiment design approach for advanced oxidation of Direct Red azo-dye by photo-Fenton treatment. J Hazard Mater 162:230–236

Buitron G, Quezada M, Moreno G (2004) Aerobic degradation of the azo dye acid red 151 in a sequencing batch bio filter. Bioresource Technol 92:143–149

Chung KT, Cerniglia CE (1992) Mutagenicity of azo dyes: structure-activity relationships. Mutat Res 277:201–220

Clesceri LS, Greenberg AE, Eaton AD (2005) Standard methods for the examination of water and wastewater, 21st edn. American Public Health Association, Baltimore, pp. 5(14)–5(19).

Edrissi M, Asl NR, Madjidi B (2008) Interaction of mefenamic acid with cobalt (II) ions in aqueous media: evaluation via classic and response surface methods. Turk J Chem 32:505–519

El-Gohary F, Tawfik A (2009) Decolourization and COD reduction of disperse and reactive dyes wastewater using chemical-coagulation followed by sequential batch reactor (SBR) process. Desalination 249:1159–1164

Erkurt EA, Unyayar A, Kumbur H (2007) Decolorization of synthetic dyes by white rot fungi, involving laccase enzyme in the process. Process Biochem 42:1429–1435

Ferreira SLC, Bruns RE, Ferreira HS, Matos GD, David JM, Brandao GC, da Silva EGP, Portugal LA, dos Reis PS, Souza AS, dos Santos WNL (2007) Box-Behnken design: an alternative for the optimization of analytical methods. Anal Chim Acta 597:179–186.

Gupta AK, Pal A, Sahoo C (2006) Photocatalytic degradation of a mixture of crystal violet (C.I. Basic Violet 3) and methyl red dye in aqueous suspensions using Ag^+ doped TiO_2. Dye Pigment 69:224–232

Kazmi AA, Thul R (2007) Colour and COD removal from pulp and paper mill effluent by q Fenton's oxidation. J Environ Sci Eng 49:189–194

Khadhraoui M, Trabelsi H, Ksibi M, Bouguerra S, Elleuch B (2009) Discoloration and detoxification of a Congo red dye solution by means of ozone treatment for a possible water reuse. J Hazard Mater 161:974–981

Khataee AR, Vatanpour V, Amani Ghadim AR (2009) Decolorization of C.I. Acid Blue 9 solution by UV/Nano-TiO_2, Fenton, Fenton-like, electro-Fenton and electrocoagulation processes: a comparative study. J Hazard Mater 161:1225–1233

Li Q, Yue QY, Su Y, Gao BY, Sun HJ (2010). Equilibrium, thermodynamics and process design to minimize adsorbent amount for the adsorption of acid dyes onto cationic polymer-loaded bentonite. Chem Eng J 158:489–497

Mahajan SP (1998). Pollution control in process industries. Tata McGraw-Hill, New Delhi.

Montgomery DC (2004) Design and analysis of experiments, 5th edn. Wiley, Singapore, pp 427–448.

Pang YL, Abdullah AZ, Bhatia S (2011) Optimization of sonocatalytic degradation of Rhodamine B in aqueous solution in the presence of TiO_2 nanotubes using response surface methodology. Chem Eng J 166:873–880

Ravikumar K, Krishnan S, Ramalingam S, Balu K (2007) Optimization of process variables by the application of response surface methodology for dye removal using a novel adsorbent. Dye Pigment 72:66–74

Sahoo C, Gupta AK (2012) Optimization of photocatalytic degradation of methyl blue using silver ion doped titanium dioxide by combination of experimental design and response surface approach. J Hazard Mater 215–216:302–310

Sahoo C, Gupta AK, Pal A (2005a) Photocatalytic degradation of Crystal Violet (C.I. Basic Violet 3) on silver ion doped TiO_2. Dye Pigment 66:189–196

Sahoo C, Gupta AK, Pal A (2005b) Photocatalytic degradation of methyl red dye in aqueous solutions under UV irradiation using Ag^+doped TiO_2. Desalination 181:91–100

Sahoo C, Gupta AK, Pillai IMS (2012a) Heterogeneous photocatalysis of real textile wastewater: evaluation of reaction kinetics and characterization. J Environ sci Health A 47(13):2109–2119

Sahoo C, Gupta AK, Pillai IMS (2012b) Photocatalytic degradation of methylene blue dye from aqueous solution using silver ion doped TIO_2 and its application to the degradation of real textile wastewater. J Environ Sci Health A 47(10):1428–1438

Sahu JN, Acharya J, Meikap BC (2009) Response surface modelling and optimization of chromium (VI) removal from aqueous solution using tamarind wood activated carbon in batch process. J Hazard Mater 172:818–825

Sathishkumar P, Anandan S, Maruthamuthu P, Swaminathan T, Zhou M, Ashokkumar M (2011) Synthesis of Fe^{3+}doped TiO_2 photocatalysts for the visible assisted degradation of an azo dye. Colloids Surf A: Physicochem Eng Aspects 375:231–236.

Sen R, Swaminathan T (2004) Response surface modelling and optimization to elucidate and analyse the effects of inoculums age and size on surfactin production. Biochem Eng J 21:141–148

Sharma MA (October 1989) Treatment of cotton mill effluent—A case study. Colourage, 6–31.

Singh KP, Gupta S, Singh AK, Sinha S (2011) Optimizing adsorption of crystal violet dye from water by magnetic nanocomposite using response surface modelling approach. J Hazard Mater 186:1462–1473

Sleiman M, Vildozo D, Ferronato C, Chovelon JM (2007) Photocatalytic degradation of azo dye Metanil Yellow: optimization and kinetic modelling using a chemometric approach. Appl Catal B: Environ 77:1–11

Subba Rao KV, Rachel A, Subrahmanyam M, Boule P (2003) Immobilization of TiO_2 on pumice stone for the photocatalytic degradation of dyes and dye industry pollutants. Appl Catal B: Environ 46:77–85

Tavares APM, Cristovao RO, Loureiro JM, Boaventura RAR, Macedo EA (2009) Application of statistical experimental methodology to optimize reactive dye decolourization by commercial laccase. J Hazard Mater 162:1255–1260

Tayade RJ, Natarajan TS, Bajaj HC (2009) Photocatalytic degradation of methylene blue dye using ultraviolet light emitting diodes. Ind Eng Chem Res 48:10262–10267

van der Zee FP, Villaverde S (2005) Combined anaerobic-aerobic treatment of azo dyes—A short review of bioreactor studies. Water Res 39:1425–1440

Vinu R, Madras G (2008) Synthesis and photoactivity of Pd substituted nano-TiO_2. J Mol Catal A: Chem 291:5–11

Zhang X, Sun DD, Li G, Wang Y (2008) Investigation of the roles of active oxygen species in photodegradation of azo dye AO7 in TiO_2 photocatalysis illuminated by microwave electrode less lamp, J. Photochem. Photobiol. A. Chem 199:311–315

Zhang J, Fu D, Xu Y, Liu C (2010) Optimization of parameters on photocatalytic degradation of chloramphenicol using TiO_2 photocatalyst by response surface methodology. J Environ Sci 22(8):1281–1289

Environmental Effects of the Fukushima Daiichi Nuclear Power Plant Accident

Katsumi Hirose

Abstract A gigantic earthquake and resulting tsunami, occurred on March 11, 2011, causing the Fukushima Daiichi Nuclear Power Plant (FDNPP) accident. Large amounts of radionuclides emitted in the atmosphere (^{131}I: 160 PBq, ^{137}Cs: 15 PBq) and predominantly deposited on the Japanese Island and the western North Pacific, and a considerable part of them were dispersed in the Northern Hemisphere. After the accident, the Japanese Government started emergency monitoring of environmental radioactivity. Worldwide radioactivity measurements including CTBTO in the Northern Hemisphere had detected the Fukushima-derived radionuclides in the atmosphere. In this chapter, environmental radioactivity monitoring results in the early stage of the Fukushima accident are described.

Keywords Fukushima earthquake · Atmospheric radioactivity · Radioactivity contamination mapping · Ambient dose rate measurement

1 Introduction

On March 11, 2011, a big earthquake (The Great East Japan Earthquake), which was a moment magnitude of 9.0 with a hypocentral region about 500 km long and 200 km wide, attacked the northeast Honshu island, Japan. After the big earthquake, a gigantic tsunami hit east coast of the northeast Honshu island, whose wave height was more than 15 m. Just after the earthquake, fission reaction in all of reactors constructed in the northeast Honshu Island was safely stopped. The location of nuclear power plants and nuclear facilities in the northeast Honshu Island is shown in Fig. 1. However, gigantic tsunami in addition to the earthquake seriously damaged the electric system of the Fukushima Daiichi Nuclear Power Plant (FDNPP) (37.42 °N, 141.03° E). As a result, there was break down in the cooling system of nuclear reactors in the FDNPP. This was a cause of meltdowns of nuclear fuels and

K. Hirose (✉)
Department of Materials and Life Sciences, Faculty of Science and Technology, Sophia University, Tokyo, Japan
e-mail: hirose45037@mail2.accsnet.ne.jp

Geosphere Research Institute, Saitama University, Saitama, Japan

D. Sengupta (ed.), *Recent Trends in Modelling of Environmental Contaminants*,
DOI 10.1007/978-81-322-1783-1_7,

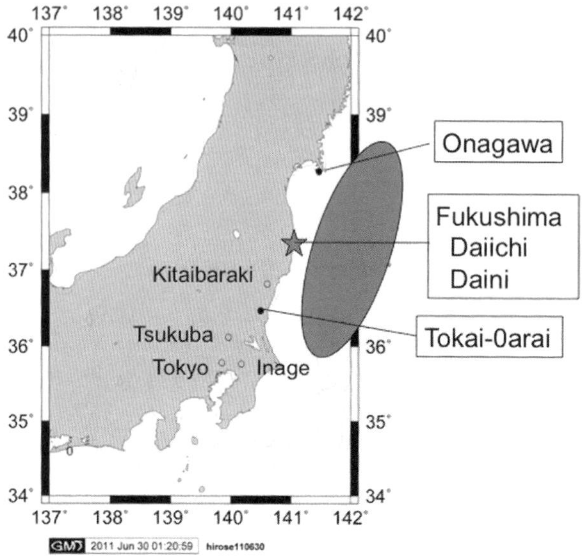

Fig. 1 Location of the FDNPP and several monitoring stations. The *red* area shows a rough area of the hypocentral region of "The Great East Japan Earthquake". The *closed circle* shows nuclear power plant. The *open circle* shows environmental radioactivity monitoring station discussed in this chapter

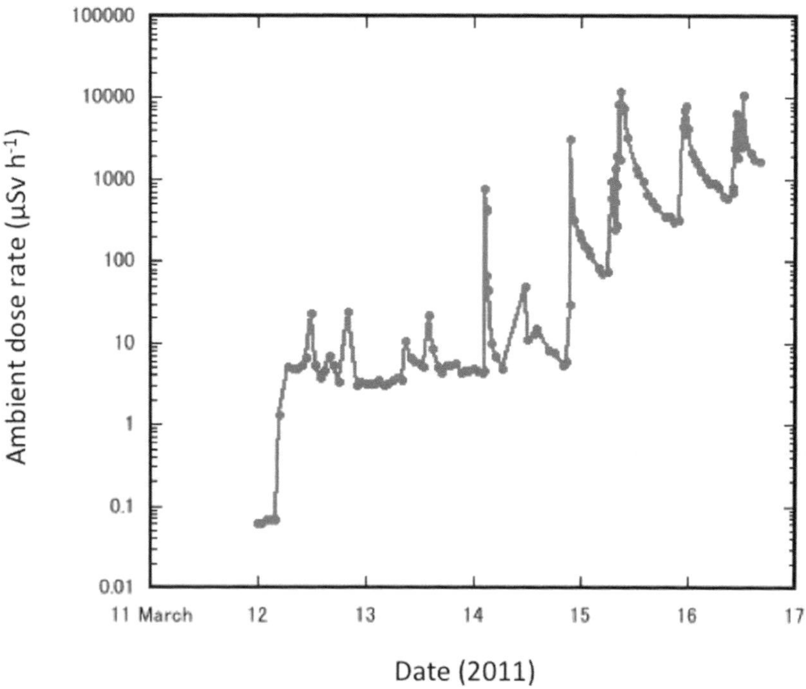

Fig. 2 Temporal change of ambient dose rate measured at the front gate of the FDNPP. This figure depicts data from TEPCO (2011). Peaks of the ambient dose rate almost corresponded to events such as explosions, vent, and others

resulting hydrogen explosions of No 1 Reactor and No 3 Reactor buildings in the FDNPP, which happened on March 12 and 14, respectively. In addition to explosions of nuclear reactors, an explosion occurred in the No 4 Reactor building on March 15, for which the sources of flammable gas are unknown because spent fuel in the No 4 reactor in spent fuel pool appeared to have been covered with water. An explosion occurred in the No 2 Reactor on March 15, suffered to lower suppression area (RJG 2011). On the early morning of March 12, ambient dose rate in the reactor site increased rapidly (Fig. 2), which means that atmospheric emission of radionuclides from the FDNPP was started as a result of meltdowns of nuclear fuel. On March 15–16, 2011, after the serious damage of the No. 2 Reactor (TEPCO 2011), the ambient dose rate at the FDNPP site showed pronounced peaks (Fig. 2), which implies occurrence of dominant radioactivity emissions. Especially, explosions of the FDNPP lead to serious radioactivity release in the atmosphere. However, exact emission history of the FDNPP–derived radionuclides in the atmosphere is the further study, although Chino et al. (2011) tried to reproduce the emission sequence of each radionuclide from the FDNPP.

Just after the accident, the Japanese Government and Prefectural Governments started emergency radioactivity monitoring to assess radiological and environmental effects of radioactivity emitted from the FDNPP accident (MEXT 2012). Preliminary airborne monitoring conducted by the US DOE (Lyons and Colton 2012) revealed that highly radioactive contamination area spread a narrow region from the FDNPP to about 60 km northwest region in the Fukushima Prefecture. The Ibaraki Prefectural Government has constructed a new monitoring system at sites (Kita-ibaraki and Takahagi) near Fukushima Prefecture. Temporal variations of the ambient dose rate observed at monitoring stations in Ibaraki Prefecture (Fig. 3) indicated that the radioactive plume first arrived in the Kanto Plain on March 15, 2011 (IPG 2011). On March 15–23, 2011, the effect of local fallout of the FDNPP-derived radionuclides spread over the Kanto Plain, which includes Tokyo and other major cities in the Kanto area. As did governmental monitoring, many research institutes and universities in Japan have measured FDNPP-derived radioactivity in environmental samples (Amano et al. 2012; Doi et al. 2013; Furuta et al. 2011; Kanai 2012). Furthermore, the Comprehensive Nuclear-Test-Ban Treaty (CTBTO) network (CTBTO, http://www.ctbto.org/press-centre/highlights/2011/fukushima-related-measurements-by-the-ctbto/) and regional monitoring stations in many countries have monitored the FDNPP-derived radioactivity in surface air and deposition.

In this chapter, monitoring results of atmospheric radioactivity derived from the FDNPP were summarized based on MEXT data (Chino et al. 2011) and published papers.

2 Total Atmospheric Release

The most important interest regarding nuclear reactor accidents is what kinds of radionuclides and what amount of each radionuclide was released in the environment. Japanese Government has conducted to estimate total atmospheric releases

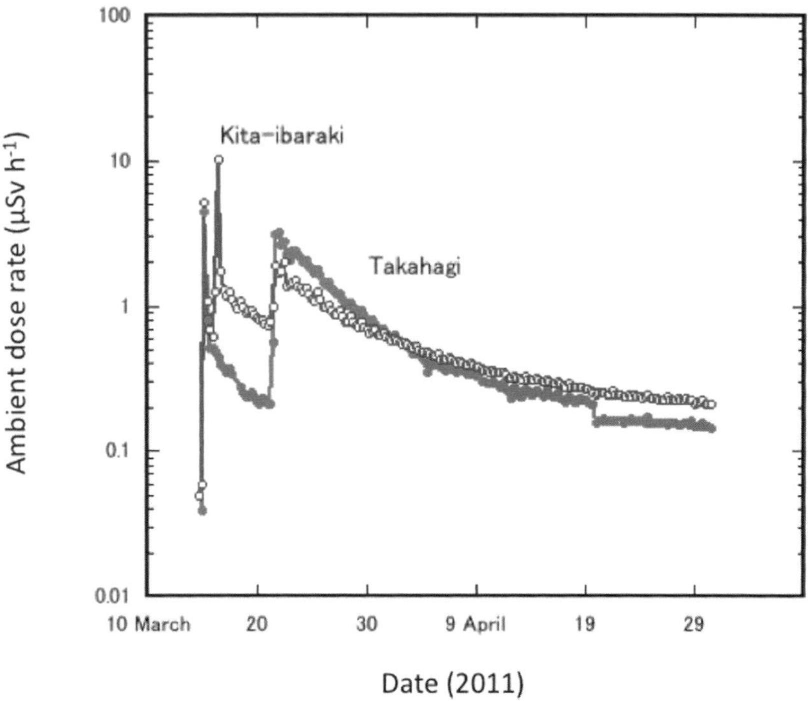

Fig. 3 Temporal variations of ambient dose rate observed in Ibaraki Prefecture. Data were cited by IPG (2011). Peaks on March 15 corresponded to the passage of radioactive plume. *Open circle*: Kita-ibaraki, *closed circle*: Takahagi

of radioactivity from the FDNPP based on the analysis of reactor status (RJG 2011); total atmospheric releases of ^{131}I and ^{137}Cs were 160 PBq and 15 PBq, respectively, which are about one order of magnitude less than the Chernobyl accident (IAEA 1986; UNSCEAR 2000). The total release of each radionuclide from the FDNPP is shown in Table 1. On the other hand, based on the inverse modeling method using monitoring data, Chino et al. (2011) estimated 150 PBq (=10^{15} Bq) of ^{131}I and 12 PBq of ^{137}Cs as the total atmospheric releases of radioactivity from the FDNPP. Stohl et al. (2011) estimated 16.7 EBq (=10^{18} Bq) of ^{133}Xe and 35.8 PBq of ^{137}Cs as the total atmospheric releases of radioactivity from the FDNPP, in which the ^{133}Xe emission is the largest radioactive noble gas release in the radioactivity emission history not associated with nuclear-weapons testing. There was no significant difference of the total release of ^{133}Xe between estimation methodologies, because all ^{133}Xe inventory, at cease of nuclear fission in three nuclear reactors of the FDNPP, was emitted in the atmosphere. On the other hand, there is rather large difference in total ^{137}Cs release between estimation methodologies. One of major causes is that monitoring data was limited in several stations of land area, although in the initial stage of the FDNPP accident the radioactive plume flowed toward ocean

Table 1 Radionuclides released from the FDNPP and total emission amount of each radionuclide. (RJG 2011)

Radionuclide	Half-life	Reactor I	Reactor II	Reactor III	Total release
		PBq	PBq	PBq	PBq
Xe-133	5.2 d	3,400	3,500	4,400	11,000
Cs-134	2.06 y	0.71	16	0.82	18
Cs-137	30.0 y	0.59	14	0.71	15
Sr-89	50.5 d	0.082	0.68	1.2	2
Sr-90	29.1 y	0.0061	0.048	0.085	0.14
Ba-140	12.7 d	0.13	1.1	1.9	3.2
Te-127m	109 d	0.25	0.77	0.069	1.1
Te-129m	33.6 d	0.72	2.4	0.21	3.3
Te-131m	30.0 h	2.2	2.3	0.45	5
Te-132	78.2 h	25	57	6.4	88
I-131	8.02 d	12	140	7	160
I-132	2.30 h	0.013		0.000037	0.013
I-133	20.8 h	12	26	4.2	42
1-135	6.57 h	2	0.074	0.19	2.3
Sb-127	3.85 d	1.7	4.2	0.45	6.4
Sb-129	4.3 h	0.14	5.6E −5	0.0023	0.14
		TBq	TBq	TBq	TBq
Ru-103	39.3 d	0.0025	0.0018	0.0032	0.0075
Ru-106	368 d	0.00074	0.00051	0.00089	0.0021
Zr-95	64.0 d	0.46	16	0.22	17
Ce-141	32.5 d	0.46	17	0.22	18
Ce-144	284 d	0.31	11	0.14	11
Np-239	2.36 d	3.7	71	1.4	76
Pu-238	87.7 y	0.00058	0.018	0.00025	0.019
Pu-239	24,065 y	0.000086	0.0031	0.00004	0.0032
Pu-240	6,537 y	0.000088	0.003	0.00004	0.0032
Pu-241	14.4 y	0.035	1.2	0.016	1.2
Y-91	58.5 d	0.31	2.7	0.44	3.4
Pr-143	13.6 d	0.36	3.2	0.52	4.1
Nd-147	11.0 d	0.15	1.3	0.22	1.6
Cm-242	162 d	0.011	0.077	0.014	0.1
Mo-99	66.0 h	0.0026	0.0012	0.0029	0.0067

side (Takemura et al. 2011). Aoyama et al. (2012) estimated 10.5±2.5 PBq as the total amount of atmospheric ^{137}Cs input in the North Pacific, based on radiocesium (^{137}Cs, ^{134}Cs) data in the North Pacific waters. In order to improve estimation of the total released radionuclides from the FDNPP, it is necessary to have information about more precious monitoring data, as did physical and chemical properties of released radionuclides.

3 Surface Air Monitoring

Radioactivity measurement in surface air is one of the most important issues in emergency environmental monitoring. The Japanese Government, including Prefectural Governments and Tokyo Electric Power Co. (TEPCO), started measurements of radioactivity in surface air using monitoring car as an emergency monitoring (TEPCO 2011; MEXT 2012). However, continuous sampling of air from both accidental reactor sites and areas from 20 to 60 km of the FDNPP has not been conducted. On the other hand, some research institutes (Amano et al. 2012; Doi et al. 2013; Furuta et al. 2011; Kanai 2012) and universities (Momoshima et al. 2012) in Japan have carried out continuous surface air sampling and have measured radioactivity in surface air samples. They had detected the FDNPP-derived radionuclides since late March 2011 in surface air samples. Air monitoring results reveal that dominant detected radionuclides originating from the FDNPP were radioactive noble gases (^{133}Xe, ^{131m}Xe, ^{133m}Xe, ^{135}Xe), radioiodine (^{132}I, ^{131}I), and radiocesium (^{134}Cs, ^{136}Cs, ^{137}Cs), although detected radionuclides in airborne particles were ^{133}I, ^{132}I, ^{131}I, ^{134}Cs, ^{136}Cs, ^{137}Cs, ^{132}Te, ^{129m}Te (^{129}Te), ^{99}Mo(^{99m}Tc), ^{140}Ba(^{140}La), and ^{110m}Ag (Amano et al. 2012; Furuta et al. 2011; Yonezawa and Yamamoto 2011). In contrast to the Chernobyl accident, significant amounts of refractory fission products such as ^{106}Ru, ^{103}Ru, and ^{95}Zr have not been detected in environmental samples for the FDNPP accident. Surface air measurements documented that radioactive plume first arrived in the Kanto Plain on March 15, 2011, which coincided with abrupt increase of the ambient dose rate (Fig. 3). After that, the FDNPP-derived radionuclides were transported in the northern hemisphere in late March and April 2011; ^{131}I and ^{137}Cs were detected in the Asia (Bolsunovsky and Dementyev 2011), North America (Bowyer et al. 2011; Zhang et al. 2011), Europe (Lozano et al. 2011; Pittauerová et al. 2011), and Arctic (Paatero et al. 2012).

3.1 Regional and Local Monitoring

Short-lived radioiodine (^{131}I, ^{132}I, and ^{133}I) within the FDNPP-derived radionuclides is an important radionuclide to monitor in early stage of nuclear reactor accidents because of higher health risk to radiation dose due to inhalation. However, there is limited information on short-live radioiodine concentrations in surface air including gaseous and particulate forms; in the sites near Fukushima, gaseous ^{131}I have been reported at Inage, Chiba (Amano et al. 2012) and Oarai, Ibaraki (Furuta et al. 2011). Although Kanai (2012) reported ^{131}I concentrations in aerosols at Tsukuba, there was no information on early stage data in March 2011 and gaseous ^{131}I. The gaseous ^{131}I is conventionally defined as radionuclides adsorbed onto activated carbon fiber filter.

The temporal variation of the total ^{131}I concentrations (total gas plus particulate) at Oarai (about 100 km south of the FDNPP) and Tsukuba (about 170 km of the FDNPP) is shown in Fig. 4. The first peak of ^{131}I, being the highest concentrations

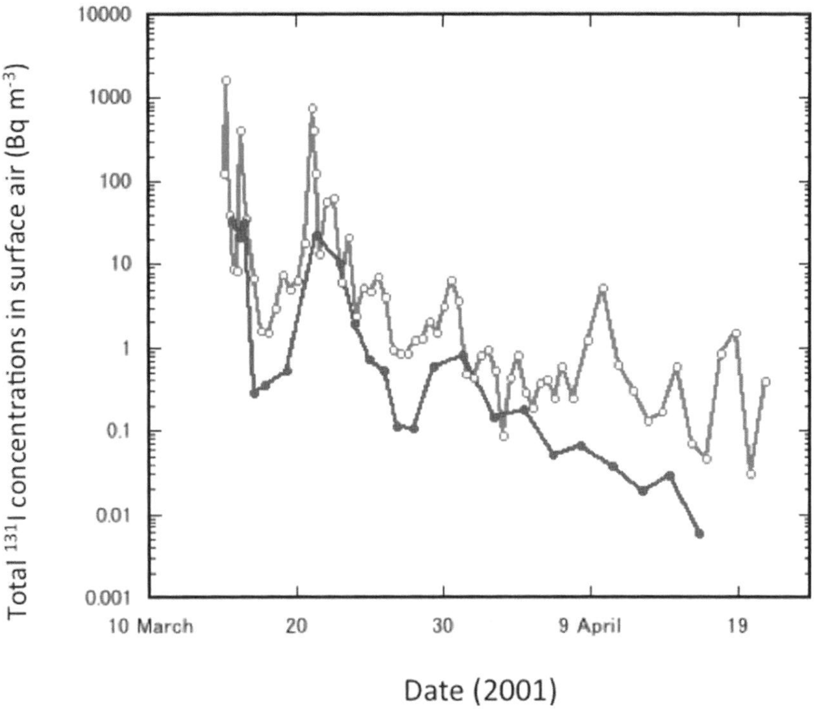

Fig. 4 Temporal variations of total ^{131}I concentrations in surface air observed at Oarai (*open circle*) and Tsukuba (*closed circle*). This figure depicts the data cited from Doi et al. (2012) and Furuta et al. (2011)

at Tsukuba (33 Bq m^{-3} (8 h mean value)) and Oarai (1.4×10^3 Bq m^{-3} (3 h mean value)) during the observation period, respectively, occurred on March 15–16, 2011, which coincided with maximum of the ambient dose rate of the corresponding site. On March 15–16, 2011, the high ^{131}I concentrations in surface air were observed in Setagaya (Tokyo) (about 240 km of the FDNPP) and Inage (about 220 km of the FDNPP), where the ^{131}I concentrations in surface air were 2.4×10^2 Bq m^{-3} (1 h mean value) and 33 Bq m^{-3} (27 h mean value) (Amano et al. 2012), respectively. The ^{131}I concentrations in surface air at four sites decreased with distance from the FDNPP. The second peak of total ^{131}I at Tsukuba (Doi et al. 2013) and Oarai (Furuta et al. 2011) appeared on March 20–22, which also coincided with the second peak of ambient dose rate at the corresponding sites. A similar peak of ^{131}I was observed in Setagaya and Inage on March 20–22. The total ^{131}I concentration in surface air at Tsukuba, which showed some peaks corresponding to periods of March 30 to April 1, April 17–19, and May 5–7, decreased as an apparent half life of 2.7 days (Doi et al. 2013), and it is difficult to detect ^{131}I in air samples of the Kanto Plain in June, 2011 (detection limit: 0.1 mBq m^{-3}).

In order to have better understanding of the atmospheric behaviors of ^{131}I, the temporal variation of gaseous ^{131}I/total ^{131}I ratio observed at Tsukuba was examined,

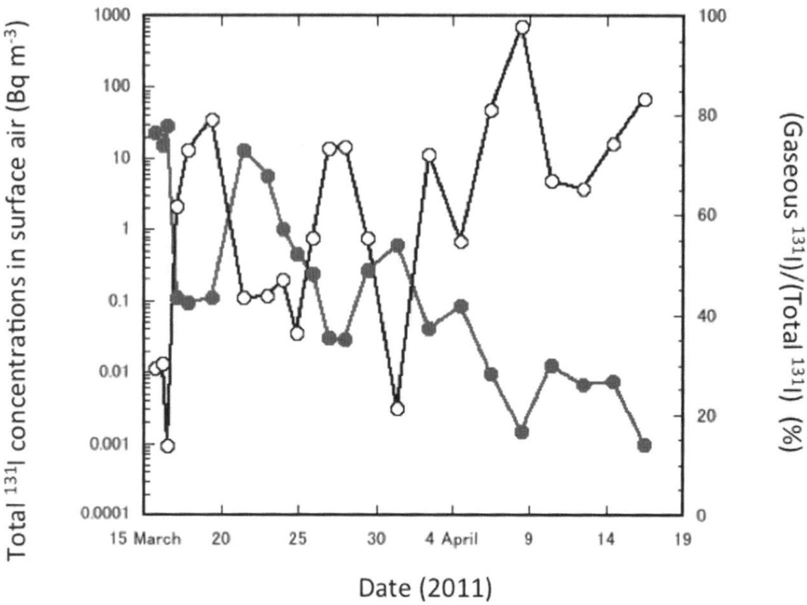

Fig. 5. Temporal variation of percentage of gaseous ^{131}I to total ^{131}I observed at Tsukuba. *Open circle* and *closed circle* show the percentage of gaseous ^{131}I to total ^{131}I and the concentration of total ^{131}I, respectively. This figure depicts the data cited from Doi et al. (2013)

in which the gaseous radioiodine (^{131}I, ^{133}I) was captured with the activated carbon fiber filter (Doi et al. 2013). The result is show in Fig. 5. A lower portion of the gaseous ^{131}I almost corresponded to a peak of the total ^{131}I concentration in surface air; the lowest percentage of the gaseous ^{131}I to the total ^{131}I (13 %) appeared in occurrence of the highest total ^{131}I concentration on March 16, 2011. The percentage of the gaseous ^{131}I to the total ^{131}I in each peak increased with time. The similar trend of increasing percentage of the gaseous ^{131}I was observed at Oarai (Furuta et al. 2011). These findings suggest that, as a possible process, the gaseous ^{131}I remained in the atmosphere longer than the particulate ^{131}I, and, as another possible process, the gaseous ^{131}I was preferentially emitted into atmosphere comparing with the particulate ^{131}I due to change of the radioactivity-released conditions in the accidental reactors such as decline of temperature. Other possible cause of increasing gaseous ^{131}I may be volatilization of deposited ^{131}I due to biological processes (Amachi et al. 2011; Muramatsu, and Yoshida 1995).

^{133}I was detected in surface air samples collected at Oarai, Tsukuba, and Inage as did ^{131}I, on March 15, 2011 (Amano et al. 2012; Doi et al. 2013; Furuta et al. 2011; Kanai 2012). The maximum concentrations of ^{133}I in surface air at Oarai, Tsukuba, and Inage on March 15, 2011 were 18, 2.6, and 1.9 Bq m^{-3}, respectively. The ^{133}I concentrations in surface air rapidly decreased and ^{133}I could not be detected in the end of March 2011. The temporal change of $^{133}I/^{131}I$ ratio in surface air at each site, which was 0.1 on March 15, 2011, decreased exponentially as shown

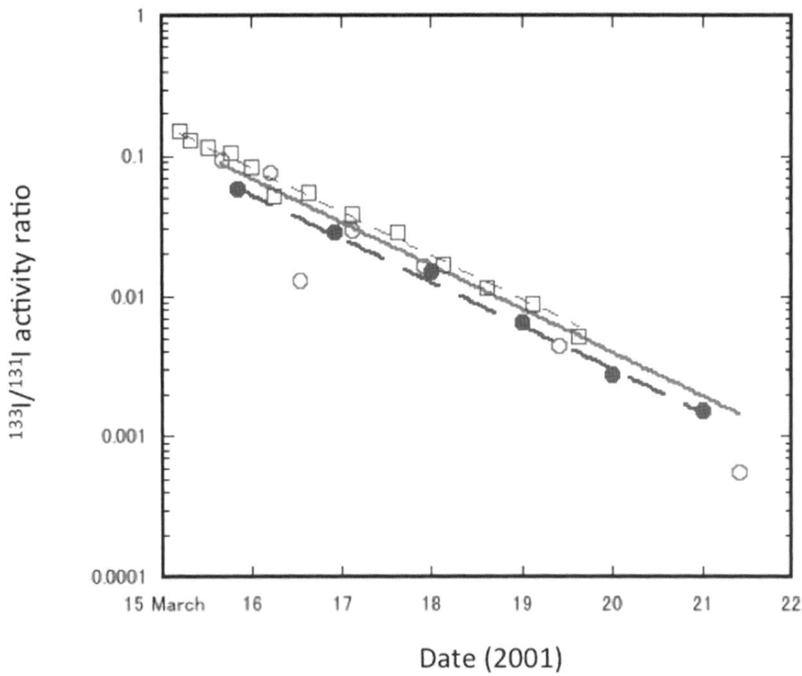

Fig. 6 Temporal variations of $^{133}I/^{131}I$ ratios in surface air observed at Tsukuba. *Open circle*: Tsukuba (Doi et al. 2013), *closed circle*: Inage (Amano et al. 2012), *open square*: Oarai (Furuta et al. 2011). The initial $^{133}I/^{131}I$ ratios on March 11 at Tsukuba, Inage, and Oarai were 1.5, 1.1, and 1.8, respectively

in Fig. 6. There was no geographical variation of temporal changes of the $^{133}I/^{131}I$ ratios observed at Tsukuba, Oarai, and Inage (Amano et al. 2012; Doi et al. 2013; Furuta et al. 2011; Kanai 2012), although lower $^{133}I/^{131}I$ ratios occurred at Tsukuba. A similar result has been obtained in US CTBTO network data (CTBTO, http://www.ctbto.org/press-centre/highlights/2011/fukushima-related-measurements-by-the-ctbto/), in which the temporal change of $^{133}I/^{131}I$ ratios measured in the US sites excellently correlated with the ORIGEN model (Biegalski et al. 2011). The temporal change of the $^{133}I/^{131}I$ ratio coincided with a decay curve of a decay constant ($\lambda_{133}-\lambda_{131}$). This suggests that the $^{133}I/^{131}I$ ratio at cease of nuclear fission in the reactors (March 11, 2011) was 1.4, which largely differed from that in estimated total release (0.26) as shown in Table 1 (RJG 2011).

The ^{134}Cs, ^{136}Cs (half-life: 13.16 d), and ^{137}Cs were detected in surface air at Oarai, Tsukuba, and Inage (Amano et al. 2012; Doi et al. 2013; Furuta et al. 2011; Kanai 2012). The temporal variation of the ^{137}Cs concentrations in surface air at Oarai and Tsukuba is shown in Fig. 7. The first peak of surface ^{137}Cs in the Kanto Plain occurred on March 15–16, 2011, which coincided with that of the total ^{131}I concentration and the ambient dose rate. On March 15, 2011, high radiocesium concentrations in surface air were observed at Oarai, Tsukuba, and Inage, where the

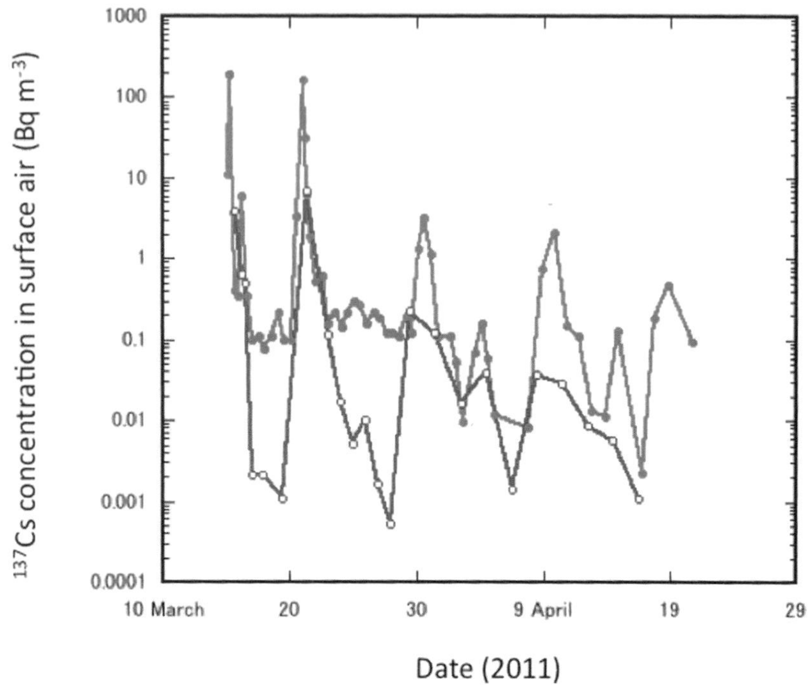

Fig. 7 Temporal variations of ^{137}Cs concentrations in surface air observed at Oarai and Tsukuba. This figure depicts the data from Doi et al. (2013) and Furuta et al. (2011)

^{137}Cs concentrations in surface air were 1.9×10^2 Bq m^{-3} (3 h mean value) (Furuta et al. 2011), 3.8 Bq m^{-3} (3 h mean value) (Doi et al. 2013), and 0.87 Bq m^{-3} (27 h mean value) (Amano et al. 2012), respectively. The ^{137}Cs concentrations in surface air at three sites decreased with increasing distance from the FDNPP, although a simple comparison between data cannot be carried out because of different air sampling time span. On March 18–20, the ^{137}Cs concentrations in surface air rapidly decreased to about three orders of magnitude lower than previous value in contrast of ^{131}I. The second peak of the surface ^{137}Cs in the Kanto Plain appeared on March 20–22, which corresponded to occurrence of the highest concentration (4.6 Bq m^{-3} (48 h mean value)) within the observation period. The observed maximum concentrations of ^{137}Cs in surface air at Oarai, Tsukuba, and Inage were 160 Bq m^{-3} (9 h mean value), 4.6 Bq m^{-3} (48 h mean value), and 6.1 Bq m^{-3} (26 h mean value), respectively. The ^{137}Cs concentration in surface air in the Kanto Plain, which showed some peaks in April—September, decreased, and ^{137}Cs can be still detected in air samples in early September, 2011 (detection limit: 0.05 mBq m^{-3}). It is noteworthy that radiocesium in surface air of Fukushima City has still been observed in the early 2012, whose concentrations were 0.1 to several mBq m^{-3} (MEXT 2012).

Isotope signature is an important tool to have better understanding of the environmental behaviors of the FDNPP-derived radionuclides. As it is well known that

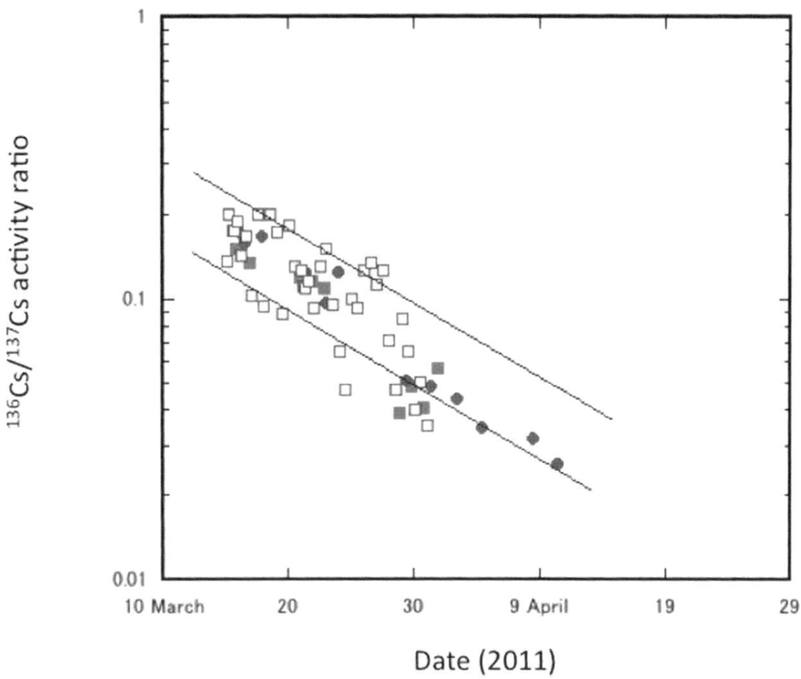

Fig. 8 Temporal variations of ^{136}Cs/^{137}Cs ratio in surface air. *Closed circle*: Tsukuba (Doi et al. 2013), *closed square*: Inage (Amano et al. 2012), *open square*: Oarai (Furuta et al. 2011). Solid lines show the radioactive decay curve of ^{136}Cs

^{134}Cs is not a direct fission product, ^{134}Cs is produced by neutron activation of ^{133}Cs, which is the decay product of ^{133}Xe present in nuclear reactors. ^{134}Cs/^{137}Cs ratios increase with burnup time of nuclear fuel. Mutual relationships between radiocesium in surface air were examined (Amano et al. 2012; Doi et al. 2013; Biegalski et al. 2011). The ^{134}Cs/^{137}Cs activity ratio in surface air at all sites including Japanese and the US stations, ranged from 0.9–1.2 as an average of 1.0±0.1, was fairly stable over the time of observation, which coincides with that of deposition (Amano et al. 2012; Hirose 2012). It must be noted that the ^{134}Cs/^{137}Cs activity ratio in the FDNPP-derived radionuclides is higher than that in the Chernobyl fallout (^{134}Cs/^{137}Cs ratio: 0.5) (IAEA 1986; UNSCEAR 2000). Temporal change of the ^{136}Cs/^{137}Cs activity ratio is shown in Fig. 8. In contrast to the ^{134}Cs/^{137}Cs ratios, there was rather large variability of the ^{136}Cs/^{137}Cs ratio in surface air. The similar different nature of the ^{136}Cs/^{137}Cs ratio from the ^{134}Cs/^{137}Cs ratio was observed in the US monitoring sites (Biegalski et al. 2011). Although the ^{136}Cs/^{137}Cs ratio decreased exponentially, the ^{136}Cs/^{137}Cs ratios at three sites in Japan seemed to not fit in the simple decay curve of ^{136}Cs; the ^{136}Cs/^{137}Cs ratios in late March-April 2011 were slightly lower than that in the period of the first and second peaks (March 15–22). The ^{136}Cs/^{137}Cs ratio was calculated to extrapolate at cease of reactors (March 11, 2011) from a decay curve of ^{136}Cs. The initial ^{136}Cs/^{137}Cs ratio ranged from 0.09 and 0.27. These findings

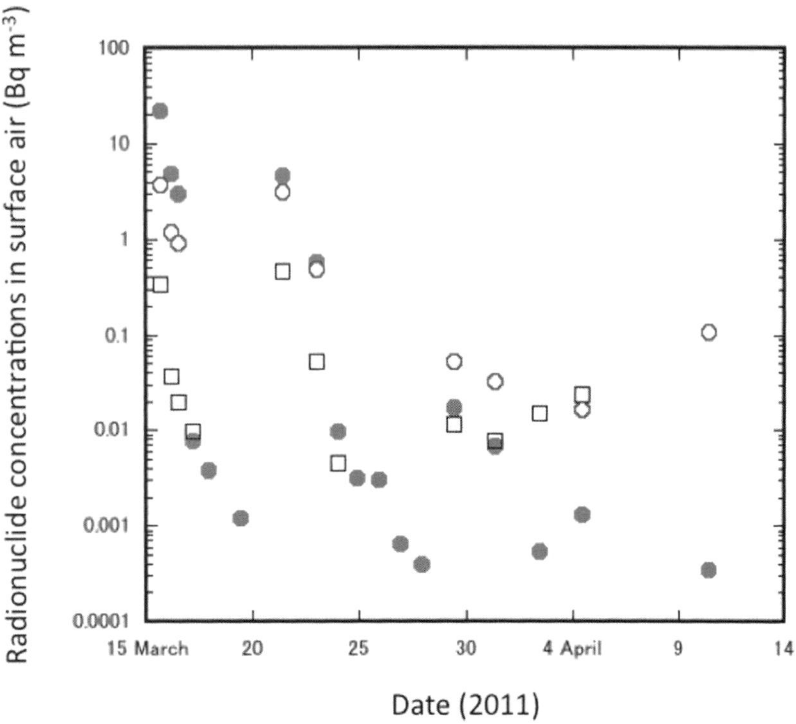

Fig. 9 Temporal variations of ^{132}Te (*closed circle*), ^{129m}Te (*open circle*), and ^{99m}Tc (*open square*) in surface air at Tsukuba. This figure depicts data from Doi et al. (2013)

suggest that there is no isotopic fractionation between ^{134}Cs and ^{137}Cs, whereas the ^{136}Cs/^{137}Cs ratio may change between major emission and following minor emission, although its cause is unknown. In this connection, Biegalski et al. (2012) mentioned that excess ^{137}Cs could be present in the samples due to the release of a damaged fuel in the Fukushima Daiichi spent nuclear fuel pools, taking into account the fact that temporal change of the ^{136}Cs/^{137}Cs ratios appear to be generally below the ORIGEN model (ORNL 2004). However, there is no report of the radioactivity release from the Fukushima Daiichi spent nuclear fuel pools. It is noteworthy that, determination of ^{136}Cs is rather complicated in the initial stage because of interference of short-lived radionuclides.

As other radionuclides, ^{132}Te (^{132}I), ^{129m}Te (^{129}Te), and ^{99}Mo(^{99m}Tc) were detected in the surface air samples at Tsukuba (Doi et al. 2013; Kanai 2012). Temporal variations of ^{132}Te, ^{129m}Te, and ^{99}Mo(^{99m}Tc) in surface air at Tsukuba are shown in Fig. 9. ^{132}Te, ^{129m}Te, and ^{99}Mo(^{99m}Tc) were detected on March 15, 2011, which coincided with the first detection of radioiodine and radiocesium. The maximum concentrations of ^{132}Te and ^{129m}Te, which occurred on March 15, were 23 and 3.8 Bq m^{-3}, respectively, which corresponded to the maximum of radioiodine, whereas the maximum ^{99}Mo(^{99m}Tc) concentration (0.46 Bq m^{-3}) occurred on March 20–22, corresponding to the maximum of radiocesium. The surface air concentrations of ^{132}Te, ^{129m}Te, and ^{99}Mo(^{99m}Tc) exhibited three major peaks, which corresponded with that of ^{131}I and ^{137}Cs, and decreased to less than detection limit after mid-April 2011.

In order to elucidate difference of atmospheric behaviors between radionuclides, mutual relationships between ^{132}Te, ^{129m}Te, and ^{99}Mo(^{99m}Tc) were examined, in which activity concentrations of short-lived radionuclides were corrected radioactive decay on March 11, 2011 (Doi et al. 2013). The decay-corrected ^{132}Te concentration in the surface air were correlated with that of the decay corrected ^{129m}Te, in which the ^{129m}Te/^{132}Te ratio was 0.086 ± 0.005. It is noteworthy that RJG (2011) documented the ^{129m}Te/^{132}Te ratio of 0.04 (Table 1). The ratios of the decay-corrected ^{132}Te, ^{129m}Te, and ^{99m}Tc concentrations to the ^{137}Cs concentration in the surface air were calculated. The decay-corrected ^{132}Te/^{137}Cs and ^{129m}Te/^{137}Cs ratios were in the ranges of 4–65 and 0.3–5.4, respectively. Peaks of the decay-corrected ^{132}Te/^{137}Cs and ^{129m}Te/^{137}Cs ratios appeared on March 22–23 after occurrence of the maximum concentrations of ^{137}Cs on March 20–22. There was rather large variation of ^{132}Te/^{137}Cs and ^{129m}Te/^{137}Cs ratios, suggesting that atmospheric behaviors of radiotellurium differed from that of radiocesium. The decay-corrected ^{99}Mo(^{99m}Tc)/^{137}Cs, ranging from 0.16–22, showed larger variation than that of the decay-corrected ^{132}Te/^{137}Cs and ^{129m}Te/^{137}Cs. The decay-corrected ^{99}Mo(^{99m}Tc)/^{137}Cs increased with time. These findings suggest that the atmospheric behavior of ^{99}Mo(^{99m}Tc) was significantly different from that of ^{137}Cs and ^{132}Te.

Atmospheric behaviors of the FDNPP-derived radionuclides depend on physical and chemical properties of radionuclide-bearing particles. Particle size distributions of the FDNPP-derived radionuclide-bearing particles collected at Tsukuba were determined during the periods of April 4–11 and April 14–21, 2011 (Doi et al. 2013) and April 29–May 12 and May 12–26, 2011 (Kaneyasu et al. 2012). The activity median aerodynamic diameters (AMAD) of ^{131}I bearing particles were calculated to be 0.7 and 0.7 μm in April 4–11, 2011 and in April 14–21, 2011, respectively, whereas the AMADs of ^{134}Cs and ^{137}Cs bearing particles were 1.8 and 1.5 μm in April 4–11, 2011, and 1.0 and 1.0 μm in April 14–21, 2011, respectively. In May, measurements of particle size distributions of radiocesium bearing particles (Kaneyasu et al. 2012) revealed that ^{137}Cs attached on sub-micrometer particles as did ^{134}Cs, are typically sulfate particles. These findings suggest that the particle size of the ^{131}I bearing particles differed from those of radiocesium, which implies that the dispersion and deposition behaviors of ^{131}I differed from that of ^{134}Cs and ^{137}Cs and that the particle size of the radio-cesium bearing particles changed with time. The result reveals that the FDNPP-derived ^{134}Cs and ^{137}Cs bearing particles observed in April 2011 were larger than that of the Chernobyl ^{137}Cs, which existed as sub-micrometer particles observed in Japan (Aoyama et al. 1992). These findings suggest that the FDNPP-derived radiocesium observed in April was preferentially removed from atmosphere by dry and wet deposition processes, and that, as a result, the FDNPP-derived radiocesium observed in May, existing as sub-micrometer particles, remained in the atmosphere (Kaneyasu et al. 2012). Paatero et al. (2012) estimated that a significant part of the FDNPP-derived radioactivity

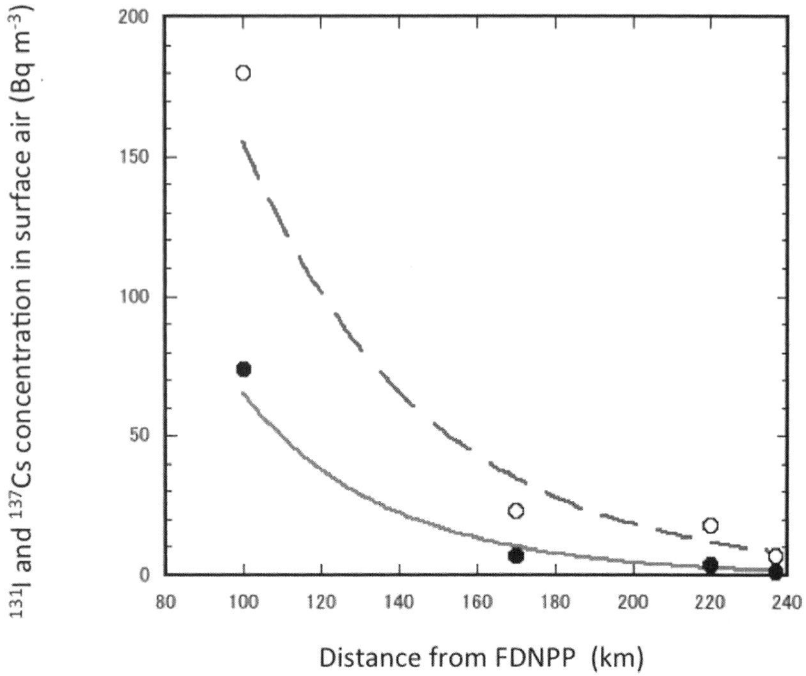

Fig. 10 Relationship between ^{131}I and ^{137}Cs surface air concentrations and distance from FDNPP. The 48 h mean values of ^{131}I and ^{137}Cs surface air concentrations during the period of March 20–21 were used. *Open circle*: ^{131}I, *closed circle*: ^{137}Cs. Used data were Oarai (Furuta et al. 2011), Tsukuba (Doi et al. 2013), Inage (Amano et al. 2012) and Setagaya (Tokyo: 237 km from the FDNPP http://www.sangyo-rodo.metro.tokyo,jp/whats-new/measurement.html)

is in hot particles from autoradiogram of a filter sample from 1–4 April 2011 at Mt. Zeppelin, Ny-Ålesund, Svalbard. It must be noted that there is no information about particle size of radionuclide-bearing particles in major radioactive plume occurring in the late March.

The surface air concentrations of the FDNPP-derived radionuclide decreased with distance from the FDNPP. The model simulation (Morino et al. 2011) revealed that the radioactive plume originating from the FDNPP flowed southwestward, in March 20–21, according to northeast wind. In order to elucidate the difference of the atmospheric behavior between ^{131}I and ^{137}Cs, the surface air concentrations of ^{131}I and ^{137}Cs at four sites corresponding to downstream of the FDNPP were potted as a function of distance form the FDNPP as shown in Fig. 10. The result reveals that the surface air concentrations of ^{131}I and ^{137}Cs decreased exponentially with distance shown in Fig. 10. The Apparent Half Decrease Distances (AHDD) of ^{131}I and ^{137}Cs can be calculated to be 32 and 26 km from the fitting curves, respectively. The AHDD of ^{131}I is longer than that of ^{137}Cs, which implies that ^{137}Cs emitted from the FDNPP is easily removed from atmosphere comparing with ^{131}I.

3.2 Global Monitoring

The Comprehensive Nuclear-Test-Ban Treaty (CTBT) (CTBTO, http://www.ctbto.org/press-centre/highlights/2011/fukushima-related-measurements-by-the-ctbto/) was adopted by the United Nations General Assembly on September 10, 1996 (CTBTO 2011). Within the CTBT, the International Monitoring System (IMS) was defined to monitor the world for nuclear explosions. The IMS comprises of four primary monitoring technologies: radionuclide, seismic, hydroacoustic, and infrasound. The FDNPP-derived radionuclides were first determined by global CTBTO network (CTBTO, http://www.ctbto.org/press-centre/highlights/2011/fukushima-related-measurements-by-the-ctbto/). The FDNPP-derived radionuclides were transported in the northern hemisphere in late March and April 2011; ^{131}I and ^{137}Cs were detected in the Asia (Bolsunovsky and Dementyev 2011; Kim et al. 2012), North America (Bowyer et al. 2011; Zhang et al. 2011), Europe (Lozano et al. 2011; Pittauerová et al. 2011), and Arctic (Paatero et al. 2012). The FDNPP-derived radionuclides had been detected around the northern Hemisphere in about two weeks and also in the southern Hemisphere about one month later.

In North America, on March 16, the FDNPP-derived gaseous radionuclides (^{133}Xe, ^{133m}Xe, ^{131m}Xe) were first detected by the Pacific Northwest Laboratory (46° 16′47″N, 119° 16′53″W), located more than 7,000 km from the FDNPP (Bowyer et al. 2011). The maximum ^{133}Xe concentration (41 Bq m^{-3}) occurred on March 20. The second-highest values of radioactive xenon were observed on March 25. The radioactive cloud containing the FDNPP-derived radionuclides had crossed the Pacific Ocean and reached Canada (Sidney: 49.3° N, 123.2° W) on March 17, 6 days after the first emission from the accidental reactor (Zhang et al. 2011). In surface air, several volatile radionuclides, such as ^{131}I, ^{132}I, ^{132}Te, ^{134}Cs, ^{136}Cs, and ^{137}Cs, have been detected. On March 19, the surface air concentrations of ^{131}I, ^{132}I, and ^{132}Te reached their maximum values. Then on March 24, 5 days later the second-highest values of ^{131}I, ^{132}I, and ^{132}Te were observed as did maximum concentrations of ^{134}Cs, ^{136}Cs, and ^{137}Cs, which implies that the second high radioactive plume had reached Sidney BC. The temporal waves of radioiodine and radiocesium were larger than radioactive xenon.

In Europe, first the FDNPP-derived radionuclides were observed by the CTBTO station in Reykjavik, Iceland, on 20 March 2011 (CTBTO, http://www.ctbto.org/press-centre/highlights/2011/fukushima-related-measurements-by-the-ctbto/). From late March to early May, many air monitoring stations across Europe including the Arctic have documented traces of the FDNPP-derived radionuclides above the usual background levels, which were supported by resuspension of deposited radionuclides from the atmospheric nuclear testing and the Chernobyl accident for ^{137}Cs and emissions from radiotherapeutic facilities for ^{131}I. The elevated levels of radionuclides derived from the Fukushima NPP were detected at the surface air sampling stations in Portugal (Carvalho et al. 2012), Spain (Lozano et al. 2011), England (Beresford et al. 2012), Monaco (Pham et al. 2012), Italy (Clemenza et al. 2012), Germany (Pittauerová et al. 2011), Greece (Manolopoulou et al. 2011), and

Lithuania (Lujanienė et al. 2012). In Arctic, particle-bound ^{131}I was first detected at Mt. Zeppelin (78° 58′N, 11° 53′E) from a sample collected between 25 March 2011 10:11 UTC and 28 March 2011 13:04UTC (Paatero et al. 2012). The detected radionuclides were ^{133}Xe, ^{131}I, ^{134}Cs, ^{136}Cs, ^{137}Cs, and ^{132}Te. The model simulations (Takemura et al. 2011; Manolopoulou et al. 2011) revealed complicated long-range air mass transport from the FDNPP (Japan) across the North Pacific, the North America, and the North Atlantic Ocean to Europe. The most comprehensive radionuclide data over the Europe has been compiled by Masson et al. (2011). Trajectory analysis and comparison with ^{7}Be, which is a cosmogenic radionuclide produced in the upper troposphere and stratosphere, suggest that the FDNPP-derived radionuclides observed in the Europe were transported via upper troposphere (Lujanienė et al. 2012).

In Asia, the elevated levels of radionuclides derived from the FDNPP were detected at the surface air sampling stations in South Korea (Kim et al. 2012), Taiwan (Huh et al. 2012), and Vietnam (Long et al. 2012). For radioactivity monitoring stations in the Asia, the FDNPP-derived radionuclides (^{131}I, ^{134}Cs, and ^{137}Cs) were initially detected in the surface air on March 27–30. The highest concentrations of ^{131}I, ^{134}Cs, and ^{137}Cs appeared in April 6–7. The model simulation (Huh et al. 2012) revealed that the FDNPP-derived radioactive plume was predominantly transported toward the southwest under phases of northern-easterly winds in April 6–7. As a result, the highest radionuclide concentrations occurred in the southwest Japan (Kyushu) (Momoshima et al. 2012), the southern region of Korea, Taiwan, and Vietnam.

4 Radioactivity Deposition

4.1 Daily Deposition at Local and Regional Stations

Measurements of radionuclides in daily deposition samples, which include wet and dry depositions, started on March 18 at the Japanese Government monitoring stations (MEXT 2012). Daily radioactive deposition rates in 44 stations in Japan were recorded until the end of 2011. High radioactive deposition derived from the FDNPP accident occurred in wide area of the Kanto Plain and South Tohoku area from March 21–23, 2011. On the other hand, in the area adjacent to the FDNPP, high radioactive deposition occurred on March 15–16, although there was no daily radioactivity deposition record during the period of March 2011.

The temporal changes of the daily ^{131}I and ^{137}Cs deposition rates during the period of March to April are shown in Fig. 11a, b. Earliest daily deposition sampling in the Kanto Plain was carried out in Japan Chemical Analysis Center (Inage, Chiba Prefecture) (Amano et al. 2012). The first deposition of the FDNPP-derived radionuclides appeared on March 15, which coincided with the first arrival of the radioactive plume. The first peaks of ^{131}I and ^{137}Cs depositions, which were 2,400

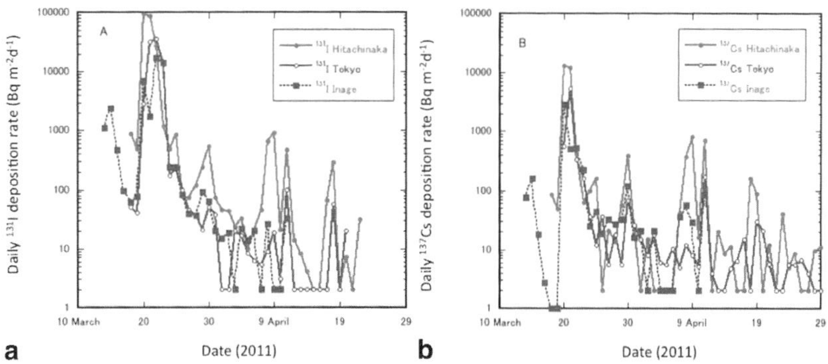

Fig. 11 Temporal changes of daily ^{131}I (**a**) and ^{137}Cs (**b**) deposition observed in the Kanto Plain. This figure depicts data from MEXT (2012). The value described as ND (not detected) is tentatively shown at 2 Bq m^{-2} in this figure

and 76 Bq m^{-2}d^{-1}, respectively, were observed at Inage on March 16. It must be noted that at the emergency stage of the accident, Government Monitoring documented only ^{131}I and ^{137}Cs as dominant radionuclides in deposition samples from the FDNPP accident by MEXT (2012). On March 20, the level of ^{131}I, a dominant radionuclide, at Hitachinaka (36.40° N, 140.54° E), which is located about 120 km south of the FDNPP, was 91 kBq m^{-2}d^{-1} as a maximum value, and ^{137}Cs deposition at Hitachinaka was 13 kBq m^{-2}d^{-1}. High radionuclide deposition rates (^{131}I: 58 kBq m^{-2}d^{-1}, ^{137}Cs: 4.3 kBq m^{-2}d^{-1}) were observed at Yamagata (38.25° N, 140.33° E), which is about 110 km northwest of the FDNPP. On March 21, high deposition rates were observed in all of the monitoring stations of the Kanto Plain. In Inage, the highest daily deposition rates of ^{131}I and ^{137}Cs were measured on March 23 (^{131}I: 17 kBq m^{-2}d^{-1}, ^{137}Cs: 2.9 kBq m^{-2}d^{-1}) (Amano et al. 2012). The high radioactive deposition rates accompanied with rainfall continued until March 23. On March 24, the radioactive deposition rates decreased dramatically due to fine weather with coverage of high-pressure system. Most of the FDNPP-derived radionuclides, which were transported from the FDNPP by northeast wind, were deposited on land surface by rainfall during March 20–23. In April 2011, the monitoring stations in most of the Japan main islands had observed the FDNPP-derived radionuclides in the daily radioactive deposition samples. On May 2011, however, the daily deposition rate of the FDNPP-derived radionuclides in most of the Japanese monitoring stations decreased below detection limit except Fukushima City (37.75° N, 140.47° E), which is located about 60 km northwest of the FDNPP. In Fukushima City, in which radioactivity measurement in daily deposition samples started on April 1, 2011, ^{137}Cs was still detected in the daily deposition samples in the end of 2011 (Fig. 12).

The ^{131}I/^{137}Cs activity ratio in the daily deposition samples, ranged from 0.3–230, varied temporally and spatially. The ^{131}I/^{137}Cs activity ratio in total atmospheric release was estimated to be 11 (NSC 2011). Compared with the ^{131}I/^{137}Cs activity ratio at the initial release of radionuclides from the FDNPP, radioactive decay of ^{131}I was

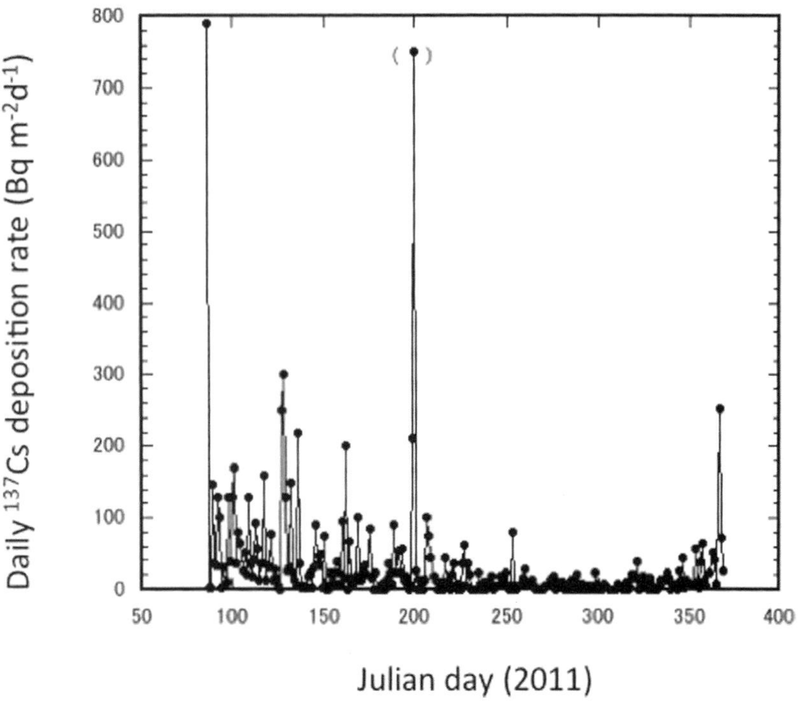

Fig. 12 Temporal changes of daily ^{137}Cs deposition observed at Fukushima City. This figure depicts data from MEXT (2012). The high value in July 2011 (shown with a parenthesis) contained large uncertainty due to measurement problems

corrected on March 11 when shutdown of nuclear reactors was conducted. The decay-corrected $^{131}I/^{137}Cs$ activity ratios ranged from 3.5–550, with a median value of 15. High decay-corrected $^{131}I/^{137}Cs$ activity ratios occurred on March 22–23, which corresponded to time just after occurrence of maximum deposition rates (March 20–21). Higher $^{131}I/^{137}Cs$ activity ratios appeared at the inland site (Utsunomiya: 36.60° N, 139.94° E), which means that the depositional behavior between ^{131}I and ^{137}Cs largely differed from each other. The $^{134}Cs/^{137}Cs$ activity ratios in the daily deposition samples were 1.0 (Amano et al. 2012).

4.2 Monthly Deposition at Local and Regional Stations

The Monitoring Centers in Local Governments have determined the FDNPP-derived radionuclides in monthly deposition samples. Iodine − 131, ^{134}Cs, and ^{137}Cs were measured as dominant radionuclides in the monthly deposition samples in all monitoring stations of Japan, whereas ^{129m}Te, ^{129}Te, ^{136}Cs, ^{110m}Ag, ^{95}Zr, ^{95}Nb (half-life: 35 d), ^{140}Ba, and ^{140}La (half-life: 1.68 days) were detected as minor radionuclides in the monthly deposition samples at the monitoring stations within 300 km of the FDNPP. For the monthly deposition, long-lived radionuclides (^{134}Cs, ^{137}Cs, and ^{110m}Ag) are meaningful because the monthly deposition rates of the short-lived radionuclides are not well defined. Activity ratios of ^{134}Cs/^{137}Cs observed on March, ranged 1.0–1.1 as an average of 1.03, were constant in case when the monthly ^{137}Cs deposition was exceed 100 Bq m^{-2}. The ^{134}Cs/^{137}Cs activity ratios on April and May at MEXT sites were from 0.99–1.09 with an average of 1.03 and from 0.93 to 1.04 with an average of 0.99, respectively (Hirose 2011). The result suggests that there were no spatial and temporal variations of the ^{134}Cs/^{137}Cs activity ratios in the monthly deposition samples during the period from March to May 2011. Activity ratios of ^{110m}Ag/^{137}Cs observed on March ranged 0.0009–0.006 as an average of 0.0019 (Hirose 2011).

The spatial distributions of the monthly ^{137}Cs deposition in March, April, and May are depicted in Fig. 13a, b, and c, respectively. The highest monthly ^{137}Cs deposition (3,340 kBq m^{-2}) was observed in March at Futaba (Fukushima Prefecture) about 5 km of the FDNPP. The monthly ^{137}Cs depositions at the stations within 300 km of the FDNPP except Kofu (35.65° N, 138.57° E; inland site) and the Japan Sea side sites (Niigata: 37.91° N, 139.04° E and Akita: 39.72° N, 140.10° E) were in the range from 1.1 kBq m^{-2} to 17 kBq m^{-2}, which are greater than the maximum monthly ^{137}Cs deposition (0.55 kBq m^{-2}) derived from the 1961–62 large-scale atmospheric nuclear testing observed at Koenji (Tokyo) in 1963 (Hirose et al. 2008; Igarashi et al. 2005). The results reveal that the high ^{137}Cs-deposited areas, comparable to the cumulative amount of the ^{137}Cs deposition at Tokyo until mid-1960 (about 7 kBq m^{-2}), appeared within a regional band from 100 to 300 km of the FDNPP. The spatial distribution of the monthly ^{137}Cs deposition on March 2011 revealed that the major deposition of the FDNPP-derived radionuclides occurred in the North Pacific coast and inland area of the east Honshu Island, whereas there was less contribution of the FDNPP-derived radionuclides in the Japan Sea side sites of the east Honshu Island. These findings suggest that the transport of the radioactive plume is strongly affected by land topography, and that most of the FDNPP-derived radionuclides might be injected in the boundary layer (about 1,000 m). Model simulation (Morino et al. 2011) revealed that land topography controls the transport of the FDNPP-derived radioactive plume.

The higher monthly ^{137}Cs deposition (0.17 Bq m^{-2}), which was one order of magnitude higher than pre-Fukushima levels (Igarashi et al. 2005), was observed on March at Fukuoka (33.51° N, 130.50° E) and Uruma (26.31° N, 127.90° E, Okinawa), which are located about 1,050 and 1,750 km southwest of the FDNPP, respectively. Detection of ^{131}I and ^{134}Cs in the same sample revealed that the FDNPP-derived radionuclides were transported to Fukuoka and Uruma in late March. The highest monthly deposition rates of radiocesium occurred at Fukuoka and Uruma in April 2011; the monthly ^{137}Cs depositions at Fukuoka and Uruma were 0.5 and 3.7 Bq m^{-2}. It must be noted that the ^{137}Cs deposition at Uruma is higher than that at Fukuoka. Model simulation (Takemura et al. 2011) suggested that the FDNPP-derived radioactivity plume spread far eastern Siberia on March 24. Another model simulation (Masson et al. 2011) revealed that the FDNPP-derived radioactive plume

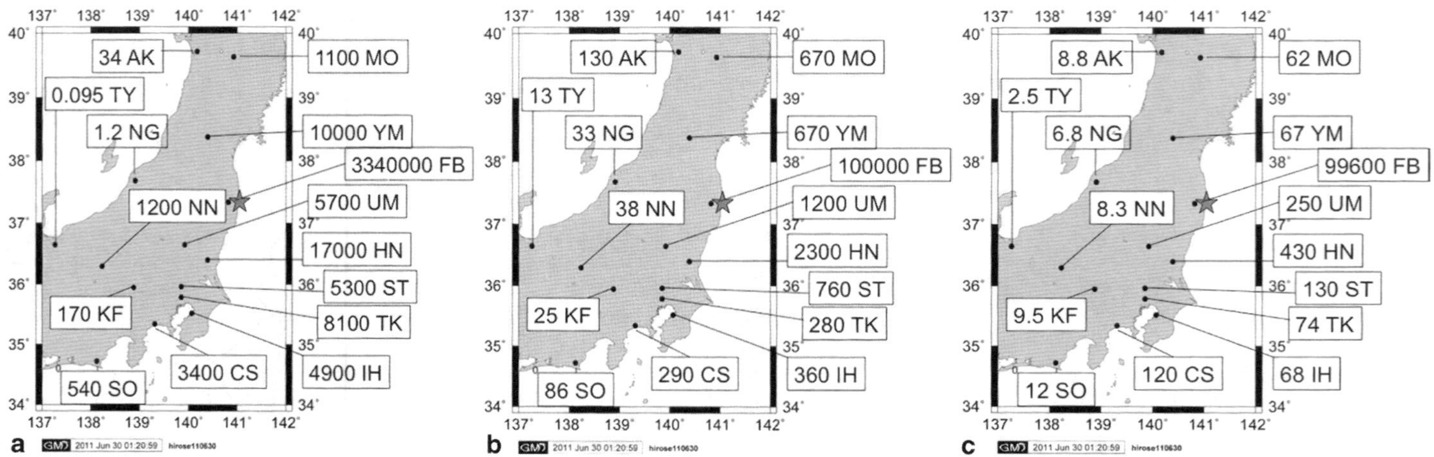

Fig. 13 Spatial distributions of monthly ^{137}Cs deposition observed in Japan. (Unit: Bq m^{-2}) **a** March, **b** April, **c** May. *AK* Akita, *MO* Morioka, *YM* Yamagata, *FB* Futaba, *NG* Niigata, *UM* Utsunomiya, *HN* Hitachinaka, *ST* Saitama, *TK* Tokyo, *IH* Ichihara, *CS* Chigasaki, *SO* Shizuoka, *KF* Kofu, *NN* Nagano, *TY* Toyama. Data were cited from MEXT (2012)

was predominantly transported toward the southwest under phases of northern-easterly winds on April 6–7. It is likely that, the FDNPP-derived radionuclides observed at Fukuoka and Uruma were transported via the east Siberia or the North Pacific southwest of Honshu Islands rather than round the globe.

On April, higher monthly ^{137}Cs depositions were observed at the North Pacific side stations and East Japan inland stations, although the levels decreased markedly. This suggests that the atmospheric emission of radionuclides from the FDNPP continued on April 2011, although the release rate decreased dramatically. On the other hand, the monthly ^{137}Cs deposition increased at southwest sites in Japan and at the Japan Sea side sites comparing with that on March, suggesting that the FDNPP-derived radioactive cloud dominantly affected north part of the Northern Hemisphere atmosphere. On May, the monthly ^{137}Cs depositions decreased at all of the monitoring stations of Japan (Fig. 13c), although higher ^{137}Cs depositions were observed within 300 km of the FDNPP. On August, the monthly ^{137}Cs depositions at most of the monitoring stations in southwest Japan decreased below detection limit.

4.3 Atmospheric Half-Life of FDNPP-Derived ^{137}Cs

In order to elucidate depositional behaviors of the FDNPP-derived radionuclides, it is important to examine the temporal change of the monthly ^{137}Cs deposition. The monthly ^{137}Cs deposition at the sites in 300 km of the FDNPP except Futaba exhibited exponential decrease during the period of March to June as shown in Fig. 14. Apparent half-lives of the FDNPP-derived atmospheric ^{137}Cs ranged from 8.8 days for Tokyo to 14 days for Morioka (39.70° N, 141.16° E) (Hirose 2011). The apparent atmospheric half-life of the FDNPP-derived ^{137}Cs is similar to the residence time (8 days) estimated for ^{210}Pb over the west central United States, based on radioactive equilibrium of ^{210}Pb with its short-lived progenies (Moore et al. 1973). Lambert et al. (1982) estimated a global mean aerosol residence time of 6.5 days by using atmospheric inventories of ^{222}Rn and ^{210}Pb extrapolated from the observations and computing the ^{210}Pb deposition sink to balance ^{222}Rn decay. The residence time of tropospheric aerosols inferred from a global three-dimensional simulation of ^{210}Pb was 5–10 days, which depends on season and latitude (Balkanski et al. 1993). On the other hand, longer troposphere residence time of aerosols (about 30 days) was obtained from radioactive debris of atmospheric nuclear explosions (Katsuragi 1983). A similar long residence time was estimated for the Chernobyl ^{137}Cs (25 days) (Aoyama 1988), which was emitted as sub-micrometer particles (Hirose 1995). The residence times of aerosols in troposphere, which are in the range of 5–30 days, have been determined by natural and anthropogenic radionuclides, which depend on particles size and altitude (Ehhalt 1973). The shorter half-life of the FDNPP-derived ^{137}Cs suggests that most of the FDNPP-derived radionuclides were injected into the lower layer of the troposphere and/or emitted as larger particles in the atmosphere. As another possible process, the apparent half-life of the FDNPP-derived ^{137}Cs may partly reflect a history of the atmospheric

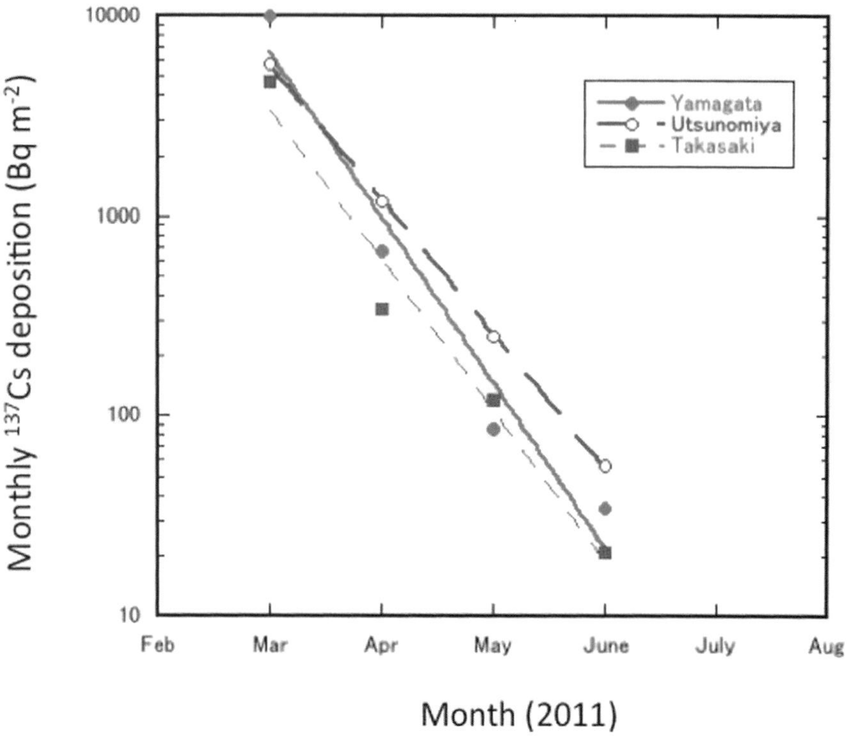

Fig. 14 Temporal changes of monthly ^{137}Cs deposition observed at several sites of east Honshu Island. Data were cited from MEXT (2012). The apparent half-lives of atmospheric ^{137}Cs at Yamagata, Utsunomiya, and Takasaki were 11, 13.5, and 12 days, respectively

emission because, the radioactive emission from the FDNPP continued as of the end of 2011, although the emission rate of radioactivity dramatically decreased (TEPCO 2011). In Futaba, the monthly ^{137}Cs deposition showed different history from other monitoring stations as shown in Fig. 15; rapid decrease of the monthly ^{137}Cs deposition occurred from March to April, after that, the decrease rate of the monthly ^{137}Cs deposition was slower than that of other stations in the corresponding period. It is likely that, the monthly ^{137}Cs deposition at Futaba was strongly affected by the atmospheric emission from the FDNPP.

The monthly ^{137}Cs deposition after July, 2011 at the monitoring stations in the Kanto Plain exhibited slower decrease rates as shown in Fig. 15, although ^{137}Cs cannot be detected in the monthly deposition samples for most of the monitoring stations in the south western region of Japan. Most of the monitoring stations in the Kanto Plain have reported meaningful amounts of the monthly radiocesium deposition in the end of 2011 (MEXT 2012). The relatively high levels of the radiocesium deposition rates in summer—winter 2011 may be supported by re-suspension of radiocesium deposited on land surface (Hirose et al. 2012; Igarashi et al. 2003, 2009). Another possible source may be subsequent emission of radiocesium from the FDNPP.

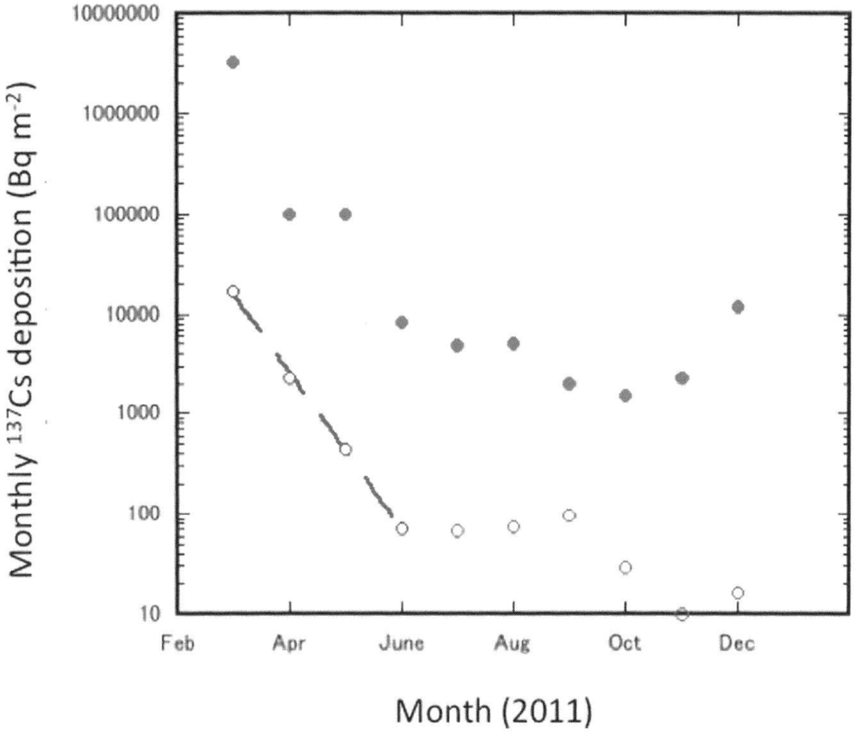

Fig. 15 Temporal changes of monthly ^{137}Cs deposition observed at sites near the FDNPP. *Closed circle*: Futaba, *open circle*: Hitachinaka. The apparent half-life of atmospheric ^{137}Cs at Hitachinaka in March—June was 11.5 days

4.4 Deposition Behavior of FDNPP-Dearived Radionuclides

Deposition velocity is a useful tool to have better understanding of the atmospheric behavior of the FDNPP-derived radionuclide: deposition velocity V_d (m s^{-1})=deposition flux (Bq $m^{-2}s^{-1}$)/surface air concentration (Bq m^{-3}). Amano et al. (2012) (Amano et al. 2012) have determined the deposition velocities of the FDNPP-derived ^{131}I and ^{137}Cs during the periods March 14–17 and March 21–24, corresponding to the first arrival (no rainfall) and the second arrival (heavier rainfall: 38 mm), respectively. The deposition velocities in the first arrival of the radioactive plume, corresponding to dry deposition velocity, were around 0.002–0.003 m s^{-1} for ^{134}Cs and ^{137}Cs and 0.001–0.003 m s^{-1} for ^{131}I. The dry deposition velocities of the Chernobyl fallout were determined to be 0.00035–0.0058 m s^{-1} for ^{134}Cs and ^{137}Cs and 0.00023–0.00095 m s^{-1} for ^{131}I (Aoyama et al. 1992). The dry deposition velocities of the FDNPP-derived radionuclides are of the same order of magnitude as those of the Chernobyl fallout, although the dry deposition velocities depend on size and chemical characteristics of radionuclide-bearing particles. The deposition velocities in the second arrival of the radioactive plume, corresponding to wet deposition velocity, were around 0.01–0.14 m s^{-1} for ^{134}Cs and ^{137}Cs and 0.004–0.03 m s^{-1}

for ^{131}I. The wet deposition velocities of the Chernobyl fallout were determined to be 0.0026–0.11 m s^{-1} for ^{137}Cs (Hirose et al. 1993). The wet deposition velocities of the FDNPP-derived radiocesium are of the same order of magnitude as those of the Chernobyl fallout. It is noteworthy that, the wet deposition velocities depend on size and chemical characteristics of radionuclide-bearing particles and meteorological conditions such as rainfall intensity (Hirose et al. 1993). The wet deposition velocities of the FDNPP-derived radionuclides were more than one order of magnitude greater than their dry deposition velocities, which implies that the FDNPP-derived radionuclides were deposited on land surface by wet deposition processes rather than dry deposition. Furthermore, these findings suggest that the FDNPP-derived radiocesium is slightly easily removed from the atmosphere by wet and dry deposition processes comparing with the FDNPP-derived radioiodine.

5 Mapping of Radioactivity Deposition

Serious radioactivity deposited area due to the FDNPP fallout spread about 300 km of the FDNPP, which suggested with deposition data (Hirose 2011) and model simulation (Morino et al. 2011). In order to effectively operate radiation protection to the FDNPP-derived radioactivity, it is essential to construct detail radioactivity contamination map of land surface. The mapping of radioactivity-contaminated area has been performed by radioactivity measurements based on high dense soil sampling in Fukushima Prefecture (Yoshida and Takahashi 2012) and airborne monitoring with NaI (Tl) detectors and helicopter. The airborne monitoring of radiation has been conducted by MEXT with cooperation of the US DOE since April 6, 2011 (Lyons and Colton 2012). Comparison between airborne monitoring data and high dense soil data provides the most adequate conversion factors to estimate the FDNPP-derived radionuclide deposition from airborne gamma-ray measurement. The airborne monitoring can be realized in the radiocesium-contaminated area with deposition density of more than 10 kBq m^{-2}. The map of the FDNPP-derived ^{137}Cs deposition is shown in Fig. 16. The high radiocesium-contaminated area due to the Fukushima fallout spread narrow area northwestward of the FDNPP. Another typical future is that higher contaminated areas are present in Fukushima Basin besides the northern and southern mountains, the mountainside of the north Kanto Plain, and the central part of the Kanto Plain. This finding suggests that the FDNPP fallout was strongly governed by land topography and geographical distribution of rainfall.

6 Conclusion

The radioactive plume was emitted from the FDNPP since March 12, 2011. Dominant radioactive emission in atmosphere occurred in March 15–16, 2011 (Chino et al. 2011). The FDNPP-derived radionuclides were detected in surface air of land

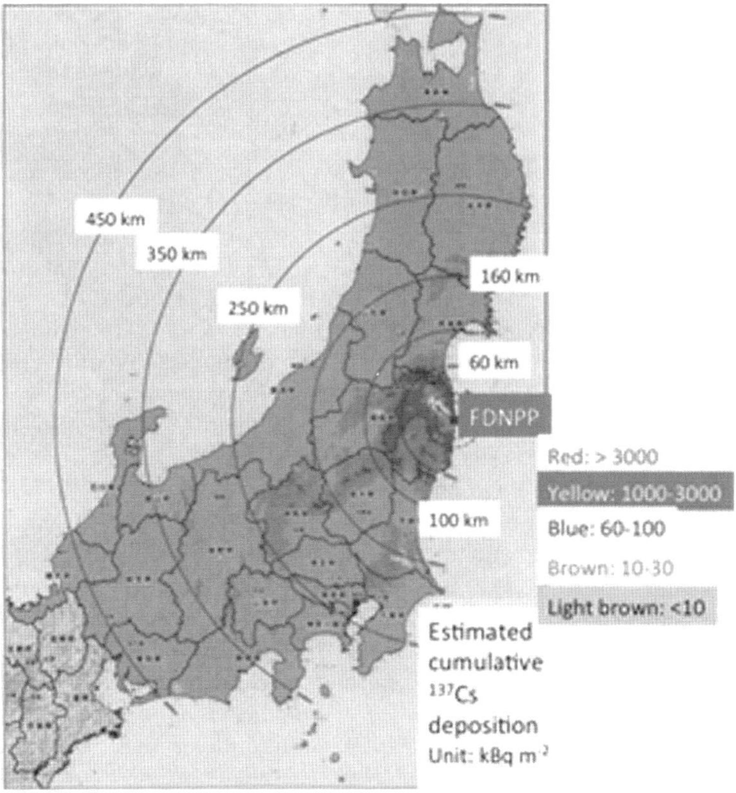

Fig. 16 Spatial distribution of FDNPP-derived ^{137}Cs deposition measured by airborne monitoring. This figure is depicted on basis of figure in MEXT (2012)

monitoring stations (especially in the Kanto Plain) since March 15, 2011, which was consistent with elevated ambient dose rate. The highest concentrations of the FDNPP-derived radionuclides in surface air at the monitoring stations of the Kanto Plain occurred in March 15–16 and March 20–22 as a result of transport of the radioactive plume from the FDNPP. Detected radionuclides were ^{133}I, ^{132}I, ^{131}I, ^{134}Cs, ^{136}Cs, ^{137}Cs, ^{129m}Te (^{129}Te), ^{99}Mo (^{99m}Tc), ^{110m}Ag, ^{95}Nb, and ^{140}Ba (^{140}La). In late March and April, global network of radioactivity monitoring in the northern hemisphere detected the FDNPP-derived radioactivity fallout. The surface air concentrations of the FDNPP-derived radionuclides decreased with distance from the FDNPP. When radioactive decay is corrected at cease of nuclear fission in the FDNPP, the ratios between isotopes are almost constant during the sampling period; the $^{133}I/^{131}I$, $^{134}Cs/^{137}Cs$, $^{136}Cs/^{137}Cs$, $^{129m}Te/^{132}Te$ ratios were 1.4, 1.0, 0.09–0.27, and 0.09, respectively. On the other hand, there were larger variations between elements; the decay-corrected $^{131}I/^{137}Cs$, $^{132}Te/^{137}Cs$, and $^{99m}Tc/^{137}Cs$ at Tsukuba were in the ranges of $8–1.0 \times 10^3$, 4 to 65, and 0.16 to 22, respectively. Large variations of the $^{131}I/^{137}Cs$, $^{132}Te/^{137}Cs$, and $^{99m}Tc/^{137}Cs$ ratios suggest that atmospheric emission of the FDNPP-derived radionuclides and removal processes of these radionuclides in the atmosphere were affected largely by the chemical and physical properties of elements. In fact, significant amounts of ^{131}I and ^{133}I existed as gaseous form, whereas radiocesium in surface air on April attached relatively larger particles (> 1 μm).

As a result of snowfall and southwest wind on March 15–16, serious radioactive contaminated area appeared in Fukushima Prefecture (MEXT 2012). The Kanto plain was mainly contaminated by the FDNPP-derived radionuclides due to the movement of radioactive cloud by northeast wind and following rainfall during March 20 to 23. The airborne monitoring is an effective tool to realize the mapping of radioactivity-contaminated area. Recent model simulation can reproduce high radioactive deposition area and global dispersion of the FDNPP-derived radionuclides. The apparent residence time of the FDNPP-derived radionuclides observed in the Kanto plain and northeast Honshu Island is about 12 days, which suggests that major radioactive materials emitted from the FDNPP accident were injected into the lower layer of the troposphere.

The FDNPP accident is still ongoing at the end of 2011 (TEPCO 2011), although major atmospheric emission of radionuclides has been ceased in the end of March 2011 apparently. To have better understanding of the environmental behaviors of the FDNPP-derived radionuclides, including factors controlling variability of temporal change between radionuclides, it is necessary to know physical and chemical properties of the FDNPP-derived radionuclides in addition to detail time course of radioactivity emission at accidental reactors. Now, there is only limited information on the environmental impact of the FDNPP-derived radioactivity, although huge amounts of the environmental radioactivity data related to the FDNPP accident have been documented. In order to do adequate response to public concern, further investigation is required.

Acknowledgment Author thanks monitoring staffs and officers of MEXT for great efforts to measure environmental radioactivity.

References

Amachi S, Kamagata Y, Kanagawa T, Muramatsu Y (2011) Bacteria mediate methylation of iodine in marine and terrestrial environment. Appl Environ Microbiol 67:2918–2722

Amano H, Akiyama M, Chunlei B, Kawamura T, Kishimoto T, Kuroda T et al (2012) Radiation measurements in the Chiba Metropolitan area and radiological aspects of fallout from the Fukushima Daiichi Nuclear Power Plants accident. J Environ Radioact 111:42–52

Aoyama M (1988) Evidence of stratospheric fallout of caesium isotopes from the chernobyl accident. Geophys Res Lett 15:327–330

Aoyama M, Hirose K, Takatani S (1992) Particle size dependent dry deposition velocity of the Chernobyl radioactivity. In Hemisphere Precipitation scavenging and atmospheric exchange processes; Fifth International Conference, vol 3, pp 1581–1593

Aoyama M, Tsumune D, Hamajima Y (2012) Budgets and temporal change of radiocesium distribution released from Fukushima NPP accidents in the North Pacific Ocean, 2012 Ocean Science Meeting, abstract ID10092

Balkanski YJ, Jacob DJ, Gardner GM, Graustein WC, Turekian KK (1993) Transport and residence time of tropospheric aerosols inferred from a global three-dimensional simulation of ^{210}Pb. J Geophys Res 98:20573–20586

Beresford NA, Barnett CL, Howard BJ, Howard DC, Wells C, Tyler AN, Bradley S, Copplestone D (2012) Observations of Fukushima fallout in Great Britain. J Environ Radioact 114:48–53. doi:10.1016/j.jenvrad.2011.12.008

Biegalski SR, Bowyer TW, Eslinger PW, Friese JA, Greenwood LR, Haas DA, Hayes JC, Hoffman I, Keillor M, Miley HS, Moring M (2011) Analysis of data from sensitive U.S. monitoring stations for the Fukushima Dai-ichi nuclear reactor accident. J Environ Radioact 114:15–21

Bolsunovsky A, Dementyev D (2011) Evidence of the radioactive fallout in the center of Asia (Russia) following the Fukushima Nuclear Accident. J Environ Radioact 102:1062–1064

Bowyer TW, Biegalski SR, Cooper M, Eslinger PW, Haas D, Hayes JC et al (2011) Elevated radioxenon detected remotely following the Fukushima nuclear accident. J Environ Radioact 102:681–687

Carvalho FP, Reis MC, Oliveira JM, Malta M, Silva L (2012) Radioactivity from Fukushima nuclear accident detected in Lisbon, Portugal. J Environ Radioact 114:152–156

Chino M, Nakayama H, Nagai H, Terada H, Katata G, Yamazawa H (2011) Preliminary estimation of released amounts of ^{131}I and ^{137}Cs accidentally discharged from the Fukushima Daiichi nuclear power plant into the atmosphere. J Nucl Sci Technol 48:1129–1134

Clemenza M, Fiorini E, Previtali E, Sala E (2012) Measurement of airborne ^{131}I, ^{134}Cs and ^{137}Cs due to the Fukushima reactor incident in Milan (Italy). J Environ Radioact 114:152–156

CTBTO (Comprehensive Nuclear-Test-Ban Treaty Organization) (2011). http://www.ctbto.org/press-centre/highlights/2011/fukushima-related-measurements-by-the-ctbto/

Doi T, Masumoto K, Toyoda A, Tanaka A, Shibata Y, Hirose K (2013) Anthropogenic radionuclides in the atmosphere observed at Tsukuba: Characteristics of the radionuclides derived from Fukushima. J Environ Radioact 122:55–62

Ehhalt DH (1973) Turnover tomes of ^{137}Cs and HTO in the troposphere and removal rates of natural particles and vapor. J Geophys Res 78:7076–7086

Furuta S, Sumiya S, Watanabe H, Nakano M, Imaizumi K, Takeyasu M et al (2011) (In Japanese) Results of the environmental radiation monitoring following the accident at the Fukushima Daiichi nuclear power plant—Interim report (ambient radiation dose rate, radioactivity concentration in the air and radioactivity concentration in the fallout). JAEA-Review 2011-035:1–84

Hirose K (1995) Geochemical studies in the chernobyl radioactivity in environmental samples. J Radioanal Nucl Chem Articles 197:331–342

Hirose K (2012) 2011 Fukushima Daiichi nuclear power plant accident: summary of regional radioactivity deposition monitoring results. J Environ Radioact 111:13–17

Hirose K, Takatani S, Aoyama M (1993) Wet deposition of long-lived radionuclides derived from the chernobyl accident. J Atmos Chem 17:61–73

Hirose K, Igarashi Y, Aoyama M (2008) Analysis of 50 years records of atmospheric deposition of long-lived radionuclides in Japan. Appl Radiat Isot 66:1675–1678

Hirose K, Kikawada Y, Igarashi Y (2012) Temporal variation and provenance of thorium deposition observed at Tsukuba, Japan. J Environ Radioact 108:24–28

Huh CA, Hsu S-C, Lin C-Y (2012) Fukushima-derived fission nuclides monitored around Taiwan: free tropospheric versus boundary layer transport. Earth Planet Sci Lett 319–320:9–14

IAEA (1986) Summary report on the post-accident review meeting on the Chernobyl' accident. Safety Series No. 75-INSAG-1

Igarashi Y, Aoyama M, Hirose K, Miyao T, Nemoto K, Tomita M, Fujikawa T (2003) Resuspension: decadal monitoring time series of the anthropogenic radioactivity deposition in Japan. J Rad Res 44:319–328

Igarashi Y, Aoyama M, Hirose K, Povinec PP, Yabuki S (2005) What anthropogenicradionuclides (^{90}Sr and ^{137}Cs) in atmospheric deposition, surface soils and Aeolian dusts suggest for dust transport over Japan. Water Air Soil Poll: Focus 5:51–69

Igarashi Y, Inomata Y, Aoyama M, Hirose K, Takahashi H, Shinoda Y, Sugimoto N, Shimizu A, Chiba M (2009) Possible change in Asian dust source suggested by atmospheric anthropogenic radionuclides during the 2000s. Atmos Environ 43:2971–2980

IPG (Ibaraki Prefectural Government) (2011) http://www.pref.ibaraki.jp/earthquake/doserate_2011.html. Accessed Mar 2014

Kanai Y (2012) Monitoring of aerosols in Tsukuba after Fukushima Nuclar Power Plant incident in 2011. J Environ Radioact 111:33–37

Kaneyasu N, Ohashi H, Suzuki F, Okuda T, Ikemori F (2012) Sulfate aerosol as a potential transport medium of radiocesium from the Fukushima nuclear accident. Environ Sci Technol 46:5720–5726

Katsuragi Y (1983) A study of ^{90}Sr fallout in Japan. Pap Meteor Geophys 33:277–291

Kim C-K, Byun J-I, Chae J-S, Choi H-Y, Choi S-W, Kim D-J et al (2012) Radiological impact in Korea following the Fukushima nuclear accident. J Environ Radioact 111:70–82

Lambert G, Polian G, Sanak J, Ardouin B, Buisson A, Jegou A, Leroulley JC (1982) Cycle du radon et de ses descendants: application a l'étude des èchanges troposphère-stratosphère. Ann Geophys 38:497–531

Long NQ, Truong Y, Hien PD, Binh NT, Sieu LN, Giap TV, Phan NT (2012) Atmospheric radionuclides from the Fukushima Dai-ichi nuclear reactor accident observed in Vietnam. J Environ Radioact 111:53–58

Lozano RL, Hernández-Ceballos MA, Adame JA, Casas-Ruíz M, Sorribas M, San Miguel EG, Bolívar JP (2011) Radioactive impact of Fukushima accident on the Iberian Peninsula: evolution and plume previous pathway. Environ Int 37:1259–1264

Lujaniené G, Bycenkiené S, Povinec PP, Gera M (2012) Radionuclides from the Fukushima accident in the air over Lithuania: measurement and modeling approaches. J Environ Radioact 114:71–80

Lyons C, Colton D (2012) Aerial measuring system in Japan. Health Phys 102:509–515

Manolopoulou M, Vagena E, Syoulos S, Loannidou A, Papastefanou C (2011) Radioiodine and radiocesium in Thessaloniki, Greece due to the Fukushima nuclear accident. J Environ Radioact 102:796–797

Masson O et al (2011) Tracking of airborne radionuclides from the damaged Fukushima Dai-ichi nuclear reactors by European networks. Environ Sci Technol 45:7670–7677

MEXT (Ministry of Education, Culture, Sports, Science and Technology) (2012) http://www.mext.go.jp/amenu/saigaijyouhou/syousai/1303856.htm. Accessed Mar 2012

Momoshima N, Sugihara S, Ichikawa R, Yokoyama H (2012) Atmospheric radionuclides transported to Fukuoka, Japan remote from the Fukushima Daiichi nuclear power complex following the nuclear accident. J Environ Radioact 111:28–32

Moore HE, Poet SE, Martell EA (1973) ^{222}Rn, ^{210}Pb, ^{210}Bi, and ^{210}Po profiles and aerosol residence times versus altitude. J Geophys Res 78:7065–7075

Morino Y, Ohara T, Nishizawa M (2011) Atmospheric behavior, deposition, and budget of radioactive materials from the Fukushima Daiichi nuclear power plant in March 2011. Geophys Res Lett L00G11

Muramatsu Y, Yoshida S (1995) Volatilization of methyl iodine from soil-plant system. Atmos Environ 29:21–25

NSC (Nuclear Safety Comission) (2011) http://www.nsc.go.jp/info/20110412.pdf. Accessed Mar 2012

ORNL (Oak Ridge National Laboratory) (2004) ORIGEN ARP 2

Paatero J, Vira J, Siitari-Kauppi. M, Hatakka J, Holmen K, Viisanen Y (2012) Airborne fission products in the high Arctic after the Fukushima nuclear accident. J Environ Radioact 114:41–47

Pham MK, Eriksson M, Levy I, Nies H, Osvath I, Betti M (2012) Detection of Fukushima Daiichi nuclear power plant accident radioactive tracers in Monaco. J Environ Radioact 114:131–137

Pittauerová D, Hettwig B, Fischer HW (2011) Fukushima fallout in Northwest German environmental media. J Environ Radioact 102:877–880

RJG (Report of Japanese Government) (2011) The accident at TEPCO's Fukushima Nuclear Power 2011. http://www.kantei.go.jp/jp/Topics/2011/iaea_houkokusho.html. (Revised total release of radionuclides: http://www.meti.go.jp/press/2011/10/20111020001.pdf). Accessed Mar 2014

Stohl A, Seibert R, Wotawa G, Arnold D, Burkhart JF, Eckhardt S et al (2011) Xenon-133 and caesium-137 releases into the atmosphere from the Fukushima Dai-ichi nuclear power plant: determination of the source term, atmospheric dispersion, and deposition. Atmos Chem Phys Discuss 11:28319–28394

Takemura T, Nakamura H, Takigawa M, Kondo H, Satomura T, Miyasaka T, Nakajima T (2011) A numerical simulation of global transport of atmospheric particles emitted from the Fukushima Daiichi Nuclear Power Plant. Sora 7:101–104

TEPCO (Tokyo Electric Power Co.) (2011) Tentative estimation of atmospheric emission rate of radionuclides from the Fukushima Daiichi NPP. http://www.tepco.co.jp/cc/press/betsu11_j/images/107194.pdf. Accessed Mar 2012

UNSCEAR (2000) Sources and effects of ionizing radiation, vol. 1: sources. United Nations, New York, pp 654

Yonezawa C, Yamamoto Y (2011) Measurements of artificial radionuclides in surface air by CTBTO network. BUNSEKI No. 8 pp 451–458 (In Japanese).

Yoshida N, Takahashi Y (2012) Land-surface contamination by radionuclides from the Fukushima Daiichi Nuclear Power Plant accident. Elements 8:201–206

Zhang W, Bean M, Benotto M, Cheung J, Ungar K, Ahier B (2011) Development of a new aerosol monitoring system and its application in Fukushima nuclear accident related aerosol radioactivity measurement at the CTBT radionuclide station in Sidney of Canada. J Environ Radioact 102:1065–1069

Applications of Remote Sensing, Geographic Information System and Geostatistics in the Study of Arsenic Contamination in Groundwater

A. R. Ghosh and Kajori Parial

Abstract The arsenic-related groundwater problem is threatening huge areas in different countries like India, Bangladesh, China, Chile, Mexico etc. Arsenic contamination in ground water affects human health through its direct and indirect consumption. In order to combat this problem, an in-depth understanding of the background process of arsenic contamination is needed. Remote Sensing and Geographic Information System (GIS) offer a suitable solution for understanding and mapping the extent of the arsenic-infected areas. Various thematic maps on geomorphology, landuse/landcover, hydrology etc. can be prepared by interpreting remote sensing data, i.e. satellite imagery of the concerned area. These thematic maps can be further utilized to explore and discover relationships between that particular theme and the distribution of arsenic. Digital elevation models prepared using photogrammetric techniques utilizing Remote Sensing data may help in understanding the nature of distribution of arsenic-contaminated areas. GIS helps capturing, retrieving and analysing geospatial data. It recreates a real-world scenario and helps in analysing and creating new relevant information. In arsenic affected areas, people are exposed to arsenic contamination through its direct consumption with drinking water as well as through consumption of crops contaminated due to irrigation with arsenic rich groundwater. Risk and vulnerability assessment of such population can be done using GIS. Web-based GIS offers a unique opportunity of crowdsourcing and development of participatory public GIS. It can help in generation and analysis of distribution and assessment of impact of arsenic contamination and evolving mitigation strategy. Geostatistical methods help in interpolating the degree of arsenic contamination in unsampled locations as well as modelling uncertainty about unknown values. Initially experimental semivariograms are constructed and are fitted with suitable models like spherical, Gaussian etc. Various techniques of kriging are employed to interpolate level of arsenic contamination in unsampled

A. R. Ghosh (✉)
Department of Science & Technology, Government of West Bengal,
Bikash Bhavan (4th Floor), Salt Lake, Kolkata 700091, India
e-mail: arghosh@rediffmail.com

K. Parial
Department of Geology and Geophysics, Indian Institute of Technology Kharagpur,
Kharagpur 721302, India
e-mail: kajorigis@gmail.com

D. Sengupta (ed.), *Recent Trends in Modelling of Environmental Contaminants*,
DOI 10.1007/978-81-322-1783-1_8,

locations using parameters obtained from variography. Zonation of arsenic contamination levels helps in the assessment of risk and vulnerability of residents of arsenic-contaminated areas and the development of efficient mitigation strategies.

Keywords Arsenic contamination · Image interpretation · Geomorphology ·Spatial analysis · Semivariography · Kriging

1 Introduction

Arsenic contamination in groundwater is one of the recent natural disasters which are slowly affecting huge populations in different countries. High concentration of arsenic in groundwater is reported in different countries like India, Bangladesh, Pakistan, China, Nepal, Vietnam, Cambodia, Taiwan, Chile, Mexico and the USA (Hossain and Piantanakulchai 2013). Groundwater is the main source of drinking water in these areas/countries, especially in rural areas. Although human beings are affected mostly through drinking of arsenic contaminated water (Smedley and Kinniburgh 2002), the water–soil–plant–human transfer pathway is also very important (Khan et al. 2009).

Arsenic is one of the most important water pollutants within the metalloids (Manahan 1994). It is a toxic, teratogenic (Reimann and Caritat 1998) and carcinogenic (Manahan 1994) contaminant. According to *WHO Guidelines (1993),* the safe limit of Arsenic in potable water is 10 µg/L (Acharyya and Shah 2007; Ahsan and Del Valls 2011). However, India, Bangladesh, China, and Nepal maintain the pre-1993 WHO Guidelines value—50 µg/L (Appelo 2006). It is well accepted that the occurrence of arsenic in the environment is ubiquitous, but its release in the groundwater is governed by various processes, most of which are poorly understood Hossain and Piantanakulchai 2013). Causes of arsenic contamination in groundwater are attributed to mainly three probable processes:

1. Oxidation of pyrite associated with sediments and rock materials due to excess pumping and water table drawdown (Chakrabarti et al. 2001; Sikdar et al. 2001).
2. Mobilization of arsenic from Fe-oxyhydroxides when the environment becomes anaerobic (Smedley and Kinniburgh 2002; Bhattacharya et al. 1997).
3. Desorption of Fe-oxyhydroxides especially at higher pH (Bhattacharya et al. 2006; Kar et al. 2010; Smedley and Kinniburgh 2002; Bhattacharya et al. 1997).

The depth of the groundwater has a strong correlation with the occurrence of the arsenic depending on the local geology (Ravenscroft et al. 2005). In West Bengal (India) and Bangladesh, it is observed that shallow tube wells (<150 m depth) are more affected compared to the tube wells with depth greater than 150 m (Hossain and Sivakumar 2005). In West Bengal it is often found that high concentration of arsenic occurs between 20 and 80 m below ground level (Das et al. 2007).

Identification of any problem essentially entails an in-depth understanding of the background processes involved, followed by designing and implementation of effective management strategies. However, whilst dealing with naturally occurring contaminant(s) (arsenic here) in natural resources (in this study groundwater), the quantification of the hazard intensity becomes a major challenge. Chemical analysis of samples collected over a large area is not only costly but also time-consuming. Remote Sensing and Geographic Information System (GIS) provide a promising solution in such situations. Experimental data coupled with various ancillary data, integrated in a common platform can efficiently provide valuable knowledge about the problem. The process of achieving such a reliable solution will be discussed in detail in the following sections. These tools and techniques have several advantages and extensive applications in various fields like engineering, environmental, geological and biological studies. In case of arsenic contamination studies, they can be employed in various ways like:

To demarcate, identify and monitor:

- Areas with safe/unsafe drinking water,
- Areas which are prone to contamination,
- Population affected or that which can be affected,
- Health status of the population exposed to the contaminants,
- Safe areas (with safe drinking water).

A systematic and well-maintained GIS database can serve in designing and planning both cost- and time-effective mitigation.

While studying environmental contamination, often, nonavailability of both qualitatively and quantitatively significant data becomes a bottleneck in assessing the situation. GIS coupled with geostatistics offers an effective solution to this problem. Using geostatistics, one can interpolate values over a larger area based on sample values determined experimentally, which are distributed throughout the area of interest. The geostatistical method of interpolation has been widely used in identifying such environmental problems.

The following section deals in detail about remote sensing, GIS and geostatistics and their applications in the study of arsenic contamination in groundwater.

2 Remote Sensing

'Remote Sensing is an art and science of obtaining information about an object, area, or phenomenon through the analysis of data acquired by a device that is not in contact with the object, area or phenomenon under investigation' (Lillesand et al. 2004). Different objects return different amounts of energy in different bands of electromagnetic energy incident upon them. However, these patterns are unique for every single object. This is called spectral signature—the response of an object in different bands of electromagnetic energy, helps in identifying objects from an image acquired through satellites. While interpreting satellite images visually, various factors like tone, texture, size, shape, shadow, association etc. are used to identify features on the earth's surface.

In the study of arsenic, remote sensing data in form of satellite imagery is used to generate various relevant thematic maps like geomorphology, landuse–landcover, elevation, hydrology etc. Landuse–landcover maps are prepared using multiseasonal satellite images. The interpretation can be done visually through on-screen digitization or using digital image classification technique. Elevation maps can be prepared through photogrammetric methods using stereo pair satellite images or aerial photographs (Lillesand 2004). Digital elevation models can be prepared through interferometric synthetic aperture radar data also (Kerle et al. 2004). Digital elevation models of comparatively coarser spatial resolution are freely available—SRTM (Shuttle Radar Topographic Mission), ASTER DEM, CartoDEM etc. Various important hydrological parameters relevant to the hydrologic processes of the area can be derived from the DEM. Along with other maps mentioned above, another important thematic map required for the study of arsenic distribution is the drainage map. The drainage patterns (rivers and canals, mostly linear features), tube wells, dug wells and natural water body distribution maps are major requirements in these studies. Geomorphological maps can be prepared using the interpretation of satellite imagery. As satellite imageries can provide synoptic view of the area, it can help in identification of geomorphic features. This reduces the field work to a great extent—however, limited field check is required to be done. These maps can be studied for the assessment of impact of geomorphological units in arsenic contamination. Digital elevation model, landuse–landcover map and soil map are used to prepare hydrological response units in GIS environments for contaminant modelling in Soil and Water Assessment Tool (SWAT) (Neitsch et al. 2009).

3 Geographic Information System

A Geographical Information System (GIS) is a computer system for capturing, storing, querying, analysing and displaying geospatial data (Chang 2011). Various kinds of geographically referenced data (Burrough and McDonnell 1998), viz., lithological, structural, soil, depth to ground water table, elevation, groundwater quality, geophysical data etc. can be stored in a GIS. GIS recreates real-world scenarios in the computer either in the form of discrete objects or in form of a continuous field. Objects are better represented in vector model while raster model is more suitable for fields. In vector model, spatial features like houses, wells, ponds, rail, road etc. are represented in terms of their spatial information (corner coordinates) along with other nonspatial information as attribute data (Ade 2004). Here, features are represented by points, lines and polygons. In raster model, the entire space of the phenomenon is subdivided into equal-sized grids or cells (pixels). Each cell has a value denoting the characteristic of the phenomenon. Continuous spatial phenomena like soil, elevation, rainfall etc. can be stored in the form of a field. (Chang 2011). Each vector or raster model representing a feature or phenomenon is termed as layers. In GIS, these layers are used singly or are integrated to derive relevant information about the problem at hand. GIS basically acts as a platform to integrate the various spatial, demographic, social, field-based and analytical data (Khan et al. 2009). In GIS, real-world objects or fields are stored and represented using projection system so that their shape, size, position and other characteristics remain intact. Different datasets registered accurately can be visualized, analysed and new relevant information can be generated using those datasets.

Spatial distribution of a contaminant and its impact on human and physical environment can be better understood using visualization capabilities of GIS (Khan et al. 2009). New probabilistic contamination layers can be generated from spatial analysis of different thematic layers (land use, land cover, drainage, vegetation, soil etc.) that share a common geographic space. Demographic and socioeconomic data can then be overlaid with probabilistic contamination layers to understand the impact of contaminant of interest on human health. Geostatistical analysis can be coupled with visualisation capabilities of a GIS for better representation. GIS-based nonpoint source pollutant model holds a promise (Corwin et al. 1997). Several authors have attempted GIS-based groundwater models to understand contamination (Watkins et al. 1996; Lasserre et al. 1999; Boutt et al. 2001; Hall et al. 2001). If capabilities of GIS are coupled with process-based models, an efficient tool for processing, storing, manipulating and visualizing data quickly and accurately can be obtained (Gogu et al. 2001).

3.1 *Applications of GIS in the Study of Arsenic Contamination*

Arsenic content analysed from samples collected from numerous wells can be stored, retrieved and analysed in a GIS system efficiently. Relationship between other information from the same geographical location can be studied rapidly and accurately. Using GIS, vulnerability and risk assessment due to consumption of high arsenic groundwater can be done. GIS can help in understanding the impact of irrigation with high arsenic groundwater on human health. Arsenic risk assessment depends upon various factors like soil arsenic, soil flooding condition, rice genotype etc. (see Chap. Arsenic Fate and Transport in the Groundwater–Soil–Plant System: An Understanding of Suitable Rice Paddy Cultivation in Arsenic-Enriched Areas). Using GIS, arsenic risk assessment can be done and mitigation strategy can be evolved.

Various kinds of spatial and aspatial data are to be brought into a GIS to assess the potential risk for human population that directly consume arsenic-contaminated water or consume crops irrigated with arsenic rich groundwater. Such data can be generated from a wide variety of sources like field surveys as well as information generated through remote sensing, maps collected from heterogeneous sources, geostatistical interpolation of known samples to unknown locations etc. All of them can be intigrated in the GIS platform. GIS has flexibility in dealing with data with different datum and projection (Khan et al. 2009).

Hossain and Piantanakulchai (2013) attempted prediction of arsenic contamination risk in groundwater using GIS and classification tree method. They prepared raster maps of soil reaction, organic matter content, iron content, geological map, surface elevation etc. Multiring buffer maps around major drainage channels were prepared through proximity analysis in GIS. The data sets were reclassified, which were used to develop classification tree using statistics toolbox of MATLAB. Finally, reclassified data of all the factors were used to create risk map in the GIS environment as per classification tree relations.

Hassan (2005) developed a public participation GIS (PPGIS) combining GIS and participatory rural appraisal for assessment of needs of the affected communities. Open Geospatial Consortium services like Web Feature Service (WFS), Web Feature Service—Transactional (WFS-T), Web Coverage Service (WCS) and Web Map Service (WMS) have been developed to make GIS more approachable and interactive for the stake-holders. With the advent of Google Maps, Bing Maps etc. the reach of spatial data has increased to a great extent. Many functionalities of GIS are available in web based GIS. These have opened the avenues for participatory evaluation of the problem of arsenic contamination and evolving suitable mitigation strategy.

4 Geostatistics

Geostatistical methods can identify the spatial patterns and interpolate the values at unsampled locations. It can model the uncertainty about estimated values at unsampled locations. In many cases, only a few data are available—thus, the resultant map provides poor estimation with high degree of corresponding uncertainty/error. Here, more closely spaced related data can be utilized to estimate the sparser data. (Goovaerts 1997). This helps in groundwater modelling, management and recharge as well as contaminant modelling.

Interpolation may be deterministic in nature, which can be estimated using mathematical formulae like spline, regression analysis etc., or it may be geostatistical in nature. Samples from natural phenomenon which appear random may have some spatial relationship. Spatial autocorrelation, (i.e. closer the sample more similar are their values) exists in natural phenomena. Using geostatistics we can try to understand the nature of variability as well as attempt to predict value at unsampled locations with an estimation of error. Nature of such variability is studied through variography or structural analysis or spatial modelling. Estimation of expected value of sample at unsampled locations is generally done through kriging. It is generally assumed that second order stationarity—a condition referred to as the intrinsic hypothesis (Journel and Huijbregts 1978) exists for covariograms and intrinsic stationarity for semivariograms between samples, i.e. it is believed that difference in value between samples located at the same distance and direction are the same, irrespective of their location. Experimental semivariograms can be fitted with various kinds of models, namely, spherical, exponential, Gaussian, hole effect, rational, quadratic etc. Here briefly spherical model of semivariogram and ordinary kriging is discussed as an example.

4.1 Variography

Experimental Semivariography

In geostatistical studies, initially, statistical properties of the dataset are studied to understand the nature of data and its distribution. If required, the data is transformed initially and after completion of computation, the same is back transformed. If distribution of data is found to be log normal in nature, log transformation of data is done before geostatistical operations. Some other transformations are box-cox and arcsine transformations. After completion of the estimation process the resultant data are suitably back-transformed.

Average squared difference of values located at a lag distance of *h* is called a semivariogram. An experimental semivariogram for a given lag *h* is computed by the following formula

$$\gamma(h) = \tfrac{1}{2}n * \sum_{1}^{n} (Z_n - Z_{n+h})^2$$

Semivariogram values are plotted against the lag interval. In the resulting experimental semivariogram, the value of the semivariogram increases with lag distance. After a certain lag distance, semivariance value does not increase any further. The semivariance value at this level is called sill. The lag distance at which value of the semivariogram flattens is known as range. Theoretically, the value of semivariogram, at zero lag distance, should be zero. Sometimes, value of semivariogram for a lag distance close to zero is more than zero. In such cases, the value of a semivariogram at the origin is called the nugget. The nugget effect occurs due to small scale variability, sampling error etc. That threshold distance of lag is called range. Lag size is decided on the basis of spatial distribution of data. Maximum lag size taken into consideration is generally half the maximum distance between samples. During computation of experimental semivariogram, some amount of lag tolerance as well as angular tolerance is allowed.

The value of covariance depends upon variation of variables around their mean. When lag is zero, i.e. when samples are very closely spaced, covariance equals the global variance of the variable. A correlogram appears as an inverted semivariogram.

These experimental semivariogram values are fitted with various models mentioned in the next section.

Semivariogram Modelling

For estimation of values at unsampled locations on the basis of surrounding known values, kriging is used. In kriging, estimation of values at any location is done using weighted average of surrounding samples. Such weights are obtained from semivariogram models. Experimental semivariograms are estimated for some discrete lag distances. The semivariogram value for any required lag distance cannot be obtained from an experimental semivariogram. For that reason, a semivariogram model prepared on the basis of certain mathematical expression is chosen which best fits the experimental semivariogram. This model provides semivariogram value at any lag distance. Some commonly used models summarized by Goovaerts (1997) are given below.

- Nugget effect model

$$g(h) = \begin{cases} 0, & h = 0 \\ 1, & otherwise \end{cases}$$

where
g (h) is semivariogram model
h is lag vector

- Spherical model with range *a*

$$g(h) = Sph\,(h/a) \begin{cases} 1.5\frac{h}{a} - 0.5\left(\frac{h}{a}\right)^3, & h \le a \\ 1, & otherwise \end{cases}$$

where
a is range

- Exponential model with practical range *a*

$$g(h) = 1 - \exp\left(\frac{-3h}{a}\right)$$

- Gaussian model with practical range *a*

$$g(h) = 1 - \exp\left(\frac{-3h^2}{a^2}\right)$$

- Power model

$$g(h) = h^{\omega} \text{ with } 0 < \omega < 2$$

4.2 Anisotropy

Sometimes it is assumed that value of a semivariogram depends solely upon distance and not on direction. Such semivariograms are referred as isotropic. Many a times, it is found that semivariogram has different parameters in different directions. In geometric anisotropy, the most common kind of anisotropy, the sill of the semivariogram in different directions is same but corresponding ranges are different. In zonal anisotropy, ranges are similar but sills are different. Anisotropy can be geometric or zonal in nature. Anisotropy ellipsoid is prepared on the basis of range values in different directions for geometric anisotropy.

To understand the nature of geometric anisotropy, experimental semivariograms are computed in different directions with some lag tolerance and angular tolerance. On the basis of the values of range and sill in different directions obtained from semivariogram models, anisotropy can be understood.

4.3 *Kriging*

Kriging is also known as best linear unbiased estimator. Kriging, like other distance-based estimators, e.g. Inverse Distance Weighting (IDW) generates weights based on distance from the samples. However, unlike IDW, kriging takes into consideration the value of the semivariogram for the distance between known and unknown samples as well as the semivariogram among all known samples for computation of weights. This is done as follows:

$$\Gamma * \lambda = g$$

where

gamma matrix, Γ is modelled semivariogram values between all pairs of sample locations

λ is matrix of weights

g contains the modelled semivariogram values between each measured location and the prediction location (Johnston et al. 2001).

or

$$\begin{pmatrix} \gamma_{11} & \cdots & \gamma_{1N} & 1 \\ \cdots & \cdots & \cdots & \cdots \\ \gamma_{N1} & \cdots & \gamma_{NN} & 1 \\ 1 & \cdots & 1 & 0 \end{pmatrix} * \begin{pmatrix} \lambda_1 \\ \cdots \\ \lambda_N \\ m \end{pmatrix} = \begin{pmatrix} \gamma_{10} \\ \cdots \\ \gamma_{N0} \\ 1 \end{pmatrix}$$

γij denotes value of model semivariogram based on the distance between two samples at i^{th} and j^{th} locations

λi denotes wights

$\gamma i0$ denotes value of model semivariogram based on the distance between i^{th} sample and prediction location

As mentioned above, for ordinary kriging, the formula $\Gamma * \lambda = g$ is used. Using matrix algebra, the weights can be obtained by the formula $\lambda = \Gamma^{-1} * g$, where Γ^{-1} is the inverse matrix of Γ. Vector g, created for the point for which estimation is to be done contains modeled semivariogram values between each measured value and the prediction point. This way, estimation for every point can be done solving the equation using linear algebra. At the same time, kriging variance and kriging standard error can be estimated for every point.

Sometimes, a significant amount of trend is observed in the dataset. In such cases, the data may be filled with a trend surface. Geostatistical analysis is done with the residual values and subsequently merged with trend surface value (Johnston et al. 2001).

Simple Kriging

For simple kriging, it is assumed that the trend component *m(u)* has known stationary mean, *m* so that

$$Z_{SK}^{*}(u) = m + \sum_{a=1}^{n(u)} \lambda_a^{SK}(u)\left[Z(u_a) - m\right]$$

The error variance $\sigma_E^2 = Var\{Z_{SK}^{*}(u) - Z(u)\}$ is minimized under the unbiasedness constraint to determine the *n(u)* weights. Simple kriging considers the mean *m(u)* to be known and constant throughout the study area *A*.

Ordinary Kriging

In ordinary kriging, it is assumed that there is an unknown mean. The linear estimator can be written as:

$$Z_{OK}^{*}(u) = m(u) + \sum_{a=1}^{n(u)} \lambda_a(u)\left[Z(u_a) - m(u)\right] = \sum_{a=1}^{n(u)} \lambda_a(u) Z(u_a) + \left[1 - \sum_{a=1}^{n(u)} \lambda a(u)\right] m(u)$$

The unknown local mean is filtered from the linear estimator by making the sum of kriging weights to one. The ordinary kriging estimator can be written as:

$$Z_{OK}^{*}(u) = \sum_{\alpha=1}^{n(u)} \lambda_\alpha^{OK}(u) Z(u_\alpha) \; with \; \sum_{\alpha=1}^{n(u)} \lambda_\alpha^{OK}(u) = 1$$

Kriging with a Trend (Universal Kriging)

Kriging with a trend (universal kriging) is like ordinary kriging, however, a linear or higher-order trend in the (*x, y*) coordinates of the data points rather than fitting a local mean in the neighbourhood of the estimation point is fitted. A local linear trend model would be given by

$$m(u) = m(x, y) = a_0 + a_1 x + a_2 y$$

Cokriging

Cokriging utilizes the relationship of one variable with another variable, which is generally more abundant. To make better predictions, the autocorrelation for the variable of interest as well as cross-correlations between that variable and other variables are used (Johnston et al. 2001).

Indicator Kriging

Indicator kriging creates binary data through the use of a threshold for continuous data if the observed data itself is not binary in nature. Like ordinary kriging, here also, there is an unknown mean and autocorrelated random component ε(s). It can be expressed as

$$\mathrm{I(s)} = \mu + \varepsilon(\mathrm{s})$$

where

I(s) is a binary variable
μ is an unknown constant
ε(s) is random autocorrelated error

4.4 Applications of Geostatistics in the Study of Arsenic Contamination

Hossain and Sivakumar (2005) proposed ordinary kriging to predict safe and unsafe zones at nonsampled locations using ground water arsenic contamination data prepared by DPHE/BGS/DFID (2000). Ordinary kriging has been used by various workers in prediction of arsenic in groundwater. Hill et al. (2009) used arsenic concentration of water samples from wells shallower than 150 m. The samples were averaged for 5 km × 5 km grid. The data were log-transformed. The transformed data has a Gaussian and symmetric distribution. They used Monte Carlo framework for the assessment of kriging estimation.

Gaus et al. (2003) attempted disjunctive kriging of arsenic concentration in groundwater in Bangladesh. They observed that taking logarithms of arsenic values the skewness can be removed, but the distribution is much flatter than a normal one. For that reason, they attempted disjunctive kriging instead of long normal kriging.

Here, data are converted to normal scores using Hermite polynomials. Simple kriging of these polynomials was done, the kriged polynomials are back-transformed in the original scale. Probabilities of values of unsampled points excluding specific thresholds are also obtained.

Lee et al. (2007) used indicator kriging for the evaluation of potential health risk of arsenic-affected groundwater in northeastern Taiwan. The experimental variogram is fitted with a theoretical model. There exists geometric anisotropy. Indicator kriging is used here to estimate the probability distribution of arsenic concentrations. Moreover, the target cancer risk (TR) and dose response function have been used to estimate the number and severity of various cancer-affected populations, respectively.

Antunes and Albuquerque (2013) used indicator kriging for the evaluation of potential arsenic contamination in abandoned mining areas in Portugal. As a standard practice, an experimental semivariogram $\hat{\gamma}(h)$ has been computed and subsequently fitted by a theoretical model $\gamma(h)$, an omnidirectional spherical model. As use of indicator geostatistics allows the mapping of probability of the attribute values to exceed as many certain cut off values, they prepared maps with two cut off values, 0.05 mg/L and 0.01 mg/L. The probability maps prepared using indicator kriging show two strong anomalies which indicate a high probability for arsenic contamination and that water should not be used for human consumption.

Andrade and Stigter (2013) have used multivariate geostatistical modelling to assess distribution of arsenic in shallow alluvial groundwater under agricultural land in central Portugal.

5 Applications of Remote Sensing, GIS and Geostatistics in Arsenic Contamination—A Case Study from West Bengal, India

Arsenic contamination in groundwater is one of the major geoenvironmental problems in the eastern part of the Gangetic plain. Several studies have already been attempted to know the status of arsenic contamination in groundwater. The arsenic-affected areas of West Bengal belong to the part of Bengal basin of the Ganga valley bordered by Himalayan fore deep zone in the north and shelf and midbasinal zone in the low-lying alluvial plain (younger alluvial plain; west and east of the Bhagirathi/Hooghly river respectively) in the south (cf. Ghosh and Majumdar 1990). Arsenic contamination in groundwater is present in the Gangetic plain–Bhagirathi/Hooghly and Ganga/Padma interfluve which is drained by a number of distributaries of Ganga–Padma river system (Chakrabarti and Sarkar 1987). Arsenic is also reported in the narrow strip of low-lying land developed in the western side of the Bhagirathi/Hooghly river, bordered by older alluvial plain in the west also—the younger alluvial plain.

Department of Science and Technology (DST), Government of West Bengal carried out projects with UNICEF and Department of Science and Technology, Government of India to identify arsenic-affected tube wells. Arsenic content of water samples collected from different government tube wells were evaluated.

A pilot scale study (Saha and Ghosh 2013) was undertaken on delineation of arsenic hazard zones using data collected from 79 arsenic affected blocks of eight districts of the state of West Bengal as a part of a study conducted by DST, Government of West Bengal. The study has been done in the western part of Malda district, West Bengal. The area is one of the most affected areas of the state in terms of arsenic contamination (Das et al. 2007). The objective of this study was to delineate safe and unsafe zones.

The total area was divided into a grid having 1 km^2 cells. Using the mean value of the data falling under each cell, the study was undertaken to identify the safe and unsafe zones. The international (WHO) and national standards of permissible limit of arsenic in drinking water are 0.01 mg/L and 0.05 mg/L respectively (Appelo 2006). In different parts of Malda district, it is found that in many water samples, arsenic contamination is above 0.01 mg/L, and even 0.05 mg/L. The study area—comprising of blocks like Manikchak, Ratua I, Ratua II, Englishbazar, Kaliachak I, Kaliachak II and Kaliachak III of Malda district—is situated between the latitude and longitude figures of 24°40′20″N to 25°32′08″N and 88°28′10″E to 88°45′50″E, respectively.

The data was stored in a GIS as point feature data with arsenic content as attribute. WGS 84 datum and UTM projection were used in this system. Geological, geomorphological, drainage and elevation maps are also brought in this GIS. Remote sensing as well as secondary data have been used to prepare these thematic layers.

Statistical analysis showed that distribution of data was log normal in nature. Hence, data was log transformed. Spherical model was fitted in the obtained experimental semivariograms. Ordinary kriging is used to interpolate the arsenic content at an unsampled location using parameters derived from semivariogram model. As already stated, ordinary kriging interpolates values by assuming a constant but unknown mean using the formula:

$$Z(s) = \mu + \varepsilon(s)$$

where

$Z(s)$	is the variable of interest
s	is location
μ	is constant mean
$\varepsilon(s)$	is random autocorrelated errors.

After ordinary kriging, cross validation of the interpolated values with real values was done. The root mean square value, average standard error, mean standardized and root mean square standardized values are found to be within acceptable limits. Three different zones of the study area indicating safe zone (below 0.01 mg/L), moderately safe zone (between 0.01 mg/L and 0.05 mg/L) and unsafe zone (above 0.05 mg/L) have been delineated (Fig. 1).

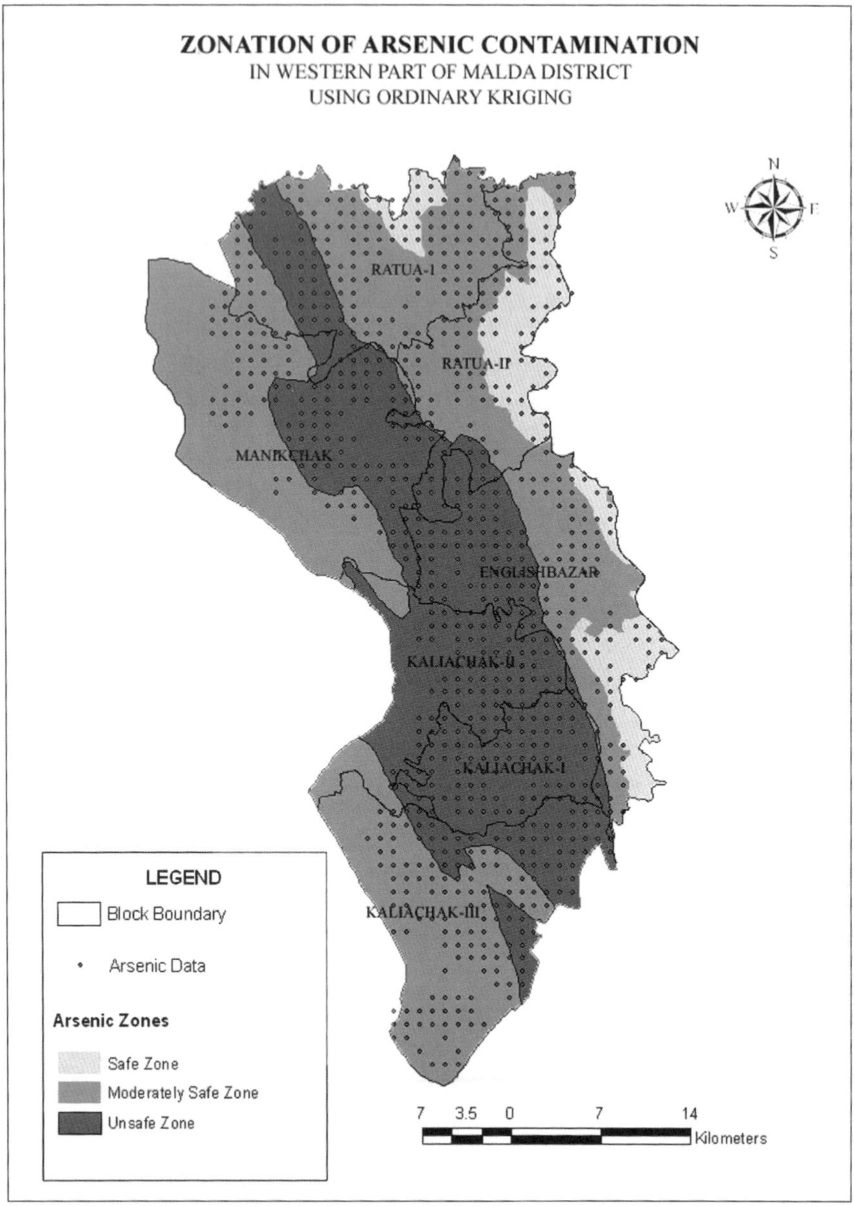

Fig. 1 Arsenic zonation map prepared using ordinary kriging in western part of Malda District (Saha and Ghosh, 2013)

Acknowledgement The first author acknowledges support provided and permission given by DST, Government of West Bengal for publishing this paper as well as for carrying out the case study on arsenic zonation.

References

Acharyya SK, Shah BA (2007) Arsenic-contaminated groundwater from parts of Damodar fan-delta and west of Bhagirathi river, West Bengal, India: influence of fluvial geomorphology and quaternary morphostratigraphy. Environ Geol 52:489–501

Ade By R (2004) Principles of Geographical Information Systems. ITC Educational textbook series 1 (3rd edn). ITC Enschede the Netherlands, p 250

Ahsan DA, Del Valls TA (2011) Impact of arsenic contaminated irrigation water in food chain: an overview from Bangladesh. Int J Environ Res 5(3):627–638

Andrade AIASS, Stigter TY (2013) The distribution of arsenic in shallow alluvial groundwater under agricultural land in central Portugal: Insights from multivariate geostatistical modelling. Sci Total Environ 449:37–51

Antunes IMHR, Albuquerque MTD (2013) Using indicator kriging for the evaluation of arsenic potential contamination in an abandoned mining area (Portugal). Sci Total Environ 442:545–552

Appelo T (ed) (2006) Arsenic in groundwater-a world problem. In: Proceedings of the Seminar, Organised by Netherlands National Committee of the International Association of Hydrologists, Utrecht

Bhattacharya P, Chatterjee D, Jacks G (1997) Occurrence of arsenic-contaminated groundwater in alluvial aquifers from delta plains, eastern India: options for safe drinking water supply. Int J Water Resour Dev 13(1):79–92

Bhattacharya P, Claesson M, Bundschuh J, Sracek O, Fagerberg J, Jacks G, Martin RA, Storniolo AR, Thir JM (2006) Distribution and mobility of arsenic in the Río Dulce alluvial aquifers in Santiago del Estero Province, Argentina. Sci Total Environ 358:97–120

Boutt DF, Hyndman DW, Pijanowski BC, Long DT (2001) Identifying potential land use-derived solute sources to stream base flow using groundwater models and GIS. Groundwater 39(1):24–34

Burrough PA, McDonnell RA (1998) Principles of Geographical Information Systems. Oxford University Press, Oxford, p 333

Chakraborti D, Basu GK, Biswas BK Chowdhury UK, Rahman MM, Paul K, Roy Chowdhury T, Chanda CR, Lodh D , Ray SL (2001) Characterization of arsenic-bearing sediments in the Gangetic delta of West Bengal, India. In: Chappell WR, Abernathy CO, Calderon RL (eds) Arsenic exposure and health effects IV. Elsevier Science, Oxford

Chakraborti P, Sarkar K (1987) Photogeomorphological appraisal of geo-environment and land use pattern around Calcutta metropolis. In: Proc Sem Earth Resources Utilization and Environmental Appraisal. Min Met Inst India pp cs-1–13

Chang KT (2011) Introduction to geographic information systems, 6th edn. McGraw Hill Higher Education, New York, p 450

Corwin DL, Vaughan PJ, Loague K (1997) Modeling nonpoint source pollutants in the vadose zone with GIS. Environ Sci Technol 31(8):2157–2175

Das D, Das T, Mukherjee A (2007) Development of map-based information system for arsenic affected areas in West Bengal. In: Ghosh, A (ed) Environment, drinking water and public health-problems and future goals. Daya, Delhi, pp 17–27

Gaus I, Kinniburgh DG, Talbot JC, Webster R (2003) Geostatistical analysis of arsenic concentration in groundwater in Bangladesh using disjunctive kriging. Environ Geol 44:939–948

Ghosh RN, Majumdar S (1990) Geology and morphostratigraphy of West Bengal—A database for archaeological exploration. In: Dutta A (ed) Studies in archaeology. Books & Books, Janakpuri, pp 7–40

Gogu R, Carabin G, Hallet V, Peters V, Dassargues A (2001) GIS-based hydrogeological databases and groundwater modeling. Hydrogeol J 9(6):555–569
Goovaerts P (1997) Geostatistics for natural resources evaluation. Oxford University Press, Oxford, p 483
Hall MD, Shaffer MJ, Waskom RM, Delgado JA (2001) Regional nitrate leaching variability: what makes a difference in northeastern Colorado. J Am Wat Res 37(1):139–150
Hassan MM (2005) Arsenic poisoning in Bangladesh: spatial mitigation planning with GIS and public participation. Health Policy 74(3):247–260
Hill J, Faisal H, Bagtzoglou AC (2009) Zonal management of arsenic contaminated ground water in northwestern Bangladesh. J Environ Manag 90:3721–3729
Hossain F, Sivakumar B (2005) Spatial pattern of arsenic contamination in shallow wells of Bangladesh: regional geology and nonlinear dynamics. Stoch Env Res Risk A 20:66–76
Hossain MM, Piantanakulchai M (2013) Groundwater arsenic contamination risk prediction using GIS and classification tree method. Eng Geol 156:37–45
International Agency Research Centre (IARC) (2012) Monographs on the evaluation of carcinogenic risks to humans: arsenic, metals, join next parafibres, and dusts: a review of human carcinogens, 100 C. International Agency for Research on Cancer, Lyon, p 41–94.
Jha MK, Chowdhury A, Chowdary VM, Peiffer S (2007) Groundwater management and development by integrated remote sensing and geographic information systems: prospects and constraints. Water Resour Manag 21:427–467
Johnston K, Ver Hoef JM, Krivoruchko K, Lucas N (2001) Using ArcGIS geostatistical analyst, ESRI, New York, p 300
Journel AG, Huijbregts CJ (1978) Mining geostatistics. Academic Press, New York, p 600
Kar S, Maity JP, Jean JS, Liu CC, Nath B, Yang HJ, Bundschuh J (2010) Arsenic enriched aquifers: occurrences and mobilization of arsenic in groundwater of Ganges delta plain, Barasat, West Bengal, India. Appl Geochem 25:1805–1814
Kerle N, Janssen LLF, Huurneman GC (2004) Principles of remote sensing. ITC, Educational textbook series 2 (3rd edn). ITC Enschede the Netherlands, p 250
Khan NI, Owens G, Bruce D, Naidu R (2009) Human arsenic exposure and risk assessment at the landscape level: a review. Environ Geochem Health 31:143–166
Lasserre F, Razack M, Banton O (1999) A GIS-linked model for the assessment of nitrate contamination in ground water. J Hydrol 224:81–90
Lee JJ, Jang CS, Wang SW, Liu CW (2007) Evaluation of potential health risk of arsenic-affected groundwater using indicator kriging and dose response model. Sci Total Environ 384:151–162
Lillesand TM, Kiefer RW, Chipman JW (2004) Remote sensing and image interpretation, 5th edn. John Wiley and Sons (Asia) Pte. Ltd, Singapore, p 763
Manahan SE (1994) Environmental chemistry, 6th edn. Lewis, Boca Raton
Neitsch SL, Arnold JG, Kiniry JR, Williams JR (2009) Soil and water assessment tool theoretical documentation, version 2000. Texas Water Resources Institute Technical report no 406, Texas, p 618
Ravenscroft P, Burgess WG, Ahmed KM, Burren M, Perrin J (2005) Arsenic in groundwater of the Bengal basin, Bangladesh: distribution, field relations, and hydrogeological setting. Hydrogeol J 13:727–751
Reimann C, Caritat P de (1998) Chemical elements in the environment: factsheets for the geochemist and environmental scientist. Springer-Verlag Berlin Heidelberg
Saha C, Ghosh AR (2013) Prediction and zonation of arsenic contamination by geostatistical method in GIS platform. In: Abstract volume of the 20th WB State Science and Technology Congress, p 418
Sahoo PK, Mukhopadhyay A (2014) Arsenic fate and transport in the groundwater-soil-plant system: an understanding of suitable rice paddy cultivation in arsenic enriched areas. In: Recent trends in modeling of environmental contaminants, D. Sengupta (Ed)
Sikdar PK, Sarkar SS, Palchoudhury S (2001) Geochemical evolution of groundwater in the quaternary aquifer of Calcutta and Howrah, India. J Asian Earth Sci 19:579–594
Smedley PL, Kinniburgh DG., 2002, A review of the source, behaviour and distribution of arsenic in natural waters. Appl Geochem 17:517–568
Watkins DW, McKinney DC, Maidment DR, Lin MD (1996) Use of geographic information systems in groundwater flow modeling. J Water Resour Plan and Manag ASCE 122(2):88–96

Modeling Radon Behavior for Characterizing and Forecasting Geophysical Variables at the Atmosphere–Soil Interface

Antonello Pasini, Roberto Salzano and Alessandro Attanasio

Abstract As well known, noble gases are often used as stable tracers in several geophysical environments, due to their basic property of being chemical noninteracting. Among these noble gases, the attention of researchers in the last decades has been focused on radon.

Here, we review the use of radon data in characterizing and forecasting geophysical variables at the atmosphere-soil interface. In particular, we analyze the applications to the study of the diffusive properties of the low atmosphere, to the research in the fields of pedology and sedimentology, where radiogenic nuclides are employed as stratigraphic markers, and to the investigations on radon exhalations in the framework of earthquake characterization and forecasting. In doing so, particular attention is paid to modeling and statistical approaches based on radon data.

In an atmospheric framework, it has been shown that radon concentration can be a good tracer of stability of the boundary layer (BL) (the lowest part of the atmosphere). In particular, this led to working out diagnostic models for the estimation of the mixing depth from radon data and prognostic neural network (NN) models for forecasting its future behavior, thus giving important information about the meteorological influences on air pollution and its possible peak events. Here, we briefly review these studies and present the structure of an improved model.

Furthermore, the interaction between radon by-products, pedosphere, and hydrosphere is commonly assumed to be an important tool for the characterization of soil and of sediment dynamics. Radon daughters are an efficient stratigraphic marker that is useful for modeling accumulation and erosion rates associated to the recent evolution of landscapes. Considering the wide range of box model approaches proposed for geomorphological purposes in the last years, a review of such applications will be illustrated.

Finally, in the framework of studies about earthquakes, we briefly review the use of statistical analyses based on radon data for characterizing and, possibly, forecasting these phenomena.

Thus, the present paper supplies a review of radon detection and the related modeling activities and statistical analyses at the atmosphere–soil interface. It clearly

A. Pasini (✉) · R. Salzano · A. Attanasio
CNR—Institute of Atmospheric Pollution Research, Rome, Italy
e-mail: pasini@iia.cnr.it

D. Sengupta (ed.), *Recent Trends in Modelling of Environmental Contaminants*,
DOI 10.1007/978-81-322-1783-1_9, © Springer India 2014

shows the importance of these studies for a better understanding of processes and phenomena that occur at this delicate and complex interface.

Keywords Radon · Modeling · Boundary layer (BL) · Soil emanation · Ground deposition · Earthquakes

1 What is Radon?

Radon is a noble gas that, from a chemical point of view, is relatively inert and which occurs in nature as intermediate by-product of the U–Th decay series. Furthermore, it is relatively mobile in the environment due to its gaseous state and it is sparingly soluble in water and appreciably more soluble in organic liquids (Appleby 2001). This element is detectable in the environment by three radioactive isotopes having atomic mass ranging between 219 and 222. The most abundant nuclide is ^{222}Rn that is the one to which we commonly refer as "*radon*". This is an isotopic daughter of ^{226}Ra and it is a member of the ^{238}U decay-series. The second most abundant nuclide is ^{220}Rn, by-product of the radioactive decay associated to the ^{224}Ra–^{232}Th chain, which is commonly known as "*thoron*". The rarest isotope is ^{219}Rn, historically known as "*actinon*", which is produced by the decay of ^{227}Ac and which is associated to the ^{235}U decay-series.

The description of the three decay-series (Fig. 1) mentioned above shows the relationship between parent and daughter nuclides. Before examining in detail the different chains, it is important to recall some basic concepts about radioactivity. The decay from a radiogenic "parent" to a "daughter" isotope occurs by emitting α or β particles. While α particles are formed by two protons and two neutrons (one Helium nucleus), β particles could be positrons (β^+) or electrons (β^-) generated by the collapse of a neutron in the nucleus. When a β decay occurs, the emission of the particle could be associated to the emission of a gamma radiation (γ). The nature of the decay is not the only feature that can distinguish one radioactive chain from another, thus the energy of each decay particle or radiation is specific for all the single nuclear structures. From this point of view it is possible to fingerprint every parent–daughter decay, which is characterized by another important feature: the "lifetime" of each isotope.

The radioactive decay follows the decay law of the type

$$\frac{\partial c}{\partial t} = -C\lambda \tag{1}$$

where the solution of the differential equation is

$$C = C_0 e^{-\lambda t} \tag{2}$$

C_0 is the initial number of atoms at $t=0$ and λ is the decay constant

$$^{238}U \Rightarrow \; ^{226}Ra \underbrace{\overbrace{\Rightarrow}^{\alpha}}_{1600y} {}^{222}Rn \underbrace{\overbrace{\Rightarrow}^{\alpha}}_{3.8d} {}^{218}Po \underbrace{\overbrace{\Rightarrow}^{\alpha}}_{3m} {}^{214}Pb \underbrace{\overbrace{\Rightarrow}^{\beta}}_{27m} {}^{214}Bi \underbrace{\overbrace{\Rightarrow}^{\beta}}_{20m} {}^{214}Po \underbrace{\overbrace{\Rightarrow}^{\alpha}}_{<1s} {}^{210}Pb \underbrace{\overbrace{\Rightarrow}^{\beta}}_{20y} {}^{210}Bi \underbrace{\overbrace{\Rightarrow}^{\beta}}_{<1s} {}^{210}Po \underbrace{\overbrace{\Rightarrow}^{\alpha}}_{140d} {}^{206}Pb$$

$$^{232}Th \Rightarrow \; ^{228}Ra \underbrace{\overbrace{\Rightarrow}^{\beta}}_{7y} {}^{228}Ac \underbrace{\overbrace{\Rightarrow}^{\beta}}_{6h} {}^{228}Th \underbrace{\overbrace{\Rightarrow}^{\alpha}}_{2y} {}^{224}Ra \underbrace{\overbrace{\Rightarrow}^{\alpha}}_{4d} {}^{220}Rn \underbrace{\overbrace{\Rightarrow}^{\alpha}}_{1m} {}^{216}Po \underbrace{\overbrace{\Rightarrow}^{\alpha}}_{<1s} {}^{212}Pb \underbrace{\overbrace{\Rightarrow}^{\beta}}_{11h} {}^{212}Bi \underbrace{\overbrace{\Rightarrow}^{\beta/\alpha}}_{1h} {}^{212}Po/{}^{208}Tl \underbrace{\overbrace{\Rightarrow}^{\alpha/\beta}}_{3m} {}^{208}Pb$$

$$^{235}U \Rightarrow \; ^{227}Ac \underbrace{\overbrace{\Rightarrow}^{\beta}}_{22y} {}^{227}Th \underbrace{\overbrace{\Rightarrow}^{\alpha}}_{20d} {}^{223}Ra \underbrace{\overbrace{\Rightarrow}^{\alpha}}_{11d} {}^{219}Rn \underbrace{\overbrace{\Rightarrow}^{\alpha}}_{4s} {}^{215}Po \underbrace{\overbrace{\Rightarrow}^{\alpha}}_{<1s} {}^{211}Pb \underbrace{\overbrace{\Rightarrow}^{\beta}}_{36m} {}^{211}Bi \underbrace{\overbrace{\Rightarrow}^{\alpha}}_{2m} {}^{207}Tl \underbrace{\overbrace{\Rightarrow}^{\beta}}_{<5m} {}^{207}Pb$$

Fig. 1 Decay series where radon nuclides are associated

$$t_{1/2} = \frac{ln2}{\lambda} \quad (3)$$

The decay constant can be measured, depending on the half-life time ($t_{1/2}$), and supports the discrimination between short- (half-life time less than days) and long-lived (half-life time more than years) isotopes. While ^{222}Rn is associated directly to a long-lived parent (^{226}Ra) with a complete alpha decay, ^{220}Rn and ^{219}Rn derive from α and β decays of short-lived parents that can be associated to the respective long-lived parents ^{228}Th and ^{227}Ac. Changing the point of view from radon as daughter of a decay-series to radiogenic parent, ^{222}Rn, ^{220}Rn, and ^{219}Rn have a half-life of about 4 days, about 1 minute, and about 4 s, respectively. While thoron and actinon by-products are short-lived isotopes which are detectable within few hours, only ^{222}Rn has a long-lived daughters that are ^{210}Pb (half-life of about 22 years).

This framework offers the opportunity of using radon as tracer of several geophysical processes at the atmosphere–ground interface. The geochemical cycle of radon is simplified by the chemical low-reactivity of this element. As a direct consequence, this gas is emitted by minerals due to U–Th decay and emanation is controlled by the lithological features (porosity, humidity, fracture, mineralogy) of the source rock. Once the gas is released, the ground emanation is influenced by the hydrological condition of the air–ground interface. The persistence of radon in the atmosphere is, subsequently, controlled by the meteo-climatological settings that create the conditions for the deposition of radon by-products in the ground. This simplified picture will be exploited in the next paragraphs and it highlights the opportunity to model geophysical processes having a time marker such as radon.

2 Relevance of Radon as a Tracer of the Low Atmosphere and Development of a Box Model for Estimation of the Stable Layer Depth

The first studies on the role of radon in characterizing the dispersion properties of the atmospheric boundary layer (BL) dated back to the late 1970s, when French researchers began to consider this noble gas, which undergoes no chemical reaction, as a perfect tracer of the BL dilution features. They found that counts of β radioactivity, coming from the decay of short-lived radon daughters, represent a simple index of the stability state of the BL (see Guedalia et al. 1980).

Here we do not discuss the instrumentation for detecting β counts from the decay of short-lived radon progeny: we just stress that the fraction attached to particulate matter is usually detected (see Allegrini et al. 1994, for details).

The typical time patterns of β counts are maxima during the night in conditions of nocturnal stability and minima during the day when the mixed layer is well developed (the more enhanced the stability of the nocturnal stable layer, then the higher the maxima in radioactivity counts). Otherwise, low quasi-constant values are found in advective situations characterized by mechanical turbulence. This qualitative analysis suggests that, in general, the number of β counts can be inversely proportional to the "degree" of stability of the lower layers.

Lopez et al. (1974) (see also Vinod Kumar et al. 1999, for more recent results) show that radon and radon daughters' concentrations are approximately homogeneous with altitude in the nocturnal stable layer and that they undergo a rapid transition to background values above the mixing height in the so-called residual layer. This fact induced Guedalia et al. (1980) to use a box model of the nocturnal stable layer endowed with a homogeneous radon concentration in the vertical. The top of this box, named equivalent mixing height h_e, has been found to be a good index of the dispersion properties of this layer in a semi-quantitative way, because low (high) values of h_e are related to low (high) dispersion power and high (low) concentrations of primary pollutants.

In Guedalia et al. (1980) the calculation of the top of the box has been performed by means of the following equation:

$$h_e(t) = \frac{\Phi \Delta t}{C(t) - C_0} \tag{4}$$

Here Φ is the radon flux at the surface, Δ*t* is the time interval from the start of accumulation, *C(t)* is the radon concentration at time *t* and C_0 is the radon concentration at the beginning of accumulation (evening).

We stress that here stable and nonadvective conditions for single nights were considered, when rain is absent, relative humidity and pressure are quite constant and the space-time interval to be analyzed is limited. In these situations, radon exhalation from the ground (its flux Φ) has been considered constant in time and spatially homogeneous, the attached fraction of radon daughters also constant and the radon concentration directly proportional to the number of β counts detected.

Allegrini et al. (1994) show that h_e can be properly identified with the height at which a parcel of air coming from the ground halts its free convection, at least in nocturnal stable situations dominated by the thermal factor.

In order to quantify the nocturnal stable layer depth over a town, they used a thermal profile from a radiosonde station in the suburbs and the surface air temperature at the radon detection site inside the town. With a potential temperature method they were able to estimate an urban mixing height, h_u, from meteorological data. A statistical analysis shows that h_e and h_u are highly correlated, so inducing to think that the box model output represents a correct estimation of the urban mixing height, at least in situations of high nocturnal stability.

As a matter of fact, the structure of the box model described above is too simple for at least two reasons: first, radon decay is neglected; secondly, entrainment of air with different radon concentrations is not allowed from the top of the box (this is critical just in situations characterized by nocturnal fluctuations of the stable layer depth). A new model structure, which includes these elements, has been developed in Pasini (2009b) and is briefly presented here.

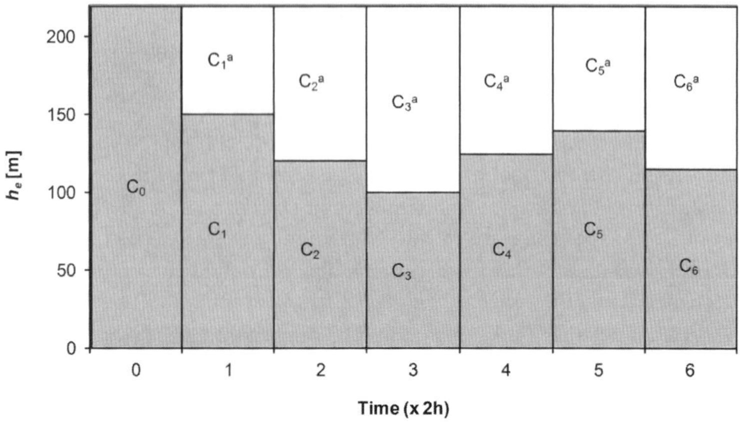

Fig. 2 Time evolution of the nocturnal stable layer (*gray boxes*). The *white* areas above represent the residual layer

Referring to Fig. 2 "compressions" are the situations in which the stable layer depth decreases (i=1, 2, 3, 6) and "expansions" are the cases when h_e increases (i=4, 5). In what follows, λ represents the decay constant of radon, Δt is our sampling rate (usually 1 or 2 hours), C^a is the calculated concentration in the residual layer and we adopt the symbolic form $\Delta h_e(\text{n, m})=h_e(\text{n})-h_e(\text{m})$ for the difference between equivalent mixing heights at time n and m, respectively.

In compression cases the generalization of Eq. 4 reads as follows:

$$h_e(i)=\frac{\Phi}{\lambda}\frac{1-\exp(-\lambda\Delta t)}{C_i-C_{i-1}\exp(-\lambda\Delta t)} \tag{5}$$

The concentration left in the residual layer after the i-th compression is:

$$C_i^a=\frac{C_{i-1}^a\exp(-\lambda\Delta t)\cdot\Delta h_e(0,i-1)+C_{i-1}\exp(-\lambda\Delta t)\cdot\Delta h_e(i-1,i)}{\Delta h_e(0,i)} \tag{6}$$

If the stable layer depth increases and overlying air is included in the box, that is, in cases of expansions, the equivalent mixing height can be calculated as:

$$h_e(i)=\frac{(\Phi/\lambda)\left[1-\exp(-\lambda\Delta t)\right]+h_e(i-1)\left(C_{i-1}-C_{i-1}^a\right)\exp(-\lambda\Delta t)}{C_i-C_{i-1}^a\exp(-\lambda\Delta t)} \tag{7}$$

and the concentration over the top of the box is of course

$$C_i^a=C_{i-1}^a\exp(-\lambda\Delta t) \tag{8}$$

This enhanced structure of the box model, tested on past results, leads to a little increase (about 1 %) in the value of the linear correlation coefficient between the total set of h_e estimated by the new model and the total set of h_u. However, a major, statistically significant improvement is obtained in critical situations such as nocturnal fluctuations of the stable layer depth, when the application of Eqs. (5–8) prevents the estimation of wide unphysical oscillations.

Even if this improvement is certainly significant, however, in order to extend the model to estimations for longer periods, we need to consider that actually the flux Φ is not constant but it depends on several variables, such as humidity of the soil, temperature, and so on. In what follows we present an improved model with variable soil flux and its application to a case study in Beijing (China).

2.1 *Soil Radon Flux Submodel*

The radon flux originated from the surface is a complex process influenced by many factors (Zhuo et al. 2008; Sun et al. 2004; Voltaggio et al. 2006). Although pedology and geology are disciplines that are not commonly involved in atmospheric modeling, the support provided by a multidisciplinary approach focused on radon emanation from soil is important. The definition of the radon-emitting source can, in fact, improve such models (Szegvary et al. 2007).

The simplest way for predicting the exhalation rate is represented by using idealized models based on the porous media transport theory. This submodel follows up the direction of Zhuo et al. (2008), which proposed a combined model where the soil radon emanation power and the soil water saturation are the main parameters that control radon flux:

$$\Phi = R\rho_b \varepsilon \left(\frac{T_S}{273} \right)^{0.75} \sqrt{\lambda D_0 p e^{-6Sp-6S^{14p}}} \tag{9}$$

In this equation Φ is the radon flux (Bq/m^2), R is the ^{226}Ra soil content (Bq/kg), ρ_b is the soil bulk density (kg/m^3), T is the soil temperature (°K), λ is the ^{222}Rn decay constant (s^{-1}), D_0 is the ^{222}Rn diffusion coefficient in air, S is the soil water saturation, p is the soil total porosity and ε is the radon emanation power. In this equation we assume for Beijing a ^{226}Ra soil content of 21.4 Bq/kg (Sun et al. 2004), a soil bulk density of 1,500 kg/m^3 (Sun et al. 2004) and a soil total porosity of 0.445 (Sun et al. 2004). λ and D_0 are constants and they are respectively $2.1*10^{-6}$ s^{-1} and $1.1*10^{-5}$ m^2/s. Soil temperature, water saturation, and emanation power are time-dependent variables that are influenced by meteorological conditions. The latter can be estimated by using the equation suggested by Zhuo et al. (2008):

$$\varepsilon = \varepsilon_0 \left[1 + a \left(1 - e^{-bS} \right) \right] \tag{10}$$

The emanation power at 25 °C (ε_0) and the two constants a and b are specific for silty soils, as the ones outcropping in the Beijing area. The former parameter, selected for the Beijing area, is 0.14 as suggested by Zhuo et al. (2008).

Considering these equations, the estimation of hourly fluxes requires the solution of heat and hydraulic balances in order to predict variations in terms of soil temperature and water saturation. The definition of a box model with a single layer (1 m height) constituted by the superficial soil represents a preliminary approach for solving these problems. At this stage several constraints are necessary for having a simple solution, but further development of the model is required for improving the prediction. Assuming that water can only infiltrate vertically (no runoff and horizontal fluxes are possible), that no heat and water gradients are present in this layer, the hydraulic balance can be defined as:

$$S = \frac{V_{water}}{V_{pores}} = \frac{V_{water}}{V_{soil} \cdot p} = \frac{P_h - ET_0}{V_{soil} \cdot p \cdot 1000} \tag{11}$$

The soil water saturation is defined as the ratio between water volume and pores volume. This relationship can be rewritten as function of soil porosity and consequently related to P_h (hourly precipitation in mm), ET_0 (hourly evapotranspiration), V_{soil} (the considered soil volume in m^3) and p as soil porosity.

Moreover the thermal balance can be expressed as follows:

$$\Delta T = \frac{G}{C_v h} \tag{12}$$

where ΔT is the soil temperature variation (°C), G is the soil heat flux (MJ m^{-2} h^{-1}), C_v is the volumetric heat capacity (J m^{-3} °K^{-1}) and h is the layer thickness (m). In detail the volumetric heat capacity can be computed as:

$$C_v = c_{soil} \rho_b + c_w \rho_w S \tag{13}$$

where c_{soil} is the average heat capacity of solid constituents in soil (J kg^{-1} °K^{-1}), ρ_b is the soil bulk density, ρ_w is the water density, c_w is the water heat capacity, and S is the soil water saturation.

The calculation of ET_0 and G were performed following the approach proposed by Alexandris and Kerkides (2003), who modified the Penman equation:

$$G = 0.1 \cdot R_n \,(\text{day}) \tag{14}$$

$$G = 0.5 \cdot R_n \,(\text{night}) \tag{15}$$

The soil heat flux in this case is related to R_n, that is the net radiation received by the surface (MJ m^{-2} h^{-1}).

$$ET_0 = \frac{\frac{\Delta}{\Delta+\gamma}(R_n - G) + 0.268\frac{\gamma}{\gamma+\Delta}(a_w + b_w u)VPD}{L} \tag{16}$$

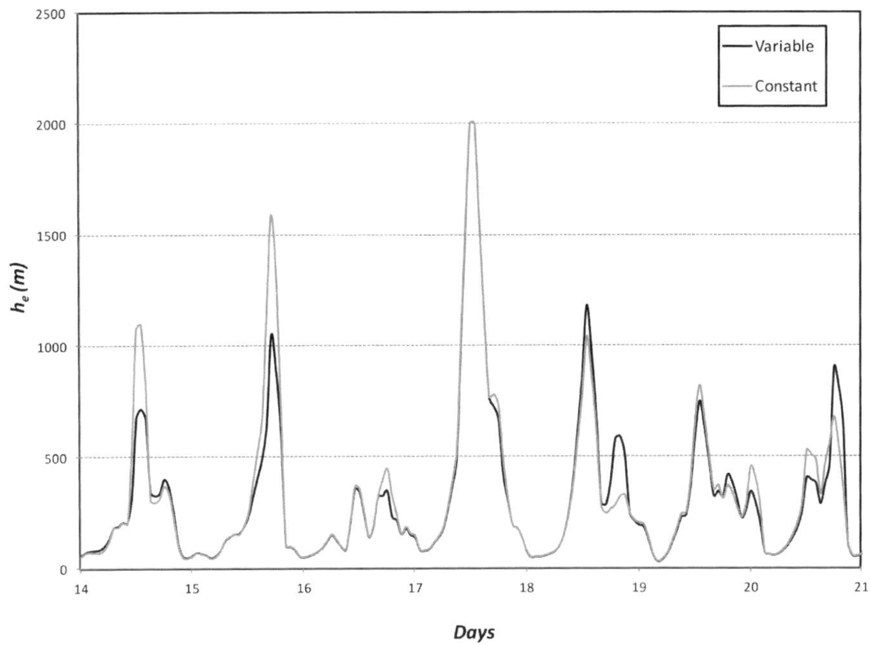

Fig. 3 Modeled equivalent mixing height (h_e) using constant and variable soil radon flux

This equation is the modified Penman equation, where ET_0 is the hourly evapotranspiration (mm), u is the wind speed (m/s), R_n is the net radiation (MJ m^{-2} h^{-1}), γ is the psychrometric constant (kPa/° C), Δ is the slope of the saturation vapour pressure curve (kPa/° C), L is the latent heat of vaporization (MJ/kg) and VPD is the vapour pressure deficit (kPa). The values of γ, VPD, and Δ depend on air temperature, pressure, and humidity and they were calculated using functions described in the same reference (Alexandris and Kerkides 2003). The net radiation was estimated using astronomical functions and considering the cloud attenuation obtained from cloudiness observations following Kasten and Czeplak (1980).

Once the soil flux is parameterized by the just-presented submodel, we are able to compare the results obtained running standard and improved models (Fig. 3). Small discrepancies were observed in concomitance of expansions. Often these anomalies are associated with major variations of soil humidity. Further validations are obviously necessary in order to confirm the performance of the improved model.

3 A Model for Forecasting the Mixing Height

The physics of the BL has been recognized as very complex and modeling in this domain represents a major challenge for physicists of the atmosphere (see, for instance, Mahrt 1998). Thus, a data-driven technique could help in a framework where dynamical modeling shows several drawbacks.

In particular, neural network (NN) modeling is able to find nonlinear links between different variables in a complex system and therefore it is natural to think of its application for finding relationships between possible predictors and predictands in the BL. In what follows we show that NNs are able to perform reliable short-term forecasts of radon concentration and that these forecast data can be used as inputs for the box model previously described, so that the final result is to achieve a prediction of future mixing layer depth.

By an NN we aim to establish a nonlinear "law" between some inputs (predictors) and one or more outputs (predictands). To do so, we have to train the network on a sample of input–output pairs, fix its free parameters (weights) and test the function found on an independent set of data. If the "law" found is able to explain the relationships between inputs and outputs even on this independent set, we may affirm that our NN has "discovered" a forecast law between predictors (in our case, past values of radon and/or meteorological parameters) and our predictand (radon concentration).

The NN model adopted in what follows has been developed in the last 15 years for analyzing environmental problems in the atmospheric environment. Here we just cite that it is built on feed-forward NNs with backpropagation training and sigmoid transfer functions, and it presents several facilities for handling short time series of data: refer to Hertz et al. (1991) and Bishop (1995) for high-quality books on this kind of networks and to Pasini (2009a, 2009b) and references therein for the specific characteristic features of our model.

In this section we will briefly describe some ways of application of an NN model to the problem of forecasting radon behavior. For more details and results (referred to the case of a constant Φ), the reader can refer to Pasini (2009b).

Among data-driven methods, in the past NNs have shown high ability of forecasting, often with better results if compared with other empirical techniques (see, for instance, Weigend and Gershenfeld 1993).

In general, forecasting through NNs may follow two distinct strategies. A time series approach could be applied, when delayed values of a variable are used as inputs in order to forecast future values of the same variable; or a quite complete description of the state of the system (through several variables' values at a certain time) can be inserted as input and the NN tries to predict the future value of a variable which is able to represent a synthetic index of its properties.

In this paper both of these approaches are presented, together with a further hybrid one.

In a time series approach, when the dynamics of the system under study is unknown an NN is often used as a black box that outputs future values of the time series, given input of a sequence of its past values. Of course, we can follow this approach even in the treatment of our forecasting problem in the BL.

Nevertheless, here a part of the dynamics is known: for instance, the influence of a day–night cycle is very clear in situations characterized by nocturnal stability and well-developed diurnal mixed layers. Thus, we can choose to preprocess our data. We can filter out the known 24-hour periodicity (by means of the so-called seasonal differencing (SD) method), subtract this signal from the detected data and model

Table 1 Performance on the test set for both the time series approach (black box (TS) and the preprocessed (SD-TS) strategies) and for the synchronous pattern approach (full set of inputs (SP) and the pruned case (SP-PR))

Period	$R(t_0+2)$	$R(t_0+4)$	$R(t_0+6)$
TS	0.812 ± 0.010	0.735 ± 0.017	0.672 ± 0.017
SD-TS	0.894 ± 0.004	0.870 ± 0.004	0.861 ± 0.005
SP	0.849 ± 0.003	0.821 ± 0.005	0.788 ± 0.006
SP-PR	0.854 ± 0.002	0.820 ± 0.005	0.773 ± 0.007

the residuals with NNs. In our application, SD consists in subtracting from the time series a replica of the series itself delayed by a 24-hour time lag.

In this manner we leave the network model the hidden (unknown) dynamics. Of course, the forecasting results of the neural modeling must be added to the preprocessing function in order to obtain final forecasts of the original time series.

Once our data is preprocessed by SD, we obtain a residual time series that looks like noise, so that a man-made analysis does not allow us to catch any additional clear dynamics in these data. In this situation, the application of an NN model to forecasting this residual time series and to evaluating its forecasting performance could lead to determining whether nonlinear dynamics is hidden in these data, or rather, if they are actually random.

As a matter of fact, in a case study on data of beta counts coming from a radon detector sited close to Rome (Italy), we find quite good performance on the test set of residuals when forecasts were extended to 2, 4 and 6 h after the most recent data in the time series. In particular, we calculate the linear correlation coefficient (modeled vs. detected) and obtain $R(t_0+2\text{h})=0.675 \pm 0.006$, $R(t_0+4\text{h})=0.500 \pm 0.008$, and $R(t_0+6\text{h})=0.431 \pm 0.009$. Here the error bars associated with the NN performance come from ensemble runs of the model with different initial random weights, so that each network is able to widely explore the landscape of its cost function, and represent ± 2 standard deviations.

Even if these values of R are not very high (remember that the time series looks like noise), they indicate a clear signal of nonlinear dynamics hidden in the residuals.

In this way the NNs show their ability in capturing this dynamics and can add information to modeling the system. In particular, they allow us to improve our forecasting performance on the system itself. In fact, Table 1 shows how the performances of SD+NN forecasts on residuals (TS-SD) are better than the results coming from a black box application from original time series (TS).

Due to the fact that β counts from detection of short-term decay radon daughters obviously depend on the physical state of the BL, it is worthwhile to test if an accurate estimation of this state (in terms of meteorological variables) could enhance the prediction of the radon behavior. With this scope, a NN model in which meteorological parameters are considered as inputs has been developed. In particular, we consider the following synchronous pattern of variables which describes the BL state at time t_0: hour of the day (expressed in two inputs as $[\sin(\pi t/12)+1]/2$ and $[\cos(\pi t/12)+1]/2$), β counts, time derivatives of β counts with respect to 2 h before, sky covering, height of the lowest cloud layer, temperature, dew point, pressure, horizontal wind speed, and visibility. Furthermore, due to

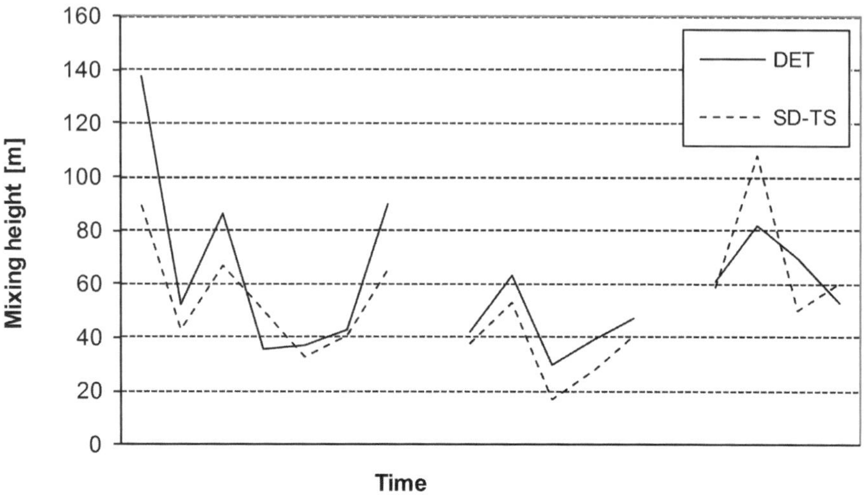

Fig. 4 The depth of the nocturnal stable layer calculated by the box model for detected data (DET) and preprocessed forecast values by time series approach (SD-TS)

the fact that, in our specific model structure (see Pasini (2009b) for further discussion), *spurious* inputs—which do not lead to important information for the future behavior of the output—tend to decrease the model performance, we prune some of them which do not show high linear and nonlinear correlation with the output. In doing so, a NN model which includes just 5 inputs (the hour of the day (two inputs), the horizontal wind speed, β counts, and time derivative of β counts) is considered.

Table 1 shows the performance results of these unpruned (SP) and pruned (SP-PR) synchronous pattern approaches in terms of the linear correlation coefficient between modeled (forecasted) and detected β counts. A comparison of the results presented in this table allows us to recognize that the synchronous pattern approach always outperforms the original time series approach (TS) in a statistically significant way. This bears witness to the relevance of meteorological information to better characterize the BL state for short range forecasting. On the other hand, recognizing periodicities and preprocessing the time series (SD-TS) leads to even better performance results, especially at t_0+4 and t_0+6 h. This bears witness to the importance of using dynamical information when available.

The final step of the modeling activities presented in this section is to join the results of these models, by introducing the forecasting β counts results into the box model. In doing so, we are able to obtain good forecasts of future equivalent mixing height. See Fig. 4 for the results on a three-night case study.

In this framework, if we consider that the day–night cycle affects the radon time series as well as time series of meteorological parameters, even if sometimes at a different degree, we can apply SD preprocessing to available meteorological data, as well. We could ask whether by a hybrid approach, it is possible to outperform previous forecasting results on residual series and, therefore, on final forecasts.

Table 2 Forecasting performance on the residual series when SD is applied also to records of meteorological variables

Method	$R\ (t_0+2\ \text{h})$
SD-TS	0.675±0.006
SD-MET	0.571±0.008
SD-TS+V	0.687±0.004
Method	$R\ (t_0+4\ h)$
SD-TS	0.500±0.008
SD-MET	0.297±0.006
SD-TS+V	0.528±0.007
Method	$R\ (t_0+6\ h)$
SD-TS	0.431±0.009
SD-MET	0.151±0.012
SD-TS+V	0.456±0.008

Therefore, first we apply SD to the time series of β counts and to the records of meteorological parameters. Then we concentrate on the following three different choices of inputs for the networks to be trained on the residual series:

SD-MET: The inputs are the residuals coming from the application of SD to several meteorological variables (wind speed, pressure, temperature, dew point, sky covering and height of the lowest cloud layer, meteorological visibility) and to values of beta counts and the time derivative of beta counts at t_0.

SD-TS+V: The inputs are the residuals coming from the application of SD to both the delayed values of the time series and the wind speed. Wind is considered very important because it characterizes the mechanical turbulence in the BL and, in particular, the situations of advection, when the periodicity in the time series (essentially due to the day–night cycle) is broken.

Further results can be found in Pasini (2009b).

The results on the test set are presented in a concise form in Table 2, where the calculation of the linear correlation coefficient between targets and outputs is reported for the forecasts on the residual series.

We can say that the information on the meteorological parameters alone in the SD-MET approach (even if data on β counts and its derivatives is included) is not able to capture a satisfying hidden dynamics or to improve forecasting results on the residual series. On the other hand, the same meteorological parameters contribute (in a statistically significant manner) to improving the results of the SD-TS approach when inserted into the input layer together with data on the time series itself. In a certain sense, one could say that the meteorological parameters act as second-order correctors to the forecast obtained by the SD-TS approach.

The increase in performance obtained through application of the SD-TS+V approach is more evident at the longest ranges, so that we can envisage a shift of the predictability horizon for forecasting radon from observations beyond 6 h.

In summary, what has been presented in this section shows that radon detection reveals a crucial role for characterization and forecasting of a BL index that summarizes its dilution properties. In this framework, two (diagnostic and prognostic) modeling strategies are presented and discussed. Their results endow us with simple but powerful tools for research and application in the complex BL environment.

4 Sediment Chrono-Stratigraphy

The deposition of radon by-products in the hydrosphere is nowadays a consistent tool for studying sedimentation in lake, marine, and river environments. Radon daughters are an efficient stratigraphic marker useful for modeling accumulation rates associated to the recent evolution of landscapes.

The timescale of geomorphological processes is a major constraint and it limits the use of radon progeny to ^{210}Pb, that is the only long-lived daughters among the three different decay series. This nuclide is member of the ^{238}U decay series and it has an half-life time ($t_{½}$) of about 22.26 years. Its daughter is ^{210}Bi while its long-lived parent nuclide is ^{226}Ra ($t_{½}$ = 1602 years). The disequilibrium between lead and radium isotopes can be interrupted by the intermediate decay of ^{222}Rn (Fig. 1). This interruption is due to high environmental mobility of this nuclide, that is, released as gas by minerals where ^{226}Ra is present. Radon is diffused, as already described, in the atmosphere and the decay to heavier and solid daughter, such as ^{210}Pb, complete the cycle with deposition in water and consequently in sediments. Lead fall-out at the end is fixed in sediments and produces an excess of ^{210}Pb that is unsupported by the amount of decayed ^{226}Ra present in minerals constituting the sediments. The disequilibrium between local ^{226}Ra and deposited ^{210}Pb is the basis on which sedimentation can be modeled. Of course, due to the contribution of decay, this method can be used only for processes involving the last 100 years, This limit value is calculated considering five times the half-life time of ^{210}Pb, time required for restoring the equilibrium between parent and daughter isotopes (Appleby and Oldfield 1992).

This chronometric tool is based on modeling sedimentation rate of ^{210}Pb falling out from the atmosphere (P). This contribution depends on the emanation flux in the atmosphere (P) and on deposition (D) and diffusion processes following the equation:

$$P = D\Pi \tag{17}$$

The ^{210}Pb activity of the depositing component is regulated by the sedimentation rate (r).

$$C_0 = \frac{p}{r} = \frac{D\Pi}{r} \tag{18}$$

The variation of ^{210}Pb activity in sediments after deposition must consider the radioactive decay and diffusion processes as Abril and Gharbi (2012) stated, where m is the accumulated dry mass and k_m is the diffusion coefficient.

$$\frac{\partial c}{\partial t} = \underbrace{-C\lambda}_{decay} + \underbrace{r\frac{\partial C}{\partial m}}_{deposition} + \underbrace{\frac{\partial}{\partial m}\left(k_m \frac{\partial A}{\partial m}\right)}_{diffusion} \tag{19}$$

Neglecting diffusion processes (k_m = 0) involving lead through the sediment core, the activity of exceeding ^{210}Pb after a time interval t, is

$$C = C_0 e^{-\lambda_{210} t} \tag{20}$$

where λ_{210} is the decay constant of ^{210}Pb, that can be estimated as

$$\frac{ln2}{t_{1/2}} = \frac{0.693}{22.26} \approx 0.03 \text{ years}^{-1} \tag{21}$$

The dating of a particular sediment level can be done using the appropriate sedimentation model. In literature several models are available and assumptions are critical for selecting the most appropriate. The simplest model is based on constant initial concentration (CIC) (Robbins 1978). The rationale behind these hypothesis is based on considering negligible the rapid variations of the atmospheric flux during the short time interval that a few millimeters of sediments represent. Fixing a constant rate (r), it is feasible to assume P as a constant value and, consequently, to associate to every sediment level the same excess of unsupported ^{210}Pb. If these assumptions are satisfied, the lead-210 concentration will vary exponentially in accumulating sediments. The rate of accumulation is obtained from the following equation and solved for the dry mass accumulation rate (r), where m is the cumulative dry mass.

$$C = C_0 e^{-\lambda_{210} \frac{m}{r}} \tag{22}$$

The resulting profile will be linear with the slope $-\frac{\lambda_{210}}{r}$. The sedimentation rate can be determined graphically from the mean slope determined by a best or least squares fit.

Another dating model, which has the most popular for 35 years, is based on assuming a constant rate of supply (CRS) (Appleby and Oldfield 1978). This assumption is applicable to most systems where the sediment supply is influenced by climatic and anthropogenic changes. In this case the hypothesis implies that the excess of ^{210}Pb during sedimentation at every moment is expressed as follows

$$C_0(t) r(t) = const \tag{23}$$

letting x be the present depth of sediment of age t, the present concentration of unsupported ^{210}Pb at depth x is

$$C(x) = C_0(t) e^{-\lambda_{210} t} \tag{24}$$

Assuming that ρ is the density of sediments at depth x, the rate of change of depth $\dot{x}$ is

$$\dot{x} = \frac{r}{\rho} \tag{25}$$

Substituting the above mentioned equations in the preliminary hypothesis it follows that the depth-age function must satisfy the differential equation

$$\rho(x)C(x)\dot{x} = \rho(0)C_0\dot{x}(0)e^{-\lambda_{210}t} \tag{26}$$

Supposing to estimate $A(x)$ as the total residual of unsupported ^{210}Pb beneath the depth x

$$A(x) = \int \rho(x)C(x)\partial x \tag{27}$$

Integrating this equation with regard to the time we obtain

$$A(x) = A(0)e^{-\lambda_{210}t} \tag{28}$$

where the age of the layer of depth x is computed by

$$t = \frac{1}{\lambda_{210}} log \frac{A(0)}{A(x)} \tag{29}$$

Appleby (2008) reviewed the use of ^{210}Pb dating over the last decades and observed that the method is unequivocal at sites with constant sedimentation rates. In different studies where sedimentation rates have varied significantly the CRS model has proved generally more reliable, but in a significant number of cases other chronostratigraphic markers have shown that neither of the two simple models is applicable over the full ^{210}Pb age range. Under these conditions accurate dates can, however, be estimated coupling such models with other chronostratigraphic dates (^{137}Cs and ^{241}Am for example) used as reference levels.

In recent years different authors (Abril and Gharbi 2012; Carrol and Lerche 2010) evidenced that diffusion processes can occur and additional data processing are required. An integrated approach that couples to the aforementioned models some marker of diffusion processes is the model proposed by Abril et al. (1992), that introduces the concept of incomplete mixing zone (IMZ). Authors describe a model that integrates ^{210}Pb output with additional and independent markers in order to trace penetration of radionuclides into the sediment–water interface. This approach supports dating on very porous sediments (about 90 %).

Figure 5 describes an example of sedimentation modeling based on a lake system located in Rome, Italy. The studied basin is an artificial lake, built in 1960 for the Olympic Games, where sediments are completely authigenic and soil erosion is negligible (Salzano 2006). Sampling was carried out in 2002 and γ-ray spectrometry was performed in order to date sediments. An independent approach, based on ^{137}Cs as marker of Chernobyl accident occurred in 1986, was considered in order to validate the above mentioned ^{210}Pb chronostratigraphy. Field evidence suggested considering a varying sedimentation rate and consequently to use a CRS model. The resulting sedimentation rate described two distinct conditions in the last 40 years.

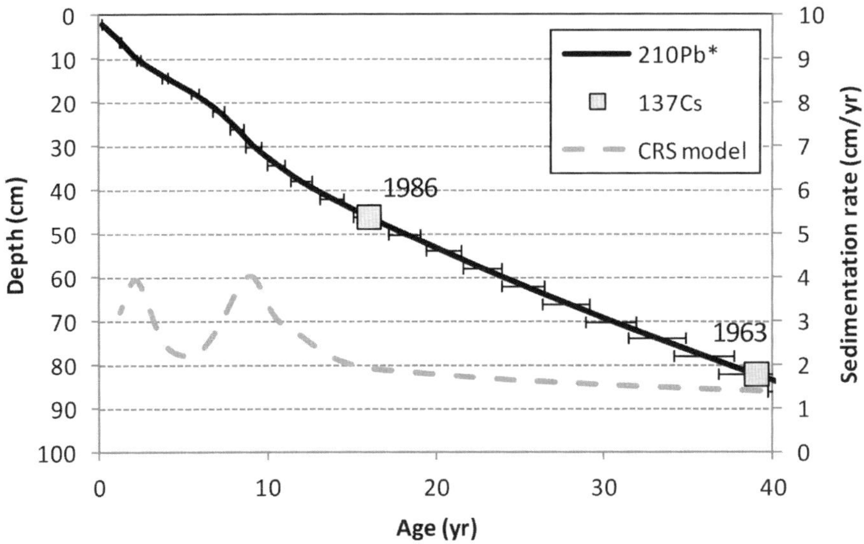

Fig. 5 Example of CRS model applied on unsupported ^{210}Pb in lake sediments sampled in the city of Rome, Italy

The first two decades were characterized by lower deposition (about 1 cm/year) while recent years presented an increasing and heavier sedimentation (above 2 cm/year) probably due to the eutrophication of the basin.

5 Soil Erosion

The sustainability of the soil resource is nowadays a major issue that stressed the scientific community for having reliable tools capable to provide information about soil erosion and degradation (Walling 2010). The sediment budget of a catchment provides a valuable summary of its internal functioning, in terms of sources, sinks, storage and output, and information on the sediment budget of a catchment must be seen as an increasingly important requirement (Walling and Collins 2008). However, such information is difficult to obtain using conventional traditional monitoring techniques (Walling 2006). Fallout radionuclides have showed the potential on filling this cognitive gap, in particular they can consider longer-term sediment mobilization and storage within catchments (Mabit et al. 2008; Walling 2006; Walling 2010; Zhang et al. 2006).

The concentration of unsupported ^{210}Pb shows generally a broad peak near the soil surface with a downward tail into the lower part of the profile. Its concentration decreases approximately exponentially with cumulative mass depth. This behavior can be described as

$$C(z) = \frac{I}{H} e^{-\frac{z}{H}} \tag{30}$$

where $C(z)$ is activity of the nuclide at mass depth z, I is the atmospheric flux and H is the relaxation cumulative mass depth, representing the depth of penetration into the soil (He and Walling 1997).

The presence of unsupported ^{210}Pb deposited from the atmosphere through the soil profile is controlled by several effects (diffusion, vertical migration, and radionuclide decay). The distribution of this fallout in the soil profile can be defined considering first of all the atmospheric input is constant over time.

$$\frac{\partial C(z,t)}{\partial t} = 0 \tag{31}$$

The resulting situation represents a steady-state condition where unsupported ^{210}Pb in an undisturbed soil profile will be achieved due to its radioactive decay (He and Walling 1997)

$$\underbrace{D\frac{\partial^2 C(z)}{\partial z^2}}_{\text{Diffusion}} - \underbrace{V\frac{\partial C(z)}{\partial z}}_{\text{Migration}} - \underbrace{\lambda C(z)}_{\text{Decay}} = 0 \tag{32}$$

where D is the effective diffusion coefficient, V is the effective migration rate and λ_{210} is the decay constant of ^{210}Pb. The solution referred to $C(z)$ is

$$C(z) = C(0)e^{-\beta z} \tag{33}$$

where $C(0)$ is the ^{210}Pb activity at the surface and β is calculated as follow

$$\beta = \frac{1}{2}\left(\sqrt{\frac{V^2}{D^2} + \frac{4\lambda_{210}}{D}} - \frac{V}{D}\right) \tag{34}$$

The value of β can be obtained by varying the values of V and D in order to minimize the sum of the squares of the deviations between the modeled and observed depth distribution.

The integration of $C(z)$ over mass depth z gives the total ^{210}Pb inventory (A) for an undisturbed site at time t

$$A(t) = \int C(z,t)dz \tag{35}$$

If erosion is induced exclusively by water runoff, the rate of soil loss R_e can be estimated from the reduction in the unsupported ^{210}Pb inventory ($\Delta^{-}A$) compared with a local reference value, in combination with information on the ^{210}Pb content of the surface soil $C(t')$.

$$\Delta^{-} A(t) = \int PR_e C(t') e^{-\lambda(t-t')} dt \qquad (36)$$

where P is a correction factor that consider the difference between the particle size in eroded sediments compared to the original soil texture. This factor generally indicates the fine particles are enriched in the mobilized material.

Alternatively, if water transport of sediments is the only cause of increasing inventories ($\Delta^{+}A$) of unsupported ^{210}Pb compared to a local reference site, the deposition rate R_d can be estimated

$$\Delta^{+} A(t) = \int R_d C_d(t') e^{-\lambda(t-t')} dt \qquad (37)$$

where $C_d(t')$ is the ^{210}Pb of accumulated soil, that can be calculated as the weighted mean unsupported ^{210}Pb concentration of soil and sediment particles mobilized from the upslope contributing area converging on the aggrading point.

$$C_d(t') = \frac{\int P'C(t')RdS}{\int RdS} \qquad (38)$$

where P' is a particle size correction factor, that consider the preferential deposition of coarser particles in areas of soil accumulation, and S is the upslope contributing area.

The potentialities of this approach can be summarized by Fig. 6 where the distribution of unsupported ^{210}Pb through the soil profile differs substantially comparing eroded or accumulated soils with a reference site. For further details on field application a lot of literature is available (Mabit et al. 2008; Dercon et al. 2012) that suppose a shift from use of the approach as a research tool to a decision support tool.

6 Approaching the Link Between Radon and Earthquakes With Statistical Tools

This section completes the picture of radon applications for Earth science purposes. We are going to briefly introduce the use of radon as earthquake precursor, topic where a large debate is present (Ghosh et al. 2009), but we will focus the attention on the available statistical approaches useful for this kind of study. Furthermore, this part will be essentially focused on devicing statistical models of complex systems in Earth sciences that do not necessarily require an exhaustive a priori physical understanding of the systems. Indeed, the model development in Earth sciences should be based on the theoretical knowledge of physical phenomena, but

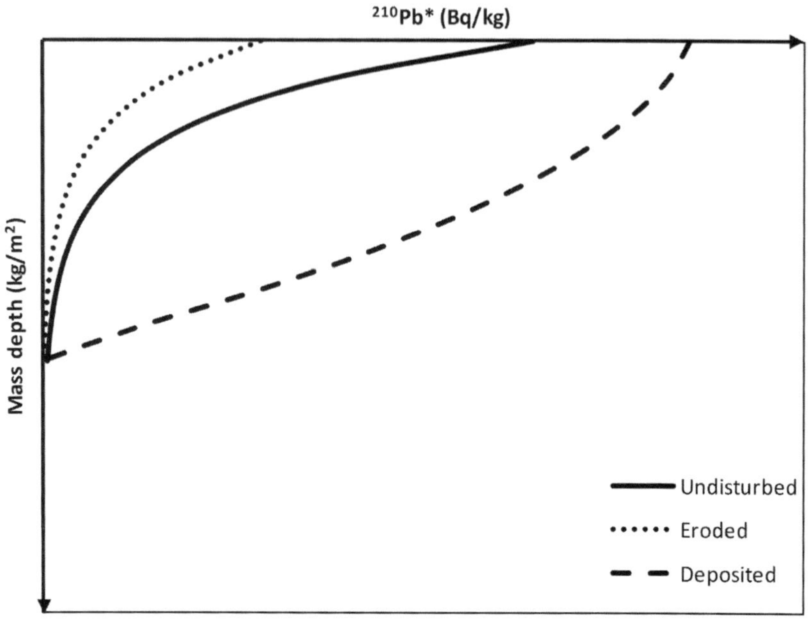

Fig. 6 Behavior of unsupported ^{210}Pb profiles for three different soil conditions.

this knowledge is often far from complete. The aim of a statistical model is often to provide predictions of the phenomena and this may be done in statistics by studying relations among the measured physical variables.

Recent years have seen an ever increasing interest in analyzing the usefulness of radon measurements in earth sciences. Radon emissions have been observed all over the world. The variations of radon exhalation from the interior of earth, as a precursory phenomenon related to earthquakes, and as an indicator of underlying geological faults, is an important field of investigation.

Many earthquakes (Sumatra, Chile, Haiti, Japan) have recently caused a disaster's massive impact in these zones, killing hundreds of thousands people. Because of these events, but more generally as a matter of human safety, a very topical theme concerns the earthquake's forecast. In particular, radon has been envisaged as a possible precursor.

Anomalous behavior of radon in soil and groundwater may be employed as a precursor for an impending earthquake. Many studies have analyzed the connection between radon emission and earthquake showing that there are often anomalies in the radon concentrations during the preearthquake activity. Igarashi et al. (1995) show that radon concentration in ground water increased for several months before the Kobe earthquake on January 17, 1995. The authors conclude these radon changes are likely to be precursory phenomena of the disastrous earthquake. Jonsson and Einarsson (1996) find interesting results about the connection between radon and earthquakes. Their main conclusion showed that: radon anomalies were observed before the 31 % of the significant events (2–5.8 magnitude); 35 % of all observed anomalies were related to seismicity; if a positive anomaly is detected at one station, the probability of a significant earthquake occurring afterwards is 38 %; 80 % of the anomalies observed before earthquakes were positive.

In Walia et al. (2003) a positive correlation is shown between the total radon emission and the microseismicity in the northwestern Himalayas. In fact the annual radon emission and number of earthquakes time series have a rising trend during the time period 1992–1999. The authors also analyze the relationship between the amplitude of radon anomaly and the earthquake magnitude showing a low positive correlation. In particular they conclude that it is not possible to define a general empirical relationship providing a reliable interpretation of radon data for all earthquake magnitudes and epicentral distances.

Yang et al. (2005) observe variations of both soil radon and thoron can serve as useful for earthquake surveillance. In fact spike-like peaks of the radon spectrum are synchronous with thoron and they occur before (few days or weeks) an earthquake event, thus can be a precursor. In order to improve their analysis, the authors suggest that more stations are necessary in order to delineate the relationship between the gas variations and the earthquakes.

Einarsson et al. (2008), by analyzing two magnitude 6.5 earthquakes in the South Iceland (June 2000), show consistent variations in radon in association with these events. Kumar et al. (2009) show that the radon anomalies observed in the Kangra valley of the northwestern Himalayas, have been correlated with the seismic events of magnitude greater than 2 within 250 km distance from the monitoring stations. They observe the radon measurements carried out for earthquake precursory studies seem to be a promising technique, but a lot of data are required to improve the models for the earthquake preparation process. Chyi et al. (2005), using continuous soil gas radon monitoring in real time with improved solid-state detector, observe spike-like anomalies which could be precursors of earthquakes. Monitoring stations located in Taiwan region showed drastic variations of radon when the terrain is stressed before the onset of earthquake. Otherwise, the spectrum recorded at other station shows no significant radon variations. With the anomalies occur before perceivable earthquakes in area within roughly a 30×100 km elliptical area; these anomalies could be used to predict earthquakes. However, an important conclusion of the authors is that they cannot determine the magnitude and precise location of the earthquake.

An important question is about the identification of the radon anomalies. Let us consider $\{r_t\}_{t=1}^{T}$ the radon time series. The very common practice for identifying possible threshold values of the anomalous radon concentration is to consider the radon mean (μ_r)$\pm k$ standard deviation (σ_r). In this way we obtain the range

$$\left(\mu_r - k\sigma_r,\ \mu_r + k\sigma_r\right) \tag{39}$$

If a value of r_t is out of this range, in the period of study, there is an anomaly. In practice it is useful to consider $k=2$. In this way, under the hypothesis of Gaussian distribution, we have a probability of about 95 % that the radon values are in the previous range. Therefore, the anomalies are the values out of the range.

An interesting alternative to identify threshold value may be the study of the structure of the radon time series in order to find possible innovative or additive outliers (Box et al. 2008). This technique is often employed to study the structural break in economic time series. This approach is based on isolating radon anomalies, that could be controlled by earthquakes, from the signal associated with analytical and environmental aspects. The first component is related to the different experimental set ups that can support the measurement of radon exhalation and we refer to Ghosh et al. (2009) and included references for more details about detectors, deployments, and performances. The second component derives from the decoupling of the background signal, associated to the deep structures (faults and microfractures), and the shallow transport of radon in the vadose zone at the air–ground interface. What we measure at the surface is, in fact, affected by environmental parameters, such as temperature, relative humidity, elevation, and air drafts. The parameterization of this component has been developed for atmospheric purposes in Sect. 2.1 and it is intuitive to understand how these factors control the exhalation process. Field evidence about the role of atmospheric parameters on radon emission, such as seasonal and daily changes in atmospheric factors, can be found in literature (Winkler et al. 2001; Singh et al. 2005; Chandrashekara et al. 2006; Groves-Kirkby et al. 2006; Sundal et al. 2008; Omori et al. 2009) and a separation of seasonal components in the radon emission is often essential in order to investigate the relationship with the seismic activity (Baykut et al. 2010). In fact, as previously noted, soil temperature variations as well as barometric pressure and precipitation may dominantly influence the soil radon emanation.

In Attanasio et al. (2012), an econometric tool is used to analyze the relationship between environmental parameters and radon. In particular the authors use the concept of Granger causality (Granger 1969), a method that has recently been employed to detect the connection between complex processes in Earth sciences such as the global warming (Attanasio et al. 2012; Pasini et al. 2012). To illustrate briefly this idea, two variables, x and y, are considered. We will say that x Granger causes y if the future values of y can be better predicted employing the past values of x and y rather than only past values of y.

The framework of radon application for seismological studies is completed by the possibility to estimate the amplitude and duration of the gaseous anomalies and the magnitude of the expected earthquake (Ghosh et al. 2009; Singh et al. 2005).

7 Conclusions

We have seen that radon and its daughters can be useful tracers for characterizing and, possibly, forecasting several geophysical variables at the interfaces of Earth's specific subsystems.

In writing this chapter we did not aim to obtain a complete and exhaustive review of the many investigations performed on these topics, but we tried to show how the use of statistical and modeling techniques applied to radon data can clarify interactions and processes which happen at the atmosphere–soil interface. In some cases the scientific literature presents consolidated and well-founded results, in other cases (such as the investigations about the connection between earthquakes and radon behavior) probably research is still in its infancy and reliable results are more difficult to obtain.

In short, this chapter wants to present concrete examples of how science is using the characteristic features of radon for synthesizing and better understanding interactions and processes in a complex system such as Earth's.

We hope that this brief review may stimulate some reader to contribute to further original applications of radon data to the study of atmosphere–soil interactions. If this should happen, the scope of this chapter will be achieved.

References

Abril JM, Garcia-Leon M, Garcia-Tenorio R, Sanchez CI (1992) Dating of Marine Sediments by an Incomplete Mixing Model. J Environ Radioactiv 15:135–151

Abril JM, Gharbi F (2012) Radiometric dating of recent sediments: beyond the boundary conditions. J Paleolimnol 48:449–460

Alexandris S, Kerkides P (2003) New empirical formula for hourly estimations of reference evapotranspiration. Agr Water Manage 60:157–180

Allegrini I, Febo A, Pasini A, Schiarini S (1994) Monitoring of the nocturnal mixed layer by means of particulateradon progeny measurement. J Geophys Res 99(D9):18765–18777

Appleby PG (2001) Chronostratigraphic techniques in recent sediments. In: Last WM, Smol JP (eds) Tracking environmental change using lake sediments. Basin analysis, coring, and chronological techniques, vol 1. Kluwer Academic, Dordrecht, pp 171–203

Appleby PG, Oldfield F (1978) The calculation of lead-210 dates assuming a constant rate of supply of unsupported 210Pb to the sediment. Catena 5:1–8

Appleby PG, Oldfield F (1992) Application of 210Pb to sedimentation studies. In: Ivanovich M, Harmon RS (eds) Uranium series disequilibrium. Oxford University Press, Oxford, pp 731–778

Attanasio A, Maravalle M, Fioravanti G (2012) Examining Granger causality between atmospheric parameters and radon. Nat Hazards 62:723–731

Attanasio A, Pasini A, Triacca U (2012) A contribution to attribution of recent global warming by out-of-sample Granger causality analysis. Atmos Sci Let 13:67–72

Baykut S, Akgul T, Inan S, Seyis S (2010) Observation and removal of daily quasi-periodic components in soil radon data. Radiat Meas 45:872–879

Bishop CM (1995) Neural networks for pattern recognition. Oxford University Press, Oxford

Box GEP, Jenkins GM, Reinsel G (2008) Time series analysis: forecasting and control. John Wiley & Sons, London

Carrol J, Lerche I (2010) Sedimentary radioactive tracers and diffusive models. J Environ Radioactiv 101:597–600

Chandrashekara MS, Sannappa J, Paramesh L (2006) Studies on atmospheric electrical conductivity related to radon and its progeny concentrations in the lower atmosphere at Mysore. Atmos Environ 40(1):87–95

Chyi LL, Quick TJ, Yang TF, Chen CH (2005) Soil gas radon spectra and earthquakes. Terr Atmos Ocean Sci 16:763–774

Dercon G, Mabit L, Hancock G, Nguyen ML, Dornhofer P, Bacchi OOS, Benmansour M, Bernard C, Froehlich W, Golosov VN, Haciyakupoglu S, Hai PS, Klik A, Li Y, Lobb DA, Onda Y, Popa N, Rafiq M, Ritchie JC, Schuller P, Shakhashiro A, Wallbrink P, Walling DE, Zapata F, Zhang X (2012) Fallout radionuclide-based techniques for assessing the impact of soil conservation measures on erosion control and soil quality: An overview of the main lessons learnt under an FAO/IAEA Coordinated Research Project. J Environ Radioactiv 107:78–85

Einarsson P, Theodorsson P, Hjartardottir A, Guojonsson GI (2008) Radon changes associated with the earthquake sequence in June 2000 in the South Iceland Seismic Zone. Pure Appl Geophys 165:63–74

Ghosh D, Deb A, Sengupta R (2009) Anomalous radon emission as precursor of earthquake. J Appl Geophys 69:67–81

Granger CWJ (1969) Investigating causal relations by econometric methods and cross-spectral methods. Econometrica 37:424–438

Groves-Kirkby CJ, Denman AR, Crockett RGM, Phillips PS, Woolridge AC, Gillmore GK (2006) Time-integrating radon gas measurements in domestic premise: comparison of short-, medium- and long-term exposures. J Environ Radioactiv 86:92–109

Guedalia D, Ntsila A, Druihlet A, Fontan J (1980) Monitoring of the atmospheric stability above an urban and suburban site using sodar and radon measurements. J Appl Meteorol 19:839–848

He Q, Walling DE (1997) The distribution of fallout 137Cs and 210Pb in undisturbed and cultivated soils. Appl Radiat Isot 48(5):677–690

Hertz J, Krogh A, Palmer RG (1991) Introduction to the theory of neural computation. Addison-Wesley, Reading.

Igarashi G, Saeki S, Takahata N, Sumikawa K, Tasaka S, Sasaki Y, Takahashi M, Sano Y (1995) Ground-water radon anomaly before the Kobe Earthquake in Japan. Science 269:60–61

Jonsson S, Einarsson P (1996) Radon anomalies and earthquakes in the South Iceland Seismic Zone 1977-1993. In: Thorkelsson B et al. (ed) Seismology in Europe. European Seismological Commission, Reykjavik, pp 247–252

Kasten F, Czeplak G (1980) Solar and terrestrial radiation dependent on the amount and type of cloud. Sol Energy 24:177–189

Kumar A, Singh S, Mahajan S, Singh Bajwa B, Kalia R, Dhar S (2009) Earthquake precursory studies in Kangra valley of North West Himalayas, India, with special emphasis on radon emission. Appl Radiat Isotopes 67:1904–1911

Lopez A, Guedalia D, Servant J, Fontan J (1974) Advantages of the use of radioactive tracers 222Rn and 212Pbfor the study of Aitken nuclei within the low troposphere. J Geophys Res 79:1243–1252

Mabit L, Benmansour M, Walling DE (2008) Comparative advantages and limitations of the fallout radionuclides 137Cs, 210Pbex and 7Be for assessing soil erosion and sedimentation. J Environ Radioactiv 99 (12):1799–1807

Mahrt L (1998) Stratified atmospheric boundary layers and breakdown of models. Theor Comp Fluid Dyn 11:263–279

Omori Y, Tohbo I, Nagahama H, Ishikawa Y, Takahashi M, Sato H, Sekine T (2009) Variation of atmospheric radon concentration with bimodal seasonality. Radiat Meas 44:1045–1050

Pasini A (2009a) Neural network modeling in climate change studies. In: Haupt SE, Pasini A, Marzban C (eds) Artificial intelligence methods in the environmental sciences. Springer, Dordrecht, pp 235–254

Pasini A (2009b) Neural networks for characterization and forecasting in the boundary layer via radon data. In: Haupt SE, Pasini A, Marzban C (eds) Artificial intelligence methods in the environmental sciences. Springer, Dordrecht, pp 255–268

Pasini A, Triacca U, Attanasio A (2012) Evidence of recent causal decoupling between solar radiation and global temperature. Environ Res Lett 7:034020. doi:10.1088/1748-9326/7/3/034020

Robbins RA (1978) Geochemical and geophysical application of radioactive lead. In: Nriagu JO (ed) The biogeochemistry of lead in the environment. Elsevier, Amsterdam, pp 286–383

Salzano R (2006) Spatial and chronological analyses of heavy metals pollution in an urban environment: the case of Roma: comparison between different monitoring techniques. Plinius 32:222–228

Singh K, Singh M, Singh S, Sahota HS, Papp Z (2005) Variation of radon (222Rn) progeny concentrations in outdoor air as a function of time, temperature and relative humidity. Radiat Meas 39:213–217

Sun K, Guo Q, Zhuo W (2004) Feasibility for mapping radon exhalation rate from soil in China. J Nucl Sci Technol 41(1):86–90

Sundal AV, Valen V, Soldan O, Strand T (2008) The influence of meteorological parameters on soil radon levels in permeable glacial sediments. Sci Total Environ 389:418–428

Szegvary T, Leuenberger MC, Conen F (2007) Predicting terrestrial 222Rn flux using gamma dose rate as a proxy. Atmos Chem Phys 7:2789–2795

Vinod Kumar A, Sitaraman V, Oza RB, Krishnamoorthy T (1999) Application of a numerical model for the planetary boundary layer to the vertical distribution of radon and its daughter products. Atmos Environ 33:4717–4726

Voltaggio M, Masi U, Spadoni M, Zampetti G (2006) A methodology for assessing the maximum expected radon flux from soils in northern Latium (central Italy). Environ Geochem Hlth 28:541–551

Walia V, Virk HS, Bajwa BS, Sharma N (2003) Relationships between radon anomalies and seismic parameters in N-W Himalaya, India. Radiat Meas 36:393–396

Walling DE (2006) Tracing versus monitoring: New challenges and opportunities in erosion and sediment delivery research. In: Owens PN, Collins AJ (eds) Soil erosion and sediment redistribution in river catchments. CABI, Oxfordshire, pp 13–27

Walling DE (2010) Using fallout radionuclides to investigate erosion and sediment delivery: some recent advances. IAHS Publ 337. IAHS Press, Wallingford, pp 3–16

Walling DE, Collins AL (2008) The catchment sediment budget as a management tool. Environ Sci Policy 11(2):136–143

Weigend AS, Gershenfeld NA (1993) Time series prediction: forecasting the future and understanding the past. Addison Wesley, New York

Winkler R, Ruckerbauer F, Bunzl K (2001) Radon concentration in soil gas: a comparison of the variability resulting from different methods, spatial heterogeneity and seasonal fluctuations. Sci Total Environ 272:273–282

Yang TF, Walia W, Chyi LL, Fu CC, Chen CH, Liu TK, Song SR, Lee CY, Lee M (2005) Variations of soil radon and thoron concentrations in a fault zone and prospective earthquakes in SW Taiwan. Radiat Meas 40:496–502

Zhang X, Qi Y, Walling DE, He X, Wen A, Fu J (2006) A preliminary assessment of the potential for using 210Pbex measurement to estimate soil redistribution rates on cultivated slopes in the Sichuan Hilly Basin of China. Catena 68(1):1–9

Zhuo W, Guo Q, Chen B, Cheng G (2008). Estimating the amount and distribution of radon flux density from the soil surface in China. J Environ Radioactiv 99:1143–1148

Index

A
Abedin, M.J., 30, 34, 35
Abril, J.M., 226, 228
Acharyya, S.K., 198
Adomako, E.E., 23, 35
Ahmed, U.Z., 29, 30, 32, 33
Ahsan, D.A., 22, 198
Akbari, A., 146
Akcil, A., 96
Åkerblom, G., 18
Aki, K., 68
Alam, Z., 147
Aldenkamp, F.J., 15
Aleboyeh, A., 147
Alexandris, S., 220, 221
Alinsafi, A., 146
Aljoe, W.W., 102
Allegrini, I., 261, 217
Alpers, C.N., 96
Amachi, S., 174
Amano, H., 169, 172–177, 180, 182–184, 189
Andrade, A.I.A.S.S., 208
Annan, A.P., 51
Aoyama, M., 171, 179, 187
Apparao, A., 51, 56
Appleby, P.G., 214, 226–228
Arabatzis, I.M., 146
Arao, T., 29
Archie, G.E., 48
Arcone, S.A., 51
Armstrong, W., 34
Arnold, D.E., 114
Attanasio, A., 234
Ay, F., 155

B
Baig, J.A., 27
Balkanski, Y.J., 187
Banerjee, K.S., 14
Baranwal, V.C., 10, 11
Bares, M., Jr., 51, 63, 81
Bauer, M., 29
Baykut, S., 234
Benson, A.K., 51
Berdichevsky, M.N., 51
Beresford, N.A., 181
Bhattacharya, C., 17
Bhattacharya, P.K., 26–29, 32, 35, 51, 67, 73, 74, 198
Bhole, A.G., 100
Bibi, M.H., 26
Biegalski, S.R., 175, 177, 178
Bishop, C.M., 222
Biswas, A., 35
Biswas, B.K., 30
Blakely, R.J., 66
Blodau, C., 29
Bockris, J., 50
Bogdan, K., 30, 34
Bolsunovsky, A., 172, 181
Boutt, D.F., 201
Bowders, J., 103
Bowyer, T.W., 172, 181
Brodie, G.A., 113
Buitron, G., 146
Burrough, P.A., 201

C
Cagniard, L., 51, 61
Campbell, J.A., 29
Campos, V., 30
Cao, X., 30
Carbonell, A.A., 35
Cardimona, S.J., 51, 84
Caritat, P., 198
Carrol, J., 228

D. Sengupta (ed.), *Recent Trends in Modelling of Environmental Contaminants*,
DOI 10.1007/978-81-322-1783-1, © Springer India 2014

Caruccio, F.T., 102
Carvalho, F.P., 181
Cerniglia, C.E., 145
Chakrabarty, A., 17
Chakraborti, D., 24, 198
Chakraborti, P., 208
Chandrashekara, M.S., 234
Chang, K.T., 200
Chaturvedi, I., 33
Chen, K.Y., 26
Chen, T.B., 27
Chen, X.P., 31
Chen, Z., 31
Cherry, J.A., 68
Chiado, E., 103
Chino, M., 169, 170, 190
Choudhury, K., 50
Chung, K.T., 145
Chyi, L.L., 233
Clemenza, M., 181
Clesceri LS, 149, 159, 162
Colton, D., 169, 190
Constable, S.C., 51, 129
Corwin, D.L., 201
Cullen, T.L., 3, 22
Cullen, W.R., 22

D

Dakhnov, V.I., 51
Das, D.K., 29, 198, 209
Das, H.K., 35
Davis, J.L., 51
Davis, S.M., 51, 82
de Meijer, R.J., 13
DelValls, T.A., 22, 198
Dementyev, D., 172, 181
Dercon, G., 231
Dittmar, J., 27
Dmitriev, V.I., 51
Doll, H.G., 51
Dowling, C.B., 26
Du, V.L., 31
Duan, G.L., 32
Dudas, M.J., 30
Duxbury, J.M., 35
Dwivedi, S., 34

E

Eaton, G.P., 68
Edrissi, M., 150
Ehhalt, D.H., 187
Eichholz, G.G., 12
El-Gohary, F., 146
Elliott, C.L., 51, 57, 78
Elliott, H.M., 51, 58
Erkurt, E.A., 146
Evangelou, V., 102

F

Faulkner, B.B., 107, 108
Ferreira, S.L.C., 150
Firestone, R.B., 8
Fisher, E.A., 51, 81, 84
Fitterman, D.V., 51, 78–80
Fitz, W.J., 28–30
Forstner, U., 99
Francesconi, K.A., 23
Frans, R., 35
Freeze, R.A., 68
Frischknecht, F.C., 51, 58, 67, 73
Frohlich, R.K., 56
Fu, Y., 29, 30
Furuta, S., 169, 172–177, 180

G

Gandhi, P.K., 99
Garelick, H., 23
Garnier, J.M., 31
Gaus, I., 207
Gediel, G., 102
Geissler, P.E., 68
Gharbi, F., 228
Ghiassi-nejad, M., 3
Ghose, R., 51, 81, 84
Ghosg, D.P., 51, 73, 74
Ghosh, A.R., 209
Ghosh, D.P., 51, 73, 74
Ghosh, R.N., 208
Glasstone, S., 50
Gogu, R.C., 201
Goovaerts, P., 202, 204
Grigorescu, E.L., 8
Grosz, A.E., 13
Guilan, D., 30, 34
Guo, H., 26
Guo, W., 34
Gupta, A.K., 147, 148, 152, 162
Gurung, J.K., 26
Guyod, H., 48

H

Haeni, F.P., 51, 63, 81
Hall, M.D., 201
Harvey, C.F., 26, 29
Hassan, M.M., 202
Hawkins, J.W., 102
Hedin, R.S., 111–113
Heikens, A., 22, 27, 30, 32, 35

Hirose, K., 177, 185, 187, 188, 190
Hohmann, G.W., 51
Hoover, D.B., 51
Horneman, A., 28
Hossain, F., 198, 207
Hossain, M.B., 27, 31, 34
Hoversten, G.M., 51
Hu, Y., 31, 34
Hu, Z.Y., 31
Hughes, L.I., 51, 63
Huh, C.A., 182
Huijbregts, Ch.J., 202
Hunag, R.Q., 29
Hussain, M., 198, 202
Hyman, D.M., 109

I
Igarashi, Y., 182, 188
Inman, J.R., 74
Islam, M.R., 30
Itoi, R., 26

J
Jahiruddin, M., 30, 34
Jensen, J.R., 51, 82
Jia, X., 51, 82
Johnston, K., 205–207
Jones, H., 23
Joseph, G., 51, 66, 82
Journel, A.G., 202

K
Kalenov, E.N., 51
Kammereck, R., 30
Kanai, Y., 169, 172, 174, 175, 178
Kaneyasu, N., 179
Kar, S., 26, 27
Katsuragi, Y., 187
Kaufman, A.A., 51, 79
Kazmi, A.A., 146
Keller, G.V., 51, 58, 67, 73, 74, 79
Kellog, O.D., 66
Kelly, W.E., 68, 69
Kerkides, P., 220, 221
Kerle, N., 200
Khadhraoui, M., 146
Khan, N.I., 198, 201
Khandaker, M.U., 6
Khataee, A.R., 146
Kim, A.G., 103
Kim, C.-K., 181, 182
Kim, K., 26
Kinniburgh, D.G., 22–24, 26, 198
Kirk, G.J.D., 31
Kleinmann, R.L.P., 102
Koefoed, O., 51, 73, 74
Koldas, S., 96
Kundu, N., 10

L
Ladwig, K., 102
Lambert, G., 187
Lasserre, F., 201
Lee, J.J., 208
Lee, J.S., 23
Lee, T., 51
Lerche, I., 228
Lewis, R., 51
Li, Q., 146
Lico, M.S., 26
Lihong, W., 30, 34
Lillesand, T.M., 199, 200
Lin, C.F., 29
Lin, H.T., 23
Liu, T.K., 26
Liu, W.J., 31
Lombi, E., 22
Long, C.L., 51
Long, N.Q., 182
Lozano, R.L., 179, 181, 182
Lucas, M., 5
Lujaniené, G., 182
Lyons, C., 169, 190

M
Ma, J.F., 35
Ma, J.M., 35
Ma, L.Q., 27, 30, 34
Mac Cary, L.M., 53
Mackie, R.L., 51, 63
Madden, T.R., 51
Madras, G., 146
Mahajan, S.P., 144
Mahimairaja, S., 28, 30
Mahrt, L., 221
Maillet, R., 56
Majumdar, S., 208
Mallick, K., 51
Manahan, S.E., 198
Mandal, B.K., 23
Mares, S., 68
Marine, A.R., 28, 29
Martinez, V.D., 22
Masson, O., 182, 185
Mayya, Y.S., 15
McArthur, J.M., 26
McDonnell, R.A., 201
Mclaren, R.G., 28

Meharg, A.A., 22, 23, 27, 28, 32, 34
Mehrotra, A., 100
Mellander, H., 18
Misra, S., 13, 14
Mittal, S., 17
Mohanty, A.K., 3, 7–9, 13
Momoshima, N., 172, 182
Montgomery, D.C., 150, 155
Mooney, H.M., 51
Moore, H.E., 187
Morino, Y., 180, 185, 190
Morrison, H.F., 51
Mukherjee, A., 29
Mulligan, C.N., 29
Muramatsu, Y., 174

N
Nabighian, M.N., 51, 57, 61, 78
Nazaroff, W.W., 16
Neitsch, S.L., 200
Nero, A.V., Jr., 16
Ng, J.C., 22
Nielson, K.K., 16
Noakes, J.E., 13
Nordstrom, D.K., 96
Norra, S., 27
Norton, G.J., 32

O
Oldenburg, D.W., 51, 63
Oldfield, F., 226–228
Omori, Y., 234
O'Neil, P., 30

P
Paatero, J., 172, 179, 181
Panaullah, G.M., 35
Pang, Y.L., 147
Parasnis, D.S., 51, 126
Pasini, A., 234
Patangay, N.S., 49, 67
Patra, H.P., 50, 51, 61, 67, 73, 74, 76
Paul, A.C., 3
Peryea, F.J., 30
Pham, M.K., 181
Piantanakulchai, M., 198, 202
Pikaray, S., 29
Pirson, S.J., 51
Pittauerová, D., 172, 181
Plumlee, G.S., 96, 97
Prasad, S.D., 99, 100

Q
Quazi, S., 29, 30

R
Raab, A., 31
Rahaman, S., 29
Rahman, M.A., 35
Rahman, M.D.M., 23, 27
Ravenscroft, P., 198
Ravikumar, K., 158
Reddy, A.K.N., 50
Reimann, C., 198
Reimer, K.J., 22
Renton, J.J., 102
Richard, P.G., 68
Richards, J.A., 51, 82
Richards, P.G., 68
Riley, R.K., 114
Robbins, R.A., 228
Roberts, L.C., 29
Rogers, V.C., 16
Rossitier, J.R., 51, 82
Routh, P.S., 63
Rowland, H.A.L., 29
Roy, A., 51, 56, 66, 74
Roy, K.K., 51, 58, 66, 74
Roychowdhury, T., 23, 27
Roymahashay, B., 55
Ryu, J., 61, 76

S
Sabnavis, M., 49, 67
Saha, C., 209
Sahoo, B.K., 14–17
Sahoo, C., 147, 148, 152, 162
Sahoo, P.K., 26
Sahu, J.N., 147
Salomons, W., 99
Salzano, R., 228
Santra, S.C., 35
Sanyal, S.K., 29
Sarkar, K., 208
Sathishkumar, P., 146
Savosin, S.I., 17
Schenk, M.K., 30, 34
Schoof, R.A., 23
Schowengerdt, R.A., 51, 82
Schreiber, M.E., 26
Scott Keys, W., 53
Seigel, B.S., 51, 82
Seigel, H.O., 51, 57, 74, 82
Selim Reza, A.H.M., 29
Sen, R., 158
Sen, S.K., 55
Sengupta, D., 13
Sengupta, M., 96
Seyfferth, A.L., 31

Shah, B.A., 198
Sharma, M.A., 146
Sharma, P.D., 96
Sharma, S.P., 10
Shastri, N.L., 51, 61, 76
Sheppard, S.C., 30
Shirley, V.S., 8
Shukla, V.K., 16
Signes-Pastor, A., 29–31
Sikdar, P.K., 24, 198
Singh, G., 109
Singh, K.P., 151, 152
Singh, V., 27
Singhal, R., 100
Sinitsyn, V.I., 17
Sivakumar, B., 198, 207
Skousen, J., 103, 107–109
Sleiman, M., 146
Slob, E.C., 51, 81, 84
Smedley, P.L., 22–24, 26, 198
Smith, A.H., 22
Smith, E., 30
Smith, H.A., Jr., 5
Sohrabi, M., 3
Stefanescu, S., 51
Stewart, M.T., 51, 79, 80
Stigter, T.Y., 208
Stohl, A., 170
Strangway, D.W., 51
Stratton, J.A., 66, 68
Streat, M., 29
Subba Rao, K.V., 146
Sugahara, T., 3
Sulekha Rao, N., 13, 14
Sumner, J.S., 51, 57
Sun, G.X., 23
Sun, K., 219, 220
Sunta, C.M., 3
Suzuki, K.T., 23
Swain, P.H., 51, 82
Swaminathan, T., 158
Syed, T., 63

T
Takahashi, Y., 28, 30, 34, 190
Takemura, T., 171, 182, 185
Talukdar, A.S.M.H.M., 29
Tavares, A.P.M., 159
Tawfik, A., 146
Tayade, R.J., 154
Thul, R., 146
Tikhnov, A.N., 51, 61
Trumm, D., 110
Tu, S., 27, 29, 34

V
Vail, W.J., 114
van Geen, A., 27, 35
Vinu, R., 146
Voegelin, A., 31
Vozoff, K., 51

W
Wait, J.R., 51, 61, 68
Walter, J.F., 22
Wang, J., 34
Wang, S., 29
Ward, S.H., 51, 124
Waters, J.C., 105
Watkins, D.W., 201
Watkins, J.S., 68
Watzlaf, G.R., 109, 113
Wei, L., 3
Welch, A.H., 26
Wenzel, W.W., 22, 28–30
Whitaker, J.H., 22, 28
Williams, M., 24
Williams, P.N., 23, 27, 28, 35
Woolson, E.A., 30
Wu, C., 33

X
Xiaboo, Li, 63
Xiaobo, L., 63
Xu, X.Y., 28

Y
Yamamoto, Y., 172
Yamashita, M., 64
Yan, W., 35
Yilmaz, O., 68, 81
Yonezawa, C., 172
Yoshida, N., 190
Yoshida, S., 174
Younger, P.L., 104

Z
Zavala, Y.J., 22, 23, 35
Zhang, J., 31, 32, 150
Zhang, W., 172, 181
Zhang, X., 146, 229
Zhao, F.J., 22, 28, 31, 34, 35
Zhdanov, M.S., 51, 74
Zhody, A.A.R., 51, 57
Zhu, Y.G., 23, 27
Zhuo, W., 219, 220
Ziemkiewicz, P.F., 107, 108, 112, 114
Zohdy, A.A.R., 68
Zonge, K.L., 51, 63

Printed by Printforce, the Netherlands